Mathematics for Technicians

Mathematics for Technicians

James E. Price

Tulsa Junior College

Richard R. Martin

John Vink Company

Larry D. Jones

Tulsa Junior College

Charles E. Merrill Publishing Company
A Bell & Howell Company
Columbus, Ohio 43216

Published by
Charles E. Merrill Publishing Co.
A Bell & Howell Company
Columbus, Ohio 43216

International Standard Book Number: 0-675-08901-8

Library of Congress Card Catalog Number: 73-86932

1 2 3 4 5 6 7 8 9 10 — 80 79 78 77 76 75 74

Printed in the United States of America

Preface

Mathematics is the tool which engineering technicians use to analyze and solve technical problems. The material contained herein is intended to be used in a two-semester sequence of technical mathematics with the first six chapters being covered the first semester. Major emphasis throughout is given to applications.

The authors believe that the outstanding features of the material will prove to be its simplicity, clarity, conciseness, and readability. Students enrolling in technical mathematics should have had one year each of high school algebra and plane geometry; however, students with less mathematics background should be able to achieve well with considerable diligence.

The algebra was divided into two chapters because the authors believe the material in chapters 4 through 7 is needed in technical courses before the material in chapter 8.

This text was tested in typed form in the classroom for two semesters at Tulsa Junior College. The authors wish to express their appreciation to the students involved for their patience and helpful evaluation of the material.

The material is a cooperative effort with the order of authorship bearing no significance. Equal responsibility is assumed for any strengths or weaknesses.

Sincere appreciation is expressed to Mr. Roger Smith of Tulsa Junior College, Mr. John Talbott, State Supervisor for Technical Education, and Dr. D. S. Phillips and Mr. R. W. Tinnell of Oklahoma State University for their constructive criticisms and very helpful suggestions. In addition, the authors wish to thank all the secretaries at Tulsa Junior College who helped type the rough draft. Special thanks goes to Linda Higgins, Martha Hood, and Yvonne Merz for typing the final draft. Finally, sincere appreciation is expressed to the State Department of Vocational Technical Education, Stillwater, Oklahoma for providing a grant to develop the material herein.

Contents

3

Geometry 89

4

Ratio, Proportion, and Variation — 133

5

Trigonometry, Part 1 — 151

6

Complex Numbers   219

7

Logarithms   239

8

9

10

Integral Calculus   345

11

Curve Fitting   363

1

Scientific Notation

Have you seen a number such as 1,230,000,000,000 or .000000783 and wondered if it might be possible to write it in some shorter form? Have you tried to find the product or quotient of two numbers such as 786,000,000 and .0002348 and wondered if there might be some way to simplify the computation?

In many scientific and technical programs numbers and problems such as these occur frequently. Scientific notation is one means of simplifying the computation.

1-2
Numbers in Scientific Notation

Let us consider the number 384,000,000 meters (which represents the distance from the earth to the moon.) This number can be written in scientific notation as 3.84×10^8 meters.

We are now ready to define scientific notation. A number is in scientific notation if it is expressed as a figure between one and ten and is multiplied by an integral power of 10. There are two ways of doing this. The following three examples illustrate both methods.

EXAMPLE 1   Express 2000 in scientific notation.

Solution:

$$
\begin{aligned}
\text{(a)} \ 2000 &= 2 \times 1000 \\
&= 2 \times 10 \times 10 \times 10 \\
&= 2 \times 10^3 \\
2000 &= 2 \times 10^3
\end{aligned}
$$

The number 2000 can be expressed as a product of the numbers 2 and 1000. The number 1000 can be expressed as $10 \times 10 \times 10$, which is normally written as 10^3. 10^3 is read "ten to the third power" or "ten cubed." The 3 is called the exponent. The number 2000 written in scientific notation is 2×10^3.

$$
\begin{aligned}
\text{(b)} \ 2000 &= 2.000 \times 10^3 \\
&= 2 \times 10^3 \\
2000 &= 2 \times 10^3
\end{aligned}
$$

The number 2000 is written as a number between one and ten, and this number, 2.000, is multiplied by ten to the third power. Notice that the power is the same as the number of places the decimal point must be moved from its place in the number 2.000 to its original position in the number 2000.

EXAMPLE 2   Express 359 in scientific notation.

Solution:

$$
\begin{aligned}
\text{(a)} \ 359 &= 3.59 \times 100 \\
&= 3.59 \times 10 \times 10 \\
&= 3.59 \times 10^2 \\
359 &= 3.59 \times 10^2
\end{aligned}
$$

$$
\text{(b)} \ 359 = 3.59 \times 10^2
$$

Again, notice that the exponent, 2, is the same as the number of places the decimal point in the number 3.59 has been moved from its original position in the number 359.

EXAMPLE 3   Express 14500 in scientific notation.

Solution:

$$
\begin{aligned}
\text{(a)} \ 14500 &= 1.4500 \times 10000 \\
&= 1.4500 \times (10 \times 10 \times 10 \times 10) \\
&= 1.4500 \times 10^4 \\
14500 &= 1.45 \times 10^4
\end{aligned}
$$

$$
\begin{aligned}
\text{(b)} \ 14500 &= 1.4500 \times 10^4 \\
&= 1.45 \times 10^4 \\
14500 &= 1.45 \times 10^4
\end{aligned}
$$

Again, you will notice that the exponent, 4, corresponds to the number of places the decimal point in 1.45 has been moved from its original position in 14500.

Let us give a more precise rule for writing a number in scientific notation and for determining the exponent, or power, of 10.

> Express the number as a figure between one and ten by placing the decimal point to the right of the first non-zero number. Multiply this number by ten to an integral power. The power can be found by counting the places the decimal point was moved from its position in the original number. If the decimal point was moved to the right the exponent will be a negative number. If it was moved to the left the exponent will be a positive number.

The following examples further illustrate the above rule.

EXAMPLE 4   Express 1346 in scientific notation.

Solution: $1346 = 1.346 \times 10^3$

EXAMPLE 5   Express .0012 in scientific notation.

Solution: $.0012 = 1.2 \times 10^{-3}$

EXAMPLE 6   Express .01780 in scientific notation.

Solution: $.01780 = 1.780 \times 10^{-2}$

EXAMPLE 7   Express 124,700,000 in scientific notation.

Solution: $124,700,000 = 1.247 \times 10^8$

Whether your interest is in electronics, mechanics, physics, or any number of other technical areas, scientific notation or "engineer's shorthand" will be of great value to you. Examples from these technical fields serve to illustrate this point.

EXAMPLE 8   Typical temperature changes of approximately 30°F occur between day and night. This causes expansion and contraction of materials. A ten foot slab of concrete pavement would expand by approximately .00005 ft. for this temperature increase. In scientific notation this would be expressed as 5×10^{-5} ft.

EXAMPLE 9 In a machine shop, machining tolerances are expressed in thousandths or ten-thousandths of an inch. A tolerance of one-thousandth (.001) written in scientific notation is 1×10^{-3} inches.

EXAMPLE 10 A current of .01 amperes will flow in an electrical circuit containing 10 volts potential difference and 1,000 ohms resistance. This current may be written as 1×10^{-2} amperes in scientific notation. The 1,000 ohms resistance may be written as 1×10^3 ohms.

1-3
Multiplication and Division

Scientific notation is convenient in multiplication and division operations. To multiply two numbers together, first express each number in scientific notation. Multiply the figures between one and ten together and add the exponents of the tens to produce a new power of ten.

EXAMPLE 11 Multiply 4000 by 600.

$$
\begin{aligned}
4000 \times 600 &= (4 \times 10^3) \times (6 \times 10^2) \\
&= (4 \times 6) \times 10^{3+2} \\
&= 24 \times 10^5 \\
&= (2.4 \times 10) \times 10^5 \\
&= 2.4 \times (10^{1+5}) \\
&= 2.4 \times 10^6 \\
&= 2,400,000
\end{aligned}
$$

EXAMPLE 12 Multiply .005 by 400

$$
\begin{aligned}
.005 \times 400 &= (5 \times 10^{-3}) \times (4 \times 10^2) \\
&= (5 \times 4) \times (10^{-3+2}) \\
&= 20 \times 10^{-1} \\
&= (2 \times 10) \times 10^{-1} \\
&= 2 \times (10^{1+(-1)}) \\
&= 2 \times 10^0 \\
&= 2 \times 1 \\
&= 2
\end{aligned}
$$

To divide two numbers, first express both the dividend and the divisor in scientific notation, then divide the figures between one and ten. Subtract the exponents of the tens to produce a new power of ten.

EXAMPLE 13   Divide 96000 by 16

$$96000 \div 16 = (9.6 \times 10^4) \div (1.6 \times 10^1)$$
$$= (9.6 \div 1.6) \times (10^{4-1})$$
$$= 6 \times 10^3$$
$$= 6000$$

EXAMPLE 14   Divide .00144 by .36

$$.00144 \div .36 = (1.44 \times 10^{-3}) \div (3.6 \times 10^{-1})$$
$$= (1.44 \div 3.6) \times 10^{-3-(-1)}$$
$$= .4 \times 10^{-2}$$
$$= (4 \times 10^{-1}) \times 10^{-2}$$
$$= 4 \times 10^{-1+(-2)}$$
$$= 4 \times 10^{-3}$$
$$= .004$$

Table 1-1 lists numbers in decimal form with their corresponding scientific notation and accepted prefix.

Table 1-1

Scientific Notation	Decimal Form	Prefix
10^{12}	1,000,000,000,000.	tera
10^{11}	100,000,000,000.	
10^{10}	10,000,000,000.	
10^9	1,000,000,000.	giga
10^8	100,000,000.	
10^7	10,000,000.	
10^6	1,000,000.	mega
10^5	100,000.	
10^4	10,000.	
10^3	1,000.	kilo
10^2	100.	hecto
10^1	10.	deka
10^0	1.	
10^{-1}	.1	deci
10^{-2}	.01	centi
10^{-3}	.001	milli
10^{-4}	.0001	
10^{-5}	.00001	
10^{-6}	.000001	micro
10^{-7}	.0000001	
10^{-8}	.00000001	
10^{-9}	.000000001	nano
10^{-10}	.0000000001	
10^{-11}	.00000000001	
10^{-12}	.000000000001	pico

Express the following numbers in scientific notation.

1. 150	2. 6,850	3. 95,240
4. 275,240	5. 5,726,371	6. .02
7. .085	8. .12734	9. .0000075
10. .00002468	11. 894,100	12. .01240
13. 521,724,398	14. .128300	15. .0029873

Express the number in each of the following statements in scientific notation.

1. One kilometer is equal to 39370 inches.
2. The velocity of light in vacuum is 299,790,000 meters per second.
3. One cubic yard contains 764,600 cubic centimeters.
4. The mean distance from the earth to the sun is 92,700,000 miles.
5. One year is equivalent to 31,600,000 seconds.
6. One degree is equivalent to .01745 radians.
7. The charge on an electron is .00000000000000000016 coulombs.
8. Seawater contains 1,290,000 micrograms per liter magnesium.
9. One angstron is a unit of measure equal to .0000000001 meters.
10. The atomic mass unit (amu) is equivalent to .000000000149 joules.
11. The mean radius of the earth is 6,371,000 meters.
12. The lowest frequency in the am commercial broadcast band is 540,000 cycles per second.
13. The mass of an electron is .0005486 atomic mass units.
14. The standard unit of length of one meter, recognized by the International Bureau of Weights and Measures, is 1,650,763.73 wavelengths of orange-red light from krypton-86.
15. The wavelength of the helium-neon gas laser beam is .0000006328 meters.

Multiply the following and express the answer in scientific notation.

1. $.01 \times 10,000 \times .0001$
2. $.0005 \times 45,000$

3. $750,000 \times .002 \times .000045$

4. $8,232,000 \times .0000469$

5. $6,724 \times .0000055 \times 3,350$

6. $(1 \times 10^6)(1 \times 10^{-3})(100)$

7. $(.002)(5,000)(1 \times 10^{10})$

8. $(100,000)(5 \times 10^{-5})(400 \times 10^3)$

9. $(729,200)(3.75 \times 10^4)(6,750 \times 10^{-6})$

10. $(4,679,000)(.0006 \times 10^{-3})(2.0 \times 10^7)$

PROBLEM SET 1-4

Perform the following division and express the answer in scientific notation.

1. $\dfrac{1,000}{.01}$

2. $\dfrac{495,000}{.00003}$

3. $\dfrac{5,732,000}{.01 \times 5,000}$

4. $\dfrac{10}{(5 \times 10^4)(3 \times 10^{-3})(8 \times 10^6)}$

5. $\dfrac{(1 \times 10^4)(7 \times 10^{-7})(5.5 \times 10^{-3})}{(85,000)(.00020)}$

6. $\dfrac{(.0000750)(.000000025)}{8,500,000,000}$

7. $\dfrac{(782,000)(234 \times 10^5)}{(649 \times 10^3)(5 \times 10^{-12})}$

8. $\dfrac{(5 \times 10^{10})(3 \times 10^{-4})(9 \times 10^3)}{625 \times 10^{12}}$

9. $\dfrac{1}{(204 \times 10^8)(780 \times 10^{-11})}$

10. $\dfrac{(9.1 \times 10^{-31})(6.2 \times 10^{19})}{24 \times 10^{11}}$

2

Algebra, Part 1

In working with numbers there are certain basic laws or rules which we know are true. These basic rules are called the fundamental laws of algebra. The following examples show how these rules are applied.

What is the sum of $2 + 3$? What is the sum of $3 + 2$? The answer is 5 in both cases. You will notice that the order in which you add two or more numbers does not affect the answer. This demonstrates the commutative law for addition. The commutative law for multiplication is very similar. For instance, $2 \times 3 = 3 \times 2$.

EXAMPLE 1   Find the sum of $(7 + 3) + 6$

$$(7 + 3) + 6 = 10 + 6$$
$$= 16$$

Now find the sum of $7 + (3 + 6)$

$$7 + (3 + 6) = 7 + 9$$
$$= 16$$

Example 1 demonstrates that grouping of elements has no effect on the answer in addition problems. This is referred to as the associative law for addition. The associative law is also true in the multiplication of three or more numbers.

9

Another basic law is the distributive law. Multiplication is distributive over addition. For example, $3(4 + 7) = (3 \times 4) + (3 \times 7)$ because $3(4 + 7) = 3(11) = 33$, and $(3 \times 4) + (3 \times 7) = 12 + 21 = 33$.

Precise statements of the above laws are:

commutative law for addition: $a + b = b + a$
associative law for addition: $(a + b) + c = a + (b + c)$
commutative law for multiplication: $a \times b = b \times a$
associative law for multiplication: $(a \times b) \times c = a \times (b \times c)$
distributive law: $a(b + c) = ab + ac$

PROBLEM SET 2-1

Identify the basic law of algebra which applies to each of the following problems.

1. $2 \times 3 = 3 \times 2$
2. $(1 + 5) + 4 = 1 + (5 + 4)$
3. $7 + 2 = 2 + 7$
4. $(2 \times 9) \times 3 = 2 \times (9 \times 3)$
5. $8(6 + 5) = 8 \times 6 + 8 \times 5$
6. $e + r = r + e$
7. $i \cdot r = r \cdot i$
8. $(s + t) + v = s + (t + v)$
9. $(\theta \cdot \phi)\alpha = \theta(\phi \cdot \alpha)$
10. $\omega(\beta + \gamma) = \omega\beta + \omega\gamma$

2-2
Real Numbers

The preceding discussion referred only to positive integers (whole numbers). It is necessary at this time to introduce other numbers.

We shall define a *positive integer* as any whole number greater than zero, such as 1, 2, 3 and so forth. Any whole number less than zero will be called a *negative integer*. Examples of negative integers are -1, -2, -3 and so forth. Subtracting a larger number from a smaller number results in a negative number. There must be negative numbers if this type of problem is to have meaning.

So far we have mentioned zero very briefly. Zero is an integer which is neither positive nor negative. If we include zero with the positive and negative integers we have the complete set of integers.

Suppose we divide a pie in such a way that four people will receive equal portions. Each person will receive one-fourth of the pie, which is normally written as $\frac{1}{4}$. This represents another division of numbers which is termed the rational numbers.

A *rational number* is any number that can be expressed as the ratio of one integer to another, that is, a numerator over a denominator. Zero can never be the denominator of a rational number because division by zero is undefined. Examples of rational numbers are $\frac{1}{3}$, $\frac{1}{2}$, $\frac{5}{7}$, $\frac{10}{4}$, $\frac{0}{3}$, 5. All integers are rational numbers because they can be expressed with a denominator of 1.

$\sqrt{4}$ and $\sqrt{16}$ are also examples of rational numbers, because they can be expressed as one integer over another, $\frac{2}{1}$ and $\frac{4}{1}$. However, $\sqrt{3}$, and π are irrational numbers because neither can be expressed as one integer (or whole number) over another. Numbers such as these are called irrational numbers. The rational and irrational numbers combined form the *real number* system.

PROBLEM SET 2-2

Identify each of the following numbers as being an integer, rational, irrational, or real. (More than one may apply.)

1. 2	2. 7/8	3. $\sqrt{9}$	4. π	5. 5/2
6. $\sqrt{5}$	7. $\sqrt{2}$	8. $7/\pi$	9. $\dfrac{1}{\sqrt{2}}$	10. $\dfrac{\sqrt{4}}{\sqrt{7}}$

2-3
Addition, Subtraction, Multiplication, and Division of Signed Numbers

The four fundamental operations of mathematics are addition, subtraction, multiplication and division. At this time we will define *absolute value* and then state the rules for working with signed (positive and negative) numbers.

The *absolute value* of a figure, denoted by vertical lines on either side of the number, is the number without a sign. It may also be defined as the distance a number is from zero without regard to direction, or it may be considered as the magnitude of the number.

EXAMPLE 2   (1) $|2| = 2$ (3) $|6| = 6$
                (2) $|-3| = 3$ (4) $|-10| = 10$

The following rules apply to operations with signed numbers.

Addition of Signed Numbers   When the signs are alike, add their absolute values and assign the same sign to the answer.

$$
\begin{array}{cccc}
(1)\ +2 & (2)\ -3 & (3)\ +5 & (4)\ -6 \\
\underline{+2} & \underline{-7} & \underline{+1} & \underline{-4} \\
+4 & -10 & +6 & -10
\end{array}
$$

When the signs are unlike, subtract the absolute value of the smaller from the absolute value of the larger and use the sign of the larger in the answer.

$$
\begin{array}{cccc}
(1)\ +7 & (2)\ -6 & (3)\ -3 & (4)\ -11 \\
\underline{-2} & \underline{+4} & \underline{+17} & \underline{+17} \\
+5 & -2 & +14 & +6
\end{array}
$$

Subtraction of Signed Numbers   To subtract two signed numbers, change the sign of the subtrahend and apply the rules for addition.

$$
\begin{array}{cc}
5 & 5 \\
\underline{-(+7)} = & \underline{-7} \\
& -2
\end{array}
$$

$$
\begin{array}{cc}
-6 & -6 \\
\underline{-(-4)} = & \underline{+4} \\
& -2
\end{array}
$$

Multiplication of Signed Numbers   If the signs are alike the product will be positive.

$$
(3)(2) = 6
$$
$$
(-3)(-4) = 12
$$

If the signs are unlike the product will be negative.

$$
(2)(-3) = -6
$$
$$
(-3)(4) = -12
$$

Division of Signed Numbers   If the signs are alike the quotient will be positive.

$$
\frac{10}{5} = 2 \qquad \frac{-28}{-7} = 4
$$

If the signs are unlike the quotient will be negative.

$$
\frac{15}{-3} = -5 \qquad \frac{-18}{2} = -9
$$

PROBLEM SET 2-3

Evaluate each of the following expressions using the laws of algebra.

1. $9 + 4$ 　　　　　　　　　　　　　　　2. $13 - 2$

3. $7 + (-5)$ 　　　　　　　　　　　　　4. $(-4) + (-4)$

5. $(11)(4)$ 　　　　　　　　　　　　　　6. $(9)(-7)$

7. $-8 - (-1)$ 　　　　　　　　　　　　8. $(-4) - (-6)$

9. $6 + (-3) - (+7)$ 　　　　　　　　10. $(9)(-2)(-5)$

11. $\dfrac{-8}{2}$ 12. $\dfrac{-15}{-5}$

13. $\dfrac{0}{-3}$ 14. $\dfrac{9(-5)}{(2)(6)}$

15. $\dfrac{(3)(0) + 7(0)}{-6}$ 16. $\dfrac{(5)(-4) + (-2)(8)}{(-3)(2)}$

17. $\dfrac{(8)(9)}{(5)(8) - (-2)(-16)}$ 18. $\dfrac{(3)(5) - (-9)(0)}{(3)(-1) - (5)(0) + (-2)(1)}$

2-4
Symbols of Inclusion

Symbols like parentheses ( ), brackets [ ], braces { }, and bar —, are called symbols of inclusion. These are used extensively in all mathematics. Basically, they are used in grouping numbers or expressions, and each group is usually simplified separately.

EXAMPLE 3

$$(2 + 3) + (4 - 1) = 5 + 3$$
$$= 8$$

$$(6 + 8 + 1) + 2 = 15 + 2$$
$$= 17$$

Sometimes two or more symbols are used in the same expression. When this is the case the general rule for simplifying is to perform the innermost inclusions first and continue until as many symbols of inclusion are removed as possible.

EXAMPLE 4   Simplify $\{2 + [3 + (1 + 4)] + 7\}$

Solution:

$\{2 + [3 + (1 + 4)] + 7\} = \{2 + [3 + 1 + 4] + 7\}$ (removing the parentheses)
$= \{2 + 3 + 1 + 4 + 7\}$ (removing the brackets)
$= 2 + 3 + 1 + 4 + 7$ (removing the braces)
$= 17$

EXAMPLE 5   Simplify $4 + [2(6 + 1)]$

Solution:

$4 + [2(6 + 1)] = 4 + [(12 + 2)]$ (distributive law)
$= 4 + [12 + 2]$ (removing the parentheses)
$= 4 + 12 + 2$ (removing the brackets)
$= 18$

You will notice from the following examples that if a parenthesis is preceded by a negative (minus) sign, each sign is changed when the parenthesis is removed.

EXAMPLE 6 Simplify $6 - (3 - 1)$

Solution: $6 - (3 - 1) = 6 - 3 + 1$
$$= 4$$

EXAMPLE 7 Simplify $10 - 2(4 - 3)$

Solution: $10 - 2(4 - 3) = 10 - 8 + 6$
$$= 8$$

PROBLEM SET 2-4

Simplify the following.

1. $1 + 4 + (3 + 2) =$
2. $(5 + 3) + 7 =$
3. $(6 + 2) + (4 + 8) =$
4. $(5 + 9) - (3 + 1) =$
5. $(2 + 7) - (4 - 3) =$
6. $3 + (2 + 5) - 4 + 7 + (-5) =$
7. $8(5 - 2) + 3(7 - 4) =$
8. $6\{3[2 + (3 - 1)] + 5(7 - 3)\} =$
9. $8\{5 - 2[4 - (7 + 3 - 6)] - 3[4 + (2) - 3(2 + 1)]\} =$
10. $\dfrac{3[2 + 9(-3 + (-2)) + 2(7 - 3)]}{5[2 + 3 + (-4) - (-3) + (-1)]} =$

2-5
Exponents

In an expression such as 3^2 the 3 is called the *base* and the 2 is called an *exponent*. An exponent tells how many times the base is used as a factor. For instance, 3^2 means 3×3. 3^2 is read "three squared" or "three to the second power." Notice the position of the exponent. It is important that it is placed to the upper right hand side of the base and that it is smaller in height than the base.

EXAMPLE 8   (1) 3^4, three to the fourth power; $3 \times 3 \times 3 \times 3$

(2) 5^3, five cubed or the third power of five; $5 \times 5 \times 5$

(3) x^2, x squared; x times x

(4) a^6, the sixth power of a or a to the sixth power; $a \cdot a \cdot a \cdot a \cdot a \cdot a$

Terms such as x^2 or a^6 are algebraic expressions. Other algebraic expressions are y^3, $4x^2$, $1/3a^4$ and $y + 3$.

In an algebraic expression such as $4x^2$, the 4 is called the numerical coefficient (or simply coefficient), the x is the base, and the 2 is the exponent.

The following operations can be performed on algebraic expressions with exponents:

(1) $a^m \cdot a^n = a^{m+n}$

(2) $\dfrac{a^m}{a^n} = a^{m-n}$ if $m > n$ and $a \neq 0$, or

$\dfrac{a^m}{a^n} = \dfrac{1}{a^{n-m}} = a^{m-n}$ if $m < n$ and $a \neq 0$

(3) $(a^m)^n = a^{m \cdot n}$

(4) $(ab)^m = a^m b^m$

(5) $(a)^0 = 1$

EXAMPLE 9

$$a^2 \cdot a^3 = a^{2+3}$$
$$= a^5, \text{ because}$$
$$a^2 \cdot a^3 = (a \cdot a)(a \cdot a \cdot a)$$
$$= a \cdot a \cdot a \cdot a \cdot a$$
$$= a^5$$

EXAMPLE 10

$$\frac{a^3}{a^2} = a^{3-2}$$

$$= a,$$
$$\text{because}$$
$$\frac{a^3}{a^2} = \frac{a \cdot a \cdot a}{a \cdot a}$$
$$= a$$

EXAMPLE 11

$$\frac{a^3}{a^5} = \frac{1}{a^{5-3}}$$

$$= \frac{1}{a^2},$$
$$\text{or}$$
$$\frac{a^3}{a^5} = a^{3-5}$$
$$= a^{-2}$$

EXAMPLE 12 $(a^3)^4 = a^{3 \cdot 4}$
$$= a^{12},$$
because
$$(a^3)^4 = a^3 \cdot a^3 \cdot a^3 \cdot a^3$$
$$= a \cdot a \cdot a \cdot a \cdot a \cdot a \cdot a \cdot a \cdot a \cdot a \cdot a \cdot a$$
$$= a^{12}$$

EXAMPLE 13 $(ab)^4 = a^4 b^4,$

because
$$(ab)^4 = ab \cdot ab \cdot ab \cdot ab$$
$$= a \cdot a \cdot a \cdot a \cdot b \cdot b \cdot b \cdot b$$
$$= a^4 b^4$$

EXAMPLE 14 $(2x^2)(3x^3) = 2 \cdot 3 \cdot x^2 \cdot x^3$
$$= 6x^5$$

EXAMPLE 15 $\dfrac{10a^7}{5a^2} = 2a^5$

EXAMPLE 16 $(a^2)^4 = a^8$

EXAMPLE 17 $(2b^3)^3 = 2^3 b^9$
$$= 8b^9$$

EXAMPLE 18 $\left(\dfrac{4x^2}{b^3}\right)^2 = \dfrac{16x^4}{b^6}$

EXAMPLE 19 $\dfrac{24b^3}{6b^5} = \dfrac{4}{b^2}$ or $4b^{-2}$

EXAMPLE 20 $5^0 = 1$

EXAMPLE 21 $(\tfrac{1}{4})^0 = 1$

Simplify the following expressions.

1. $a^2 \cdot a^5$

2. $\dfrac{b^4}{b^3}$

3. $\dfrac{x^2 \cdot x^7}{x^5}$

4. $\dfrac{4y^3 \cdot 3y^5}{6y^4}$

5. $(Z^3 \cdot Z^4)^2$

6. $(5a^2)(2b)$

7. $(10^3)^2$

8. $\sqrt{16}$

9. $\sqrt{49}$

10. $\dfrac{1}{\sqrt{9}}$

11. $\dfrac{\sqrt{25}}{\sqrt{64}}$

12. $\sqrt{50}$

13. $\dfrac{(3ab^2c)(a^4bc^3)}{a^2b}$

14. $\dfrac{5a^2 + 9a^2 + (-2a^2)}{6a^5}$

15. $\dfrac{\sqrt{81} - \sqrt{36}}{\sqrt{9}}$

16. $(.001x^5)(.0005x^2)$

17. $\dfrac{\sqrt{169}(2^2 \cdot 3)^2}{\sqrt{121} + (2 + 7^0)}$

18. $\sqrt{(5)^4}$

19. $(a^2b^3)(ab)^2$

20. $(1/3x^2)(1/2x^5)(1/8x^0)$

2-6
Algebraic Expression

Algebraic expressions such as x, $3x$, $4y$, $6y^2$, or $2ab$ can be added or subtracted in a manner very similar to operations with real numbers. Remember that in the algebraic expression $3a$, the a is called the base and the 3 is called the coefficient. An algebraic expression such as $3a$ is also a *single term* or a *monomial*. Two algebraic expressions such as $2ab + c^2$ separated by a plus or minus sign form a *binomial*. Three algebraic expressions such as $x^2 + 2x - 1$ which are separated by plus or minus signs are called *trinomials*. Any one of these can be called a *polynomial*. Thus, a monomial is a polynomial, and a trinomial is a polynomial.

Two polynomials can be added or subtracted by adding or subtracting the coefficients if the bases are the same.

EXAMPLE 22

$\quad$ (1) $2x + 5x = 7x$

$\quad$ (2) $4a^2 - 3a^2 = a^2$

$\quad$ (3) $6ab^2 + 2ab^2 - 3ab^2 = 5ab^2$

$\quad$ (4) $2x + 3y = 2x + 3y$

$\quad$ (5) $(2x + y) + (x + 2y) = 3x + 3y$

$\quad$ (6) $(2a + 3b - c) + (a - b + 2c) = 3a + 2b + c$

$\quad$ (7) $(5x + 7y) - (4x + 2y) = x + 5y$

$\quad$ (8) $(x + 4y) - (2x - y) = -x + 3y$

$\quad$ (9) $(x + 2) + (3x - 4) + (4x + 7) = 8x + 5$

$\quad$ (10) $(a + b + 2) + (2a - 3b + 8) = 3a - 2b + 10$

You cannot add coefficients when the bases are not the same.

$\quad$ (11) $(2x + y) + (a - b) = 2x + y + a - b$

$\quad$ (12) $(3\alpha + \theta) - (\gamma - \rho) = 3\alpha + \theta - \gamma + \rho$

PROBLEM SET 2-6

$\quad$ Simplify the following.

1. $3a + 5a$

2. $7x - 3x$

3. $9\alpha^2 - 4\alpha^2$

4. $5x^2y + x^2y$

5. $4\theta + 5\theta - 2\theta$

6. $a^2b^3c + a^2b^3c$

7. $5ax^2 + 3ax^2 - ax^2$

8. $3abc + 4abc + 5ab$

9. $xyz^2 + 3xyz^2 - 9xyz^2$

10. $3\alpha\beta^2 + 12\alpha^2\beta$

11. $c(ab) + a(bc) + b(ac)$

12. $3x^2y - x^2(y) + y(8x^2)$

13. $9\rho\omega + 3\rho\omega + \rho\omega$

14. $5\phi^2\lambda + 3\phi^3\lambda + \phi^2\lambda$

15. $7 + 3y + y$

16. $x(yz) + y(3xz) - z(2xy)$

17. $3(cd) + c(5d) - d(c)$

18. $4a^2b^2c - a^2b^2c - 8a^2b^2c$

19. $(x + y - 3) + (3x - 2y + 4)$

20. $(\omega^2 + \Delta + 7\pi) - (5\omega^2 - 3\Delta + 4\pi)$

2-7
Roots and Fractional Exponents

What is the square root of 25? Does the expression $\sqrt{25}$ mean the same thing? What is the value of $\sqrt{36}$, $\sqrt{49}$ or $\sqrt{64}$? Do you see that $\sqrt{100}$ indicates a figure such that the number squared will equal 100? What is $\sqrt{100}$? Was your answer 10? Could it be -10? What is 10^2? What is $(-10)^2$? The square root of a number can be both positive and negative. However, unless you are asked to find both positive and negative answers it will be understood that only the positive answer is desired. The positive square root of a number is called the *principal root*.

EXAMPLE 23 $\sqrt{25} = 5$ $\sqrt{1/4} = 1/2$

$\sqrt{100} = 10$ $\sqrt{4/25} = 2/5$

What is 3^3, 4^5 or 6^4? If 2^3 equals 8, we can say that the cube root of 8 equals 2. The expression $\sqrt[3]{8}$ is read "the cube root of 8." What is $\sqrt[3]{27}$ or $\sqrt[4]{16}$?

Another important property concerning *radicals* ($\sqrt{}$) is that the square root of a product is the product of their square roots. In other words this property states that $\sqrt{ab} = \sqrt{a} \cdot \sqrt{b}$.

EXAMPLE 24   (1) $\sqrt{18} = \sqrt{9} \cdot \sqrt{2}$

$= 3\sqrt{2}$

This property can be extended to other roots also.

EXAMPLE 25   (1) $\sqrt[3]{54} = \sqrt[3]{27} \cdot \sqrt[3]{2}$

$= 3\sqrt[3]{2}$

(2) $\sqrt[4]{32} = \sqrt[4]{16} \cdot \sqrt[4]{2}$

$= 2\sqrt[4]{2}$

Let us examine some other ways of expressing radicals.

EXAMPLE 26   (1) $\sqrt{100} = \sqrt{10} \cdot \sqrt{10}$

$= (\sqrt{10})^2$

(2) $\sqrt{81} = \sqrt{9 \cdot 9}$

$= \sqrt{9^2}$

(3) $\sqrt{5} = ?$

$\sqrt{5}$ means that $(?)(?) = 5$? Notice that each of the $(?)$ have to be identical. Therefore, we might say that $(?)^2 = 5$. What is $(5^{\frac{1}{2}})(5^{\frac{1}{2}})$? The product is 5. Then we can say $5^{\frac{1}{2}} = \sqrt{5}$.

From Example 26, (3), another important principle concerning radicals may be derived. This principle states that $\sqrt[a]{n^b} = n^{b/a}$.

EXAMPLE 27 (1) $\sqrt{x} = x^{1/2}$, because $x^{1/2} \cdot x^{1/2} = x$

(2) $\sqrt[3]{x} = x^{1/3}$, because $x^{1/3} \cdot x^{1/3} \cdot x^{1/3} = x$

(3) $\sqrt[3]{x^2} = x^{2/3}$, because $x^{2/3} \cdot x^{2/3} \cdot x^{2/3} = x^2$

Suppose we wish to find $\sqrt{\frac{3}{5}}$. One way of simplifying $\sqrt{\frac{3}{5}}$ is called *rationalizing the denominator*. This means making the denominator a rational number. As it is now the denominator is considered to be irrational.

To *rationalize the denominator*, multiply both the numerator and the denominator by some number which will make the denominator rational.

EXAMPLE 28

$$(1) \quad \sqrt{\frac{3}{5}} = \frac{\sqrt{3}}{\sqrt{5}} \cdot \frac{\sqrt{5}}{\sqrt{5}}$$

$$= \frac{\sqrt{3 \cdot 5}}{\sqrt{5 \cdot 5}}$$

$$= \frac{\sqrt{15}}{\sqrt{25}}$$

$$= \frac{\sqrt{15}}{5}$$

$$(2) \quad \sqrt{\frac{2}{3}} = \frac{\sqrt{2}}{\sqrt{3}} \cdot \frac{\sqrt{3}}{\sqrt{3}}$$

$$= \frac{\sqrt{2 \cdot 3}}{\sqrt{3 \cdot 3}}$$

$$= \frac{\sqrt{6}}{\sqrt{9}}$$

$$= \frac{\sqrt{6}}{3}$$

$$(3) \quad \sqrt[3]{\frac{2}{3}} = \frac{\sqrt[3]{2}}{\sqrt[3]{3}} \cdot \frac{\sqrt[3]{9}}{\sqrt[3]{9}}$$

$$= \frac{\sqrt[3]{18}}{\sqrt[3]{27}}$$

$$= \frac{\sqrt[3]{18}}{3} \quad \text{In this example the denominator had to be a perfect cube.}$$

$$(4) \quad \sqrt{\frac{9}{32}} = \frac{3}{\sqrt{32}}$$

$$= \frac{3}{\sqrt{32}} \cdot \frac{\sqrt{2}}{\sqrt{2}}$$

$$= \frac{3\sqrt{2}}{\sqrt{64}}$$

$$= \frac{3\sqrt{2}}{8}$$

PROBLEM SET 2-7

Solve the following.

1. $9^{1 \cdot 2}$ 2. $8^{1/3}$ 3. $16^{1/2}$ 4. $-(64)^{1/3}$

5. $(2^4 \cdot 3^2)^{1/2}$ 6. $125^{1/3}$ 7. $(3^3 \cdot 5^2)$ 8. $(4^3 \cdot 5^2)^{1/2}$

9. $49^{1/2} \cdot 81^{1/2}$ 10. $\left(\dfrac{64x^6}{y^3}\right)^{1/3}$

Simplify the following.

1. $\sqrt{\frac{1}{5}}$ 2. $\sqrt{\frac{1}{3}}$ 3. $\sqrt{\frac{5}{8}}$ 4. $\sqrt{\frac{1}{27}}$ 5. $\sqrt{\frac{8}{64}}$

6. $\sqrt{\frac{1}{2}}$ 7. $\sqrt{\frac{2}{5}}$ 8. $\sqrt{\frac{1}{8}}$ 9. $\sqrt{\dfrac{3x}{5y}}$ 10. $\dfrac{9\sqrt{6}}{\sqrt{3}}$

2-8
Fractions

At this point before stating the basic rules for working with fractions it is necessary to define a fraction and some different types of fractions.

A fraction is a comparison or ratio of two numbers. Suppose we wish to make a comparison of 1 and 2. One way of writing this comparison is $\frac{1}{2}$. The number $\frac{1}{2}$ represents a fraction. A fraction has two parts. The number above the line or bar (—) is called the numerator, the number below the line is called the denominator. Thus in the fraction $\frac{3}{4}$, 3 is the numerator and 4 is the denominator.

2-9
Types of Fractions

There are five types of fractions which will be discussed here. These are (1) proper fractions, (2) improper fractions, (3) complex fractions, (4) mixed numbers, and (5) decimal fractions.

A *proper fraction* is one in which the numerator is smaller than the denominator. For example, $\frac{1}{2}$, $\frac{2}{3}$, $\frac{7}{9}$, and $\frac{12}{15}$ are proper fractions.

An *improper fraction* is one in which the numerator is larger than the denominator. For example, $\frac{5}{3}$, $\frac{7}{4}$, $\frac{14}{5}$ are improper fractions.

A *complex fraction* is a fraction in which the numerator, the denominator, or both, are also fractional. For example, $\dfrac{\frac{1}{2}}{3}$, $\dfrac{2}{\frac{2}{3}}$, and $\dfrac{\frac{5}{6}}{\frac{3}{7}}$, are complex fractions.

A *mixed number* is the sum of a whole number and a fraction. For example, $1\frac{1}{2}$, $2\frac{3}{4}$, and $19\frac{2}{3}$ are mixed numbers.

A *decimal fraction* is a number which contains a decimal point. For example, 1.5, 2.75, .021, and 29.001 are decimal fractions.

Write a fraction for the following, and in each case identify the numerator and the denominator.

1. one-half

2. two-thirds

3. The ratio of 3 to 5.

4. The ratio of 9 to 10.

5. A comparison of 13 and 8.

6. A comparison of 8 and 13.

Identify the following fractions according to the type (proper, improper, mixed, complex, or decimal fraction.)

1. $\frac{2}{9}$ 2. $\frac{4}{9}$ 3. 1.25   4. $\dfrac{\frac{1}{2}}{3}$ 5. $1\frac{1}{3}$

6. $\frac{5}{4}$ 7. $\frac{2}{3}$ 8. $1\frac{4}{7}$ 9. 4.09   10. $\dfrac{\frac{3}{4}}{\frac{7}{8}}$

11. .006   12. $\dfrac{7}{1\frac{1}{5}}$ 13. $\frac{9}{8}$ 14. $\frac{25}{30}$ 15. $10\frac{5}{9}$

2-10
Reducing Fractions to Lowest Terms

To reduce a fraction is simply a matter of determining a number that will divide both the numerator and denominator of the fraction. For example, to reduce the fraction $\frac{12}{36}$ it is possible to divide both numbers by 4. The result is $\frac{3}{9}$. If you divide both numbers by 6, the result is $\frac{2}{6}$. There are other numbers which will divide both 12 and 36. In each case this would be reducing the fraction.

Usually, when fractions are reduced, they are reduced to *lowest terms*. A fraction is in lowest terms when it is not possible to divide the numerator or the denominator by the same number other than 1. Hence, $\frac{12}{36}$ in lowest terms would be $\frac{1}{3}$.

EXAMPLE 29   Reduce $\frac{7}{21}$ to lowest terms.

Solution: The problem here is to find a number which will divide both 7 and 21. By examination you can see the 7 will divide both.

$$\frac{7}{21} = \frac{7 \div 7}{21 \div 7}$$

$$= \frac{1}{3}, \text{ hence}$$

$$\frac{7}{21} = \frac{1}{3} \text{ in lowest terms.}$$

EXAMPLE 30 Reduce $\frac{15}{50}$ to lowest terms.

Solution: In examining both 15 and 50, we see that each is divisible by 5, therefore

$$\frac{15}{50} = \frac{15 \div 5}{50 \div 5}$$

$$= \frac{3}{10} \text{, hence}$$

$$\frac{15}{50} = \frac{3}{10} \text{, in lowest terms.}$$

EXAMPLE 31 Reduce $\dfrac{12x^2}{15x}$ to lowest terms.

Solution: In the problem we see that 3 will divide both 12 and 15, but we must also apply the rules for exponents to reduce the x^2 and x, therefore

$$\frac{12x^2}{15x} = \frac{(12 \div 3)x^2}{(15 \div 3)x}$$

$$= \frac{4x^{2-1}}{5}$$

$$= \frac{4x}{5}$$

PROBLEM SET 2-10

Reduce the following fractions to lowest terms.

1. $\frac{2}{4}$

2. $\frac{3}{9}$

3. $\frac{5}{25}$

4. $\dfrac{2x}{6}$

5. $\frac{12}{30}$

6. $\frac{10}{24}$

7. $\dfrac{30}{150}$

8. $\frac{9}{30}$

9. $\dfrac{8a^2}{12a}$

10. $\dfrac{26\alpha}{40\alpha}$

11. $\dfrac{18}{20\omega}$

12. $\frac{96}{144}$

2-11
Equivalent Fractions

Equivalent fractions are fractions which are equal. Examples of equivalent fractions are $\frac{1}{2}$ and $\frac{3}{6}$, or $\frac{3}{4}$ and $\frac{6}{8}$. The simplest way to determine whether two fractions are equivalent or not is to reduce both fractions to lowest terms. If the results are the same then the original fractions are equivalent fractions.

EXAMPLE 32 Are $\frac{1}{4}$ and $\frac{2}{8}$ equivalent?

Solution: To determine if $\frac{1}{4} = \frac{2}{8}$, reduce both fractions to lowest terms. Since $\frac{1}{4}$ is in lowest terms it is necessary only to reduce $\frac{2}{8}$ to lowst terms.

$$\frac{2}{8} = \frac{1}{4}$$

in lowest terms, hence

$$\frac{1}{4} = \frac{2}{8}$$

EXAMPLE 33 Are $\frac{4}{8}$ and $\frac{6}{12}$ equivalent?

Solution: Reduce each fraction to lowest terms.

$$\frac{4}{8} = \frac{1}{2} \quad \text{and} \quad \frac{6}{12} = \frac{1}{2}$$

Since each fraction is equal to $\frac{1}{2}$ in lowest terms they are equivalent, hence

$$\frac{4}{8} = \frac{6}{12}$$

PROBLEM SET 2-11

Determine which of the following are equivalent fractions.

1. $\frac{1}{4} = \frac{3}{12}$ 2. $\frac{2}{3} = \frac{6}{9}$ 3. $\frac{9}{13} = \frac{27}{42}$

4. $\frac{2}{9} = \frac{6}{18}$ 5. $\frac{5}{8} = \frac{25}{40}$ 6. $\frac{9}{11} = \frac{27}{44}$

7. $\frac{7}{3} = \frac{35}{18}$ 8. $\frac{5}{9} = \frac{40}{72}$ 9. $\frac{3}{8} = \frac{30}{80}$

10. $\frac{12}{20} = \frac{3}{5}$ 11. $\frac{56}{81} = \frac{7}{9}$ 12. $\frac{30}{15} = \frac{3}{2}$

2-12
Finding Equivalent Fractions

Many times it is necessary to change a fraction to another fraction of equal value. This is true very often when adding or subtracting fractions. When you perform an operation of this type you have found a fraction which is equivalent to the original fraction.

Suppose you are asked the question "One-half is equal to how many eighths?" That is, $\frac{1}{2} = \frac{?}{8}$. To solve this example, divide the 8 by 2 and multiply the result, 4, by the numerator in the fraction $\frac{1}{2}$, which is 1. This product is 4. Replace the question mark with this product, 4, and you have now changed one-half to four-eighths. This would be written as: $\frac{1}{2} = \frac{4}{8}$.

EXAMPLE 34
$$\frac{1}{3} = \frac{?}{18}$$

Solution:
$$\frac{1}{3} = \frac{}{18}$$

Divide 18 by 3. The result is 6. Multiply 6 by the numerator in $\frac{1}{3}$, which is 1. $1 \times 6 = 6$. Hence, 6 is the number which goes above the 18. Therefore:

$$\frac{1}{3} = \frac{6}{18}$$

EXAMPLE 35
$$\frac{2}{5} = \frac{?}{15}$$

Solution:
$$\frac{2}{5} = \frac{}{15}$$

Divide 15 by 5. $15 \div 5 = 3$.
Multiply 3 by 2. $2 \times 3 = 6$. Therefore:

$$\frac{2}{5} = \frac{6}{15}$$

EXAMPLE 36
$$\frac{3}{4} = \frac{?}{36}$$

Solution:
$$\frac{3}{4} = \frac{}{36}$$

Divide 36 by 4. $36 \div 4 = 9$.
Multiply 9 by 3. $3 \times 9 = 27$. Therefore:

$$\frac{3}{4} = \frac{27}{36}$$

EXAMPLE 37
$$\frac{?}{3} = \frac{16}{24}$$

Solution:
$$\frac{}{3} = \frac{16}{24}$$

This is a little different from the previous examples because it involves a reverse process. You must determine how many times larger 24 is than 3. $24 = 8 \times 3$, hence 24 is 8 times as large as 3. This means that 16 must be 8 times as large as the number that goes in the space above the 3. 16 is 8 times as large as 2. Therefore:

$$\frac{2}{3} = \frac{16}{24}$$

EXAMPLE 38
$$\frac{?}{7} = \frac{10}{35}$$

Solution:
$$\frac{-}{7} = \frac{10}{35}$$

This problem is solved exactly like *Example 37*. 35 is 5 times as large as 7 and 10 is 5 times as large as 2, therefore:

$$\frac{2}{7} = \frac{10}{35}$$

PROBLEM SET 2-12

Change each of the following to an equivalent fraction with the indicated denominator.

1. $\frac{1}{2} = \frac{}{8}$

2. $\frac{2}{3} = \frac{}{12}$

3. $\frac{4}{5} = \frac{}{20}$

4. $\frac{2x}{5} = \frac{}{10}$

5. $\frac{}{5} = \frac{15}{25}$

6. $\frac{9}{10} = \frac{}{70}$

7. $\frac{8}{3} = \frac{}{24}$

8. $\frac{12}{30} = \frac{}{5}$

9. $\frac{20}{40} = \frac{}{4}$

10. $\frac{3x}{7} = \frac{}{14}$

11. $\frac{7}{35} = \frac{}{5}$

12. $\frac{}{16} = \frac{3}{8}$

13. $\frac{}{4} = \frac{18}{24}$

14. $\frac{1}{2x} = \frac{}{6x}$

15. $\frac{}{6} = \frac{24}{48}$

16. $\frac{}{12} = \frac{3}{36}$

17. $\frac{7}{8} = \frac{}{64}$

18. $\frac{1}{x^2} = \frac{}{4x^2}$

19. $\frac{2}{(a + b)} = \frac{}{3(a + b)}$

20. $\frac{3}{x} = \frac{}{x^3}$

2-13
Signs of a Fraction

There are three signs to every fraction. The numerator, denominator, and the fraction itself have signs. For example, in the fraction $\frac{3}{4}$, the numerator is positive, the denominator is positive and the fraction itself (the sign in front of the fraction) is positive. In the fraction $-\frac{-2}{+3}$, the numerator is negative, the denominator is positive, and the sign of the fraction is negative.

EXAMPLE 39 In the following fractions, determine the sign of the (1) numerator, (2) denominator, (3) fraction.

$$\text{a.} \quad -\frac{-4}{+5} \qquad\qquad \text{b.} \quad \frac{6}{-7} \qquad\qquad \text{c.} \quad -\frac{2}{9}$$

a. (1) negative	b. (1) positive	c. (1) positive
(2) positive	(2) negative	(2) positive
(3) negative	(3) positive	(3) negative

2-14
Negative of a Fraction

The negative, or opposite, of a fraction can be found by multiplying the fraction by a negative one (-1).

EXAMPLE 40   (1) Find the negative of $\frac{3}{5}$

$$(-1)(\tfrac{3}{5}) = -\tfrac{3}{5}$$

(2) Find the negative of $\dfrac{x}{7}$

$$(-1)\left(\frac{x}{7}\right) = -\frac{x}{7}$$

(3) Find the negative of $(\frac{1}{2})$

$$(-1)(\tfrac{1}{2}) = -\tfrac{1}{2}$$

Notice, that the negative of a fraction can be found by changing one sign of the fraction.

EXAMPLE 41

(1) The negative of $\dfrac{1}{-3}$ is $\dfrac{1}{3}$

(2) The negative of $\dfrac{-2}{5}$ is $\dfrac{2}{5}$

(3) The negative of $\dfrac{4}{9}$ is $-\dfrac{4}{9}$

(4) The negative of $\dfrac{x-1}{3} = \dfrac{-(x-1)}{3}$

(5) The negative of $\dfrac{3x-2}{4} = \dfrac{-3x+2}{4}$

To *add like fractions* (fractions having a common denominator), add the numerators and place the sum over the common denominator.

EXAMPLE 42

(1) $\dfrac{1}{8} + \dfrac{3}{8} = \dfrac{1 + 3}{8} = \dfrac{4}{8}$

(2) $\dfrac{3}{7} + \dfrac{2}{7} = \dfrac{5}{7}$

(3) $\dfrac{x + 1}{5} + \dfrac{2x - 3}{5} = \dfrac{3x - 2}{5}$

(4) $\dfrac{2a - b}{c + d} + \dfrac{4a + 2b}{c + d} = \dfrac{6a + b}{c + d}$

To subtract like fractions, change the signs of the numerator in the subtrahend and add the two numerators. Place the sum over the common denominator.

EXAMPLE 43

(1) $\dfrac{2}{3} - \dfrac{1}{3} = \dfrac{2 + (-1)}{3} = \dfrac{1}{3}$

(2) $\dfrac{3}{7} - \dfrac{-5}{7} = \dfrac{3}{7} + \dfrac{+5}{7} = \dfrac{3 + 5}{7} = \dfrac{8}{7}$

(3) $\dfrac{5x - 1}{3} - \dfrac{2x + 3}{3} = \dfrac{5x - 1 + (-2x - 3)}{3}$

$= \dfrac{5x - 1 - 2x - 3}{3}$

$= \dfrac{3x - 4}{3}$

Suppose you are trying to add or subtract *unlike fractions* (fractions having unlike denominators). First you must change each fraction to a common denominator and then proceed with the above rules.

EXAMPLE 44

(1) $\dfrac{1}{2} + \dfrac{1}{4} = \dfrac{2}{4} + \dfrac{1}{4} = \dfrac{3}{4}$

(2) $\dfrac{1}{3} + \dfrac{1}{6} = \dfrac{2}{6} + \dfrac{1}{6} = \dfrac{3}{6} = \dfrac{1}{2}$

(3) $\dfrac{1}{2} - \dfrac{1}{3} = \dfrac{3}{6} - \dfrac{2}{6} = \dfrac{3 + (-2)}{6} = \dfrac{1}{6}$

(4) $\dfrac{x + 1}{3} + \dfrac{x + 2}{6} = \dfrac{2(x + 1)}{6} + \dfrac{x + 2}{6}$

$= \dfrac{2x + 2}{6} + \dfrac{x + 2}{6} = \dfrac{3x + 4}{6}$

Perform the indicated addition and subtraction.

1. $\frac{1}{4} + \frac{1}{4} =$

2. $\frac{5}{6} - \frac{1}{6} =$

3. $\frac{1}{3} + \frac{1}{2} =$

4. $\frac{4}{5} - \frac{1}{4} =$

5. $\dfrac{1}{2x} + \dfrac{2}{3x} =$

6. $\frac{5}{2} - \frac{1}{8} + \frac{1}{4} =$

7. $\dfrac{3}{16a} + \dfrac{9}{4a} - \dfrac{5}{32a} =$

8. $\dfrac{7}{8} + \dfrac{-4}{7} + \dfrac{5}{14} =$

9. $2\frac{1}{2} + 3\frac{1}{3} - 1\frac{1}{6} =$

10. $3\frac{5}{7} + 5\frac{1}{2} + 4\frac{2}{3} =$

11. $\frac{23}{40} + \frac{1}{8} + \frac{3}{5} - \frac{19}{20} =$

12. $3\frac{5}{9} + 12\frac{3}{5} - 9\frac{5}{6} =$

13. $\dfrac{3}{4e} + \dfrac{5}{8e} - \dfrac{1}{10e} =$

14. $\frac{3}{5} + \frac{19}{11} + \frac{5}{2} - \frac{1}{4} =$

15. $6\frac{7}{10} + 12\frac{1}{2} - 5\frac{7}{15} =$

2-16
Multiplication of Fractions

To multiply fractions multiply the numerators (this will be the numerator of the answer) and multiply the denominators (this will be the denominator of the answer), then reduce the fraction to lowest terms.

EXAMPLE 45   Find the product of $\frac{1}{2} \times \frac{2}{3}$.

Solution:
$$\frac{1}{2} \times \frac{2}{3} = \frac{1 \times 2}{2 \times 3}$$
$$= \frac{2}{6}$$
$$= \frac{1}{3}$$

EXAMPLE 46   Find the product of $\dfrac{2x}{3} \cdot \dfrac{4}{5x^2}$.

Solution:
$$\frac{2x}{3} \cdot \frac{4}{5x^2} = \frac{(2x)(4)}{(3)(5x^2)}$$
$$= \frac{8x}{15x^2}$$
$$= \frac{8}{15x}$$

Find the product of the following.

1. $\dfrac{a^2b^2}{a^2b} \cdot \dfrac{ab^2}{4ab}$

2. $\dfrac{6x^2y}{7b} \cdot \dfrac{21b^2d}{24xy^2}$

3. $\dfrac{7e^3iz}{9er^2} \cdot \dfrac{27er^2}{14e^2z}$

4. $\dfrac{5ab^2}{2b} \cdot \dfrac{1}{5(a-b)}$

5. $\dfrac{15x^4y^4}{48x^8y^9} \cdot \dfrac{64x^3y^3}{75x^7y^2}$

6. $\dfrac{8r^2}{5v^2} \cdot \dfrac{15w^3}{14r} \cdot \dfrac{7v^2}{24w}$

7. $\dfrac{5K1}{3K^51} \cdot \dfrac{12K^41^3}{25K1}$

8. $\dfrac{3(b-1)}{2ab} \cdot \dfrac{6a^2b^2}{4(b-1)}$

2-17
Division of Fractions

To divide two fractions invert the divisor and multiply.

EXAMPLE 47   Divide $\dfrac{8x^2y}{21ab}$ by $\dfrac{2x}{3b}$

Solution: $\dfrac{8x^2y}{21ab} \div \dfrac{2x}{3b} = \dfrac{8x^2y}{21ab} \cdot \dfrac{3b}{2x}$

$$= \dfrac{\overset{4}{\cancel{8}}x^2y}{\underset{7}{\cancel{21}}\cancel{ab}} \cdot \dfrac{\overset{1}{\cancel{3b}}}{\underset{1}{\cancel{2x}}}$$

$$= \dfrac{4xy}{7a}$$

EXAMPLE 48   Simplify $\dfrac{15xy^2}{14ac} \div \dfrac{3xy}{7c}$

Solution: $\dfrac{15xy^2}{14ac} \div \dfrac{3xy}{7c} = \dfrac{15xy^2}{14ac} \cdot \dfrac{7c}{3xy}$

$$= \dfrac{\overset{5}{\cancel{15}}\overset{y}{x}\cancel{y^2}}{\underset{2}{\cancel{14}}a\cancel{c}} \cdot \dfrac{\overset{1}{\cancel{7c}}}{\cancel{3xy}}$$

$$= \dfrac{5y}{2a}$$

Find the quotient of the following.

1. $\dfrac{4ax^2}{15b} \div \dfrac{6ax}{5b^2}$

2. $\dfrac{5}{6u^3} \div \dfrac{7}{12u^8}$

3. $\dfrac{32e^2}{70K^7} \div \dfrac{24e^6}{35K^3}$

4. $\dfrac{450e^3}{21d^2e^7} \div \dfrac{400^4}{112de^5}$

5. $\dfrac{10abc}{9a} \div \dfrac{25c^3}{45b^3}$

6. $\dfrac{z^3}{14a^5c^2} \div \dfrac{z}{7a^6c^9}$

7. $\dfrac{x^2}{2(x-1)} \div \dfrac{x^3}{x(x-1)}$

8. $\dfrac{(z-3)}{3(z+3)} \div \dfrac{(z+3)}{(z-3)(z+3)}$

2-18
Linear Equations

Sometimes we let x stand for a number or unknown quantity. For instance, let x equal the number of chairs in a room. Then $2x$ would represent 2 times the number of chairs in the room. Let b equal the number of hours that you spend in school per day. Now suppose you have a friend that spends 2 hours a day more in school than you. Then you could say that your friend spends $b + 2$ hours per day in school. These are literal representations of numbers. Give a literal representation of each of the following.

1. If John is x years old now, how old will he be in 2 years?
   Ans. $x + 2$
2. If George has y dollars in the bank and he wants to double the amount, how many dollars will he have in the bank?
   Ans. $2y$
3. If you are driving 50 m.p.h. in a speed zone in which you should be driving d m.p.h. slower, what is the m.p.h. of the speed zone?
   Ans. $50 - d$

Expressions like these and others can help us to solve problems. Suppose someone asks you a question such as "I am thinking of a number such that if you add 6 to the number the sum will be 17. What is the number?" Of course you can find the answer by subtraction, but do you see that if you let x represent the number you write $x + 6 = 17$. This is called an equation. An *equation* is an expression stating that two things are equal.

EXAMPLE 49   (1) $x = 3$ (3) $2x - 5 = x + 1$

(2) $x + 1 = 7$ (4) $\dfrac{x}{3} = 7$

The above expressions are examples of linear equations. A *linear equation* is an equation in which the exponent of the unknown is one.

An equation can be thought of as being balanced, since one side is equal to the other side. Therefore, an operation can be performed on one side of an equation providing that the same operation is performed on the other side. For example, given the equation $x + 3 = 10$ you might add (-3) to both sides resulting in the following:

$$x + 3 = 10$$
$$x + 3 + (-3) = 10 + (-3)$$

Now combine similar terms and you have $x = 7$ which solves the equation for x.

Suppose you have the equation $2x - 4 = 8$ and add 4 to both sides:

$$2x - 4 = 8$$
$$2x - 4 + (4) = 8 + 4$$

Combine similar terms and you have $2x = 12$; however the problem is not completely solved.

If you were to divide both sides by 2

$$\frac{2x}{2} = \frac{12}{2}$$

this results in $x = 6$ which solves the equation for x.

The above examples show some of the operations which can be used in solving equations. Three of the basic operations are as follows:

1. The same number may be added or subtracted from both sides of an equation.
2. Both sides of an equation may be multiplied by the same number.
3. Both sides of an equation may be divided by the same non-zero number.

EXAMPLE 50

$$(1)\ \text{Solve } x - 2 = 4$$
$$x - 2 + (2) = 4 + (2) \quad \text{(add 2 to both sides)}$$
$$x = 6$$

Check: $\qquad (6) - 2 = 4$
$$4 = 4$$

$$(2)\ \text{Solve } 2x = 10$$
$$\frac{2x}{2} = \frac{10}{2} \quad \text{(divide both sides by 2)}$$
$$x = 5$$

Check: $\qquad 2x = 10$
$$2(5) = 10$$
$$10 = 10$$

(3) Solve $2x + 4 = 16$

$$2x + 4 + (-4) = 16 + (-4) \quad \text{(add } (-4) \text{ to both sides)}$$

$$2x = 12$$

$$\frac{2x}{2} = \frac{12}{2} \quad \text{(divide both sides by 2)}$$

$$x = 6$$

Check: $\qquad 2x + 4 = 16$

$$2(6) + 4 = 16$$

$$12 + 4 = 16$$

$$16 = 16$$

(4) Solve $\dfrac{10}{x} = 2$

$$(x)\frac{10}{x} = 2(x) \quad \text{(multiply each side by } x\text{)}$$

$$10 = 2x$$

$$\frac{10}{2} = \frac{2x}{2} \quad \text{(divide each side by 2)}$$

$$5 = x$$

Check: $\qquad \dfrac{10}{x} = 2$

$$\frac{10}{5} = 2$$

(5) Solve $\dfrac{15}{x} + 1 = 4$

$$\frac{15}{x} + 1 + (-1) = 4 + (-1) \quad \text{(add } (-1) \text{ to both sides)}$$

$$\frac{15}{x} = 3$$

$$(x)\frac{15}{x} = 3(x) \quad \text{(multiply both sides by } x\text{)}$$

$$15 = 3x$$

$$\frac{15}{3} = \frac{3x}{3}$$

$$5 = x$$

Check: $\qquad \dfrac{15}{x} + 1 = 4$

$$\frac{15}{5} + 1 = 4$$

$$3 + 1 = 4$$

$$4 = 4$$

EXAMPLE 51   The sum of twice a number and 7 is 41. Find the number.

Solution:

$$x = \text{the number}$$
$$2x = \text{twice the number}$$
$$2x + 7 = 41$$
$$2x + 7 + (-7) = 41 + (-7)$$
$$2x = 34$$
$$x = 17$$

Check:

$$2x + 7 = 41$$
$$2(17) + 7 = 41$$
$$34 + 7 = 41$$
$$41 = 41$$

EXAMPLE 52   The sum of two numbers is 75. One number is 21 more than the other number. Find the numbers.

Solution:

$$x = \text{small number}$$
$$75 - x = \text{large number}$$
$$75 - x = x + 21$$

Since the large number is 21 more than the small number, 21 must be added to the small number to make it equal to the large number.

$$75 - x + (-21) = x + 21 + (-21) \quad \text{(adding } -21 \text{ to both sides)}$$
$$54 - x = x$$
$$54 = 2x$$

$$\frac{54}{2} = \frac{2x}{2} \quad \text{(dividing both sides by 2)}$$

$$27 = x$$

Check:

$$75 - x = x + 21$$
$$75 - (27) = (27) + 21$$
$$48 = 48$$

EXAMPLE 53   Bob is 8 years older than his brother Bill. In 4 years he will be twice as old. What are their ages now?

Solution:

$$\text{Let } x = \text{Bill's age now}$$
$$\text{and } x + 8 = \text{Bob's age now}$$
$$x + 4 = \text{Bill's age in 4 years}$$
$$(x + 8) + 4 = \text{Bob's age in 4 years}$$

In 4 years Bob will be twice as old as Bill, therefore multiply "Bill's age in 4 years" by 2 and it will equal "Bob's age in 4 years."

$$2(x + 4) = (x + 8) + 4$$
$$2x + 8 = x + 12$$
$$2x + 8 + (-8) = x + 12 + (-8)$$
$$2x = x + 4$$

$$2x + (-x) = x + (-x) + 4$$
$$x = 4 \quad \text{(Bill's age now)}$$
$$x + 8 = 12 \quad \text{(Bob's age now)}$$

EXAMPLE 54 Because of ice on the roads the firemen had to reduce their normal speed to a fire by 30 m.p.h. and required 10 minutes to drive 4 miles. What is the normal driving speed?

Solution:
$$\text{Let } x = \text{normal driving speed}$$
$$x - 30 = \text{speed on iced roads}$$
$$10 \text{ min.} = \tfrac{1}{6} \text{ hr.} = \text{time}$$
$$\tfrac{1}{6}(x - 30) = 4$$
$$\tfrac{1}{6}(x) - \tfrac{1}{6}(30) = 4$$

$$\frac{x}{6} - 5 = 4$$

$$\frac{x}{6} = 9$$

$$x = 54 \quad \text{(normal driving speed)}$$

EXAMPLE 55 An airplane which can travel 300 m.p.h. in still wind made a trip with a tail wind in 2 hrs. The return trip against the wind required 3 hrs. Find the speed of the wind.

Solution:
$$\text{Let } x = \text{speed of wind}$$
$$300 + x = \text{speed of plane with the wind}$$
$$300 - x = \text{speed of plane against the wind}$$
$$2(300 + x) = \text{distance traveled with the wind}$$
$$3(300 - x) = \text{distance traveled against the wind}$$

Since the distance traveled with the wind is equal to the distance traveled against the wind then:
$$2(300 + x) = 3(300 - x)$$
$$600 + 2x = 900 - 3x$$
$$600 + (-600) + 2x = 900 + (-600) - 3x$$
$$2x = 300 - 3x$$
$$2x + (3x) = 300 - 3x + (3x)$$
$$5x = 300$$

$$\frac{5x}{5} = \frac{300}{5}$$

$$x = 60 \quad \text{speed of the wind}$$

In many actual situations a linear relationship exists between two physical parameters such that if one parameter is changed by some amount the second parameter changes by a proportional amount. The following examples illustrate this point.

EXAMPLE 56 According to Newton's second law of motion, a force of 5 newtons will cause a body whose mass is 1 kilogram to be accelerated at 5 meters/second². If the force is doubled to 10 newtons by how much will the acceleration change?

Solution: Newton's 2nd law of motion states $F = ma$, therefore

$$a_1 = \frac{F_1}{m} = \frac{5 \text{ newtons}}{1 \text{ kg}} = 5 \text{ meters/sec}^2$$

$$a_2 = \frac{F_2}{m} = \frac{10 \text{ newtons}}{1 \text{ kg}} = 10 \text{ meters/sec}^2$$

EXAMPLE 57 A battery is to be used to supply current to a starter on an airplane engine. It was found that a 28 volt battery will supply only half the required current. How much will the voltage need to be increased to supply the required current?

Solution: Ohm's law states $E = IR$, therefore

$$28 \text{ volts} = (\tfrac{1}{2}I_x)(R) \text{ where } I_x \text{ is the required current.}$$

Multiplying both sides by 2 gives

$$2(28 \text{ volts}) = 2(\tfrac{1}{2})(I_x)(R)$$
$$56 \text{ volts} = I_x R, \text{ therefore the voltage must be doubled.}$$

EXAMPLE 58 A hose, such as might be used by firemen, will supply water at a rate $R = VA$. If the rate of flow, R, is to be doubled, how much must the cross-sectional area of the hose be increased?

Solution: $$R_1 = VA_1 \quad \text{or} \quad A_1 = \frac{R_1}{V}$$

R_1 is to double, therefore

$$2A_1 = \frac{2R_1}{V} \quad \text{or} \quad A_2 = \frac{2R_1}{V}$$

Therefore $A_2 = 2A_1$

Solve for the unknown in the following equations.

1. $x = 9 + 5$
2. $y - 2 = 7$
3. $\Delta + 3 = 10$
4. $-x + (-5) = 2$
5. $3(x - 2) = 15$
6. $K(3 + 5) = 24$
7. $6(\alpha - 2) = 2\alpha$
8. $\dfrac{x - 3}{4} = x$

9. $7 - 3(\beta - 1) = -2$

10. $8\rho + (2\rho - 3)(-2) = 22$

11. $5(k - 6) + (-2) = k$

12. $\dfrac{(\gamma + 3)(2 + (-4))}{2\gamma} = 5$

13. $5\omega + (-6)(\omega - 3) = 8\omega$

14. $\dfrac{(d - 3)(6)}{d + 7} = \dfrac{9 + (-3)}{2(4 + (-1))}$

15. $\dfrac{50 + (-5)}{2 + (-3)(3 + 1)} = (\theta + 5)$

PROBLEM SET 2-17

Using the equation $E = IR$ solve for the missing letter in the following problems.

1. $E = 24$ volts, $R = 6$ ohms
2. $I = 7$ amps, $R = 4$ ohms
3. $E = 400$ volts, $I = 0.1$ amps
4. $E = 1.5$ volts, $R = 300$ ohms
5. $I = .002$ amps, $R = 1000$ ohms

Using the equation $F = ma$ solve for the missing letter in the following problems.

1. $F = 80$ lbs., $m = 20$ slugs
2. $F = 10$ n, $m = 2$ Kg
3. $F = 54$ n, $a = 3$ m/sec^2
4. $m = 15$ slugs, $a = 5$ ft/sec^2
5. $F = 150$ n, $a = 0.5$ m/sec^2

Using the equation $W = Fd$ solve for the missing letter in the following problems.

1. $F = 120$ lbs, $d = 3$ ft
2. $W = 700$ ft·lbs, $F = 35$ lbs
3. $W = 500$ joule, $d = 2$ m
4. $F = 180$ n, $d = .5$ m
5. $W = 625$ in.·lbs, $d = 5$ in.

PROBLEM SET 2-18

1. Two numbers total 60. One number is 4 times the other. Determine the numbers.

2. One number is 3 times another and their difference is 50. Determine the numbers.

3. Two numbers total 98. The numbers differ by 10. Determine the numbers.

4. The product of two numbers is 128. If one is twice the other determine the numbers.

5. The sum of three numbers is 147. If the second number is twice the first and the third number is twice the second determine the numbers.

6. Two gears have a total of 84 teeth. One has $\frac{3}{4}$ as many teeth as the other. Determine the number of teeth on each.

7. Three resistors have a total resistance of 775 ohms. If the second resistor has 25 ohms more resistance than the first, and the third has three times the resistance of the first, determine the value of the resistors.

8. A hydraulic circuit contains a total of 116 inches of hose. The second hose is four times the length of $\frac{1}{3}$ the first. The third hose is five times the length of $\frac{1}{2}$ the first. Determine the length of each hose.

9. Mary is seven years older than her brother John. In three years she will be twice his age. Determine their present age.

10. Paul's father is four times his age. In fourteen years he will be twice Paul's age. Determine their present age.

11. The value of a number of pennies, nickels and dimes is $1.80. The total value of the nickels equals the value of the dimes but is four times the value of the pennies. Determine the number of each coin.

12. An airplane flies 1500 miles in 2.5 hours. Due to wind resistance the return trip takes $\frac{1}{5}$ more time. Determine the speed of the airplane on the return trip.

13. An airplane flies 1800 miles in 3 hours with a tailwind of 50 m.p.h. The return trip requires 3.6 hours. Determine the speed of the airplane with no wind.

14. The sum of the current through three parallel resistors is 18 amperes. Resistor A passes $\frac{1}{3}$ as much current as resistor C while resistor B passes twice as much current as A. Determine the current through each resistor.

15. Gear A makes 2.5 times as many revolutions per minute as gear B which makes $\frac{2}{3}$ as many revolutions per minute as gear C. The total for the three gears is 4000 r.p.m. Determine the revolutions per minute for each gear.

16. The kilogram is the unit of mass in the metric system. The total mass of three blocks of copper is 36 kilograms. The mass of block A is half that of block B and the mass of block C is 1.5 times that of block A. What is the mass of each block?

17. Two columns of a punched tape for a numerical control machine contain a total of 90 holes. One column contains 4 times as many holes as the other. How many holes are in each column?

18. In a triangle commonly used by a draftsman, angle A is $\frac{2}{3}$ of angle B, and angle C is $\frac{1}{2}$ of angle A. Find the angles of the triangle.

19. In a machine shop one press can punch the number of holes required in a stack of metal plates in two hours. Another press requires three hours. How long will the work require if both machines are used together?

20. Three transducers in a control system are intended to have identical resistance at 80°F, however in troubleshooting the system an electro-mechanical technician finds that transducer A has $\frac{13}{10}$ as much resistance as transducer B and C has $\frac{8}{5}$ more resistance than B. The total resistance of the three is 2340 ohms. If this is the correct total resistance at 80°F do any of the transducers have the correct resistance? What is the resistance of each transducer?

PROBLEM SET 2-19

The following formulas are from the technical or scientific areas listed in the right hand column. Solve each for the letter indicated.

Formula	Solve for	Description	Technical Area
1. $V = V_0 + at$	a	Speed of a uniformly accelerating body	Physics
2. $E = IR$	R	OHM's law	Electricity
3. $Q = \dfrac{P_1 - P_2}{R}$	P_2	Rate of flow in a hydraulic circuit	Fluids
4. $D = D_0 \dfrac{N}{N + 2}$	D_0	Pitch diameter of a gear wheel	Mechanics
5. $X_c = \dfrac{1}{2\pi FC}$	C	Reactance of a Capacitor	Electricity
6. $\dfrac{P_1 V_1}{T_1} = \dfrac{P_2 V_2}{T_2}$	T_2	Ideal Gas Law	Chemistry
7. $hf = eE_w + \dfrac{mv^2}{2}$	m	Photoelectric effect	Electronics
8. $R = K\dfrac{L}{A}$	A	Strain gauge resistance	Mechanics
9. $P_2 - P_1 = \left(1 + \dfrac{A_t}{A_w}\right) dh$	A_w	Differential pressure measurement using a manometer	Instrumentation

Formula	Solve for	Description	Technical Area
10. $e = 61.6 \dfrac{u - v}{u + v}$	v	Human cell potential	Biomedical Electronics
11. $S = \dfrac{8PD}{\pi d^3}\left(1 + \dfrac{d}{2D}\right)$	P	Shearing stress in a spring	Mechanics
12. $\dfrac{1}{f} = (n - 1)\left(\dfrac{1}{R_1} + \dfrac{1}{R_2}\right)$	R_1	Lensmaker's equation	Physics
13. $\dfrac{1}{R} = \dfrac{1}{r_1} + \dfrac{1}{r_2}$	r_1	Parallel resistance	Electricity
14. $C = \dfrac{D + d}{2}$	D	Center distance between meshed gears	Mechanics
15. $F = \tfrac{9}{5}C + 32$	C	Fahrenheit-centigrade temperature conversion	Physics
16. $Q = MC(P_1 - P_2)$	P_2	Heat transfer	Fluids
17. $\dfrac{e_o}{e_s} = \dfrac{A}{1 - AB}$	B	Gain of an amplifier with feedback	Electronics
18. $F = 1 \dfrac{uIi}{2\pi d}$	d	Force between parallel current-carrying conductors	Electricity
19. $f_L = f_s\left(\dfrac{V + V_L}{V - V_s}\right)$	V_L	Doppler effect	Physics
20. $f = F \sin \theta$	F	Perpendicular force acting on a lever	Mechanics
21. $e = E_m \sin wt$	E_m	Instantaneous value of a sinusoidal voltage	Electricity
22. $N = W - F \sin \theta$	F	Force normal to a surface	Physics
23. $F = qvB \sin \theta$	$\sin \theta$	Force acting on a charge in a magnetic field	Physics

Formula	Solve for	Description	Technical Area
24. $i = \dfrac{E_m \sin wt}{R_L}$	E_m	Instantaneous value of sinusoidal current	Electricity
25. $X = \dfrac{(2a + m + n)(a \sin \theta)}{(m + a) - a \sin \theta}$	n	Displacement for a quick return mechanism	Mechanics

2-19
Graphing

Consider yourself playing a game where you think of a number between 0 and 10 and someone tries to guess what number you are thinking about. Let us assume that you have played this five times and the following numbers represent your numbers and their guesses.

Numbers	1	8	0	7	6
Guesses	3	4	5	7	2

Another way in which we might pair these numbers is to plot them on a graph. Suppose we draw a graph (Figure 2-1) and let the horizontal distance represent your numbers and the vertical distance represent the guesses.

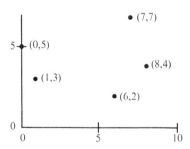

Figure 2-1

Do you see that for the point representing (1,3) you would go to the right one unit and up a distance of three units. The other points can be found in similar manner.

Normally, when a pair of numbers is written as (1,3), they are called an *ordered pair*. This means that the first number came from a particular group or set of numbers and that the second number also came from a particular group or set of numbers.

EXAMPLE 59   Plot the following pairs of numbers.

A (1,6)
B (2,3)
C (4,7)
D (1,1)

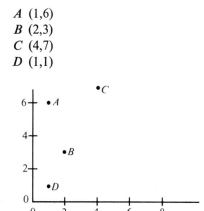

Figure 2-2

In the distance formula $d = rt$, d is said to be a function of r and t. If we let r be a constant rate such as 50, then we have $d = 50t$. Letting $t = 0, 1, 2, 3, 4$, and so forth we can find the distance traveled at 50 m.p.h. for a particular number of hours.

t	0	1	2	3	4
d	0	50	100	150	200

These points can be plotted on a graph. Let t be plotted horizontally and d be plotted vertically.

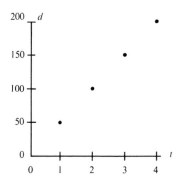

Figure 2-3

Graphing can be extended to include all integers if we make our graph as shown in Figure 2.4.

A graph like Figure 2-4 above is called a *plane rectangular coordinate system*. The two lines that divide the graph into four parts are called *axes*. The *horizontal axis* is called the *x-axis*, and the *vertical axis* is called the *y-axis*. The point where the two axes intersect is called the *origin*. The four sections of

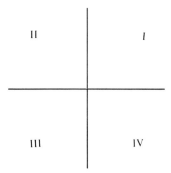

Figure 2-4

the graph are called *quadrants* and are labeled as I, II, III, IV, the first positioned in the upper right-hand corner and the others in a counterclockwise direction.

When an ordered pair is plotted on a rectangular coordinate system, the pair of numbers associated with the point are called the *coordinates of the point*. The first number, representing a value on the *x*-axis, is called the *abscissa*, and the second number, representing a value on the *y*-axis, is called the *ordinate*.

EXAMPLE 60 Draw a rectangular coordinate system and plot the following ordered pairs.

A (0,−5)	C (4,−4)	E (5,3)
B (−3,−7)	D (−4,6)	F (−8,0)

Solution:

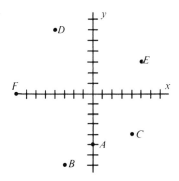

Figure 2-5

PROBLEM SET 2-20

Plot the following sets of points and indicate the quadrant in which each point is located.

1. (1,3)   2. (2,−4)   3. (−2,5)   4. (−5,−3)   5. (7,1)
6. (−6,−6)   7. (−4,1)   8. (5,1)   9. ($\frac{1}{2}$,$\frac{1}{4}$)   10. ($\frac{1}{2}$,10)

The expression $y = 2x + 3$ is a *linear function*. It differs from a linear equation like $2x = 10$. Recall that a linear equation has only one answer, whereas many numbers will satisfy a linear function. Also, the answer to a linear equation is one number, and the answers to a linear function are represented as ordered pairs.

The solution of a linear function can be found by assigning a value for x and solving for y.

EXAMPLE 61   Find the solution of $y = 2x + 3$.

Solution:

$$\begin{aligned}
\text{Let } x &= 1 \\
y &= 2x + 3 \\
&= 2(1) + 3 \\
&= 2 + 3 \\
&= 5
\end{aligned}$$
when $x = 1$, $y = 5$

$$\begin{aligned}
\text{Let } x &= 2 \\
y &= 2x + 3 \\
&= 2(2) + 3 \\
&= 4 + 3 \\
&= 7
\end{aligned}$$
when $x = 2$, $y = 7$

$$\begin{aligned}
\text{Let } x &= -3 \\
y &= 2x + 3 \\
&= 2(-3) + 3 \\
&= -6 + 3 \\
&= -3
\end{aligned}$$
when $x = -3$, $y = -3$

$$\begin{aligned}
\text{Let } x &= 0 \\
y &= 2x + 3 \\
&= 2(0) + 3 \\
&= 0 + 3 \\
&= 3
\end{aligned}$$
when $x = 0$, $y = 3$

Since x could be any real number, do you see that y could also be any real number? Do you see that this function could have an infinite number of

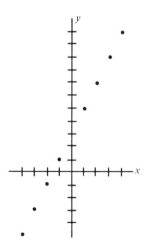

Figure 2-6

solution pairs? The following is a list of some of these pairs.

x	-4	-3	-2	-1	0	1	2	3	4
y	-5	-3	-1		1	3	5	7	9	11

Figure 2-6 shows what the graph of these points would look like. But, if you plot all the points in the solution of $y = 2x + 3$ the result would be the straight line shown in Figure 2-7. Remember that the line would continue in each direction indefinitely.

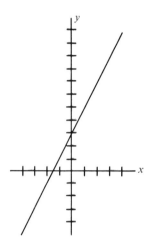

Figure 2-7

PROBLEM SET 2-21

Graph the following linear functions.

1. $y = 5x$
2. $y = x + 2$
3. $y = 4x + 3$
4. $3x - 6y = 2$
5. $4x + 3 = 6y - 1$
6. $3(2x + 1) = y$
7. $2y + 5 = 4(x + 2)$
8. $8(y + 2) = 2x + 4y$
9. $5y + 2x - 3 = 7x + 2y - 4$
10. $9y - 1 = 3(x + 5) + 2$

2-21
Slope

The steepness or inclination of a line is one of the more important aspects of graphing. This is commonly referred to as the *slope* of the line. The slope of a line can be found if the coordinates of two points on the line are known. Between any two points on a line there are two distances (vertical and horizontal) which we will use in finding the slope of a line. In Figure 2-8, a line passes through the points (1,2) and (3,6). Line segment b is the *vertical distance* and line segment a is the *horizontal distance*.

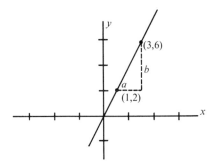

Figure 2-8

The *vertical distance* can be found by subtracting the values of the ordinates (y values). Notice that the values of y are 6 and 2. $6 - 2 = 4$. 4 is the vertical distance.

The *horizontal distance* can be found by subtracting the values of the abscissas (x values). The values of x are 1 and 3. $3 - 1 = 2$. 2 is the horizontal distance.

The slope of a line is equal to the vertical distance between two points on the line, divided by the horizontal distance between the same two points on the line. The slope of the line in Figure 2-8 is as follows:

$$\text{slope} = \tfrac{4}{2} = \tfrac{2}{1} = 2$$

Suppose we generalize what we have just said about slope. Given any two points $P(x,y)$ and $Q(x,y)$, the slope of the line passing through these two points can be found as follows:

$$\text{slope} = \frac{\text{ordinate } Q - \text{ordinate } P}{\text{abscissa } Q - \text{abscissa } P} = \frac{y_2 - y_1}{x_2 - x_1}$$

Since m is used to designate slope we have,

$$m = \frac{y_2 - y_1}{x_2 - x_1}$$

$(y_2 - y_1)$ can be considered the change in y, and $(x_2 - x_1)$ as the change in x. Therefore, the slope is sometimes referred to as:

$$m = \frac{\text{change in } y}{\text{change in } x}$$

EXAMPLE 62   Find the slope of the line joining $P(1,4)$ and $Q(7,6)$.

Solution:
$$m = \frac{y_2 - y_1}{x_2 - x_1} = \frac{6 - 4}{7 - 1}$$

$$m = \frac{2}{6} = \frac{1}{3}$$

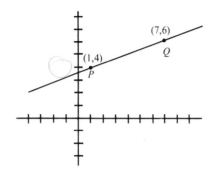

Figure 2-9

EXAMPLE 63   Find the slope of the line joining $(3,7)$ and $(-2,-3)$.

Solution:
$$m = \frac{y_2 - y_1}{x_2 - x_1}$$

$$= \frac{-3 - 7}{-2 - 3} = \frac{-10}{-5}$$

$$= 2$$

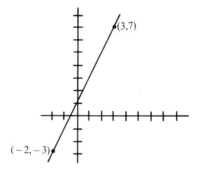

Figure 2-10

It should be made clear that it makes no difference which point is considered to be (x_1, y_1) or (x_2, y_2).

EXAMPLE 64   Find the slope of the line joining $(6,1)$ and $(-3,4)$.

Solution: (1) Let $(6,1) = (x_2,y_2)$ and $(-3,4) = (x_1,y_1)$

$$m = \frac{y_2 - y_1}{x_2 - x_1} = \frac{1 - 4}{6 - (-3)} = \frac{-3}{9} = \frac{1}{-3}$$

(2) Let $(6,1) = (x_1,y_1)$ and $(-3,4) = (x_2,y_2)$

$$m = \frac{y_2 - y_1}{x_2 - x_1} = \frac{4 - 1}{-3 - 6} = \frac{3}{-9} = \frac{1}{-3}$$

PROBLEM SET 2-22

Determine the horizontal and vertical distance between the following sets of points.

1. $(0,3)$, $(5,3)$
2. $(0,0)$, $(0,3)$
3. $(2,0)$, $(2,6)$
4. $(1,-3)$, $(-3,-3)$
5. $(-2,-3)$, $(-4,-3)$
6. $(-1,1)$, $(-1,2)$
7. $(-2,-1)$, $(-2,3)$
8. $(6,4)$, $(0,4)$
9. $(-1,3)$, $(-4,3)$
10. $(1,-4)$, $(1,3)$

PROBLEM SET 2-23

Using the equation for slope determine the slope for the following sets of points.

1. $(0,0)$, $(5,5)$
2. $(7,2)$, $(1,-1)$
3. $(0,4)$, $(2,1)$
4. $(-5,5)$, $(5,-5)$
5. $(3,4)$, $(0,-7)$
6. $(1,6)$, $(-1,-1)$
7. $(-1,3)$, $(-2,-5)$
8. $(-2,0)$, $(-4,-3)$
9. $(-4,1)$, $(6,-3)$
10. $(5,-2)$, $(-2,5)$

2-22
Straight Lines

A linear function or a linear equation represents the equation of a straight line. By *straight line*, we mean a curve whose slope is always constant. An equation in the form of $Ax + By + C = 0$ where A, B, and C are real numbers is the *standard form* of the equation of a straight line. A is the coefficient of x, B is the coefficient of y, and C is the constant. Suppose we graph the equation $y + 3x = 4$. First find several values for x and y and plot the points.

x	-3	-2	-1	0	1	2	3
y	13	10	7	4	1	-2	-5

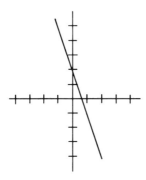

Figure 2-11

Now find the slope of the line. Let us use $(-1,7)$ as (x_2, y_2) and $(3, -5)$ as (x_1, y_1).

$$m = \frac{y_2 - y_1}{x_2 - x_1}$$

$$= \frac{7 - (-5)}{-1 - 3}$$

$$= \frac{12}{-4}$$

$$= \frac{3}{-1}$$

$$= -3$$

Solve $y + 3x = 4$ for y.

$$y + 3x = 4$$
$$y = -3x + 4$$

What is the coefficient of x? How does this compare with the slope? Notice that the constant 4 is equal to the value of y where the line crosses the y axis.

This example shows us that if an equation is in the form of $y = -\dfrac{A}{B} x + \dfrac{C}{B}$ the coefficient of x, which is $-\dfrac{A}{B}$, will equal the slope, and the value of the constant $\dfrac{C}{B}$, will equal the y-intercept. Letting $-\dfrac{A}{B} = m$ and $\dfrac{C}{B} = b$, we have the equation $y = mx + b$. The equation $y = mx + b$ is called the *slope-intercept* form of a straight line.

EXAMPLE 65   Find the slope and y-intercept of the equation $2x + y = 4$.

Solution: $2x + y = 4$ $m = -2$
                    $y = -2x + 4$ $b = 4$

Graph the equation $2x + y = 4$

x	-2	0	4
y	8	4	-4

Use $m = \dfrac{y_2 - y_1}{x_2 - x_1}$ to find the slope;

$$m = \frac{y_2 - y_1}{x_2 - x_1}$$

$$= \frac{4 - 8}{0 - (-2)}$$

$$= \frac{-4}{2}$$

$$= -2$$

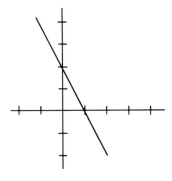

Figure 2-12

Does this compare with the answer above? What is the value of y where the graph crosses the y-axis? Does this compare with the value of b above?

When an equation is written in the form of $y = mx + b$, the coefficient m will equal the slope of the line and the constant b will equal the y-intercept. Suppose we are given two points $P(x,y)$ and $0(x_1,y_1)$, Figure 2-13, and are asked to find the slope of the line connecting these points.

$$m = \frac{y - y_1}{x - x_1}$$

multiply both sides by $(x - x_1)$

$$(x - x_1)m = y - y_1 \quad \text{or} \quad y - y_1 = m(x - x_1)$$

The formula $(y - y_1) = m(x - x_1)$ is known as the *point slope* form of a straight line. Given any point on a line and the slope of a line, this formula can be used to find the equation of the line.

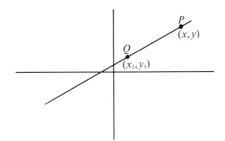

Figure 2-13

EXAMPLE 66   Find the equation of the line having slope 2 and passing through the point (1,5).

Solution:
$$y - y_1 = m(x - x_1)$$
$$y - 5 = 2(x - 1)$$
$$y - 5 = 2x - 2$$
$$y = 2x + 3$$

EXAMPLE 67   Find the equation of the line passing through the points (0,3) and (5,5).

Solution: We first find the slope of the line by

$$m = \frac{5 - 3}{5 - 0}$$

$$= \frac{2}{5}$$

Then, by using either of the points and the slope we have:

$$y - 3 = \frac{2}{5}(x - 0)$$

$$y - 3 = \frac{2}{5}x$$

$$5y - 15 = 2x$$

$$5y - 2x = 15$$

EXAMPLE 68   Graph the line $y = 4$. This means that every point will have 4 as its y coordinate, regardless the value of the x coordinate.

Solution: Since x can be any value as long as $y = 4$ then the points (0,4), (6,4) are on the graph of $y = 4$.

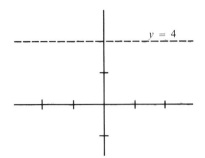

Figure 2-14

EXAMPLE 69   Find the slope of $y = 4$.

Solution:
$$m = \frac{4 - 4}{0 - 6}$$
$$= \frac{0}{-6}$$
$$= 0$$

The slope of any line parallel to the x-axis will always equal *zero* (0).

EXAMPLE 70   Graph the line $x = 3$. This means that every point will have 3 as its x coordinate regardless what the y coordinate equals.

Solution: Since y can be any value as long as $x = 3$, then the points (3,0) and (3,5) are on the graph of $x = 3$.

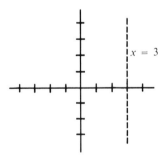

Figure 2-15

EXAMPLE 71   Find the slope of $x = 3$.

Solution:
$$m = \frac{0 - 5}{3 - 3}$$
$$= \frac{-5}{0}$$
$$= \text{undefined, Why?}$$

The slope of any line parallel to the y-axis is undefined.

Introduction to the Tangent and Derivative

Now that we have defined the slope of a line, there are other related topics that can be considered. We will define the *tangent of an angle* θ where θ is the angle shown in Figure 2-16.

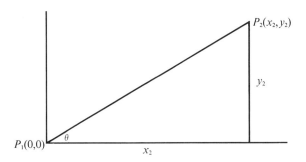

Figure 2-16

The tangent of the angle θ, written $\tan \theta$ is defined as the ordinate divided by the abscissa a.

$$\tan \theta = \frac{y_2}{x_2}$$

Now note that if we find the slope of the same line, we have

$$m = \frac{y_2 - y_1}{x_2 - x_1}$$

$$= \frac{y_2 - 0}{x_2 - 0}$$

$$= \frac{y_2}{x_2}$$

therefore;

$$\tan \theta = \text{slope} = \frac{y_2}{x_2}.$$

In other words, when we discuss the tangent of an angle in later chapters, we will be discussing nothing more than the slope of a line.

Another fundamental mathematical expression is the *derivative of a function*. The derivative of a function is the rate of change of y with respect to x at any point.

Figure 2-17 graphs the function $y = 3x + 2$. We can find the change in y with respect to x by taking two values for x, finding the corresponding values for y and then forming the ratio

$$\frac{y_2 - y_1}{x_2 - x_1}$$

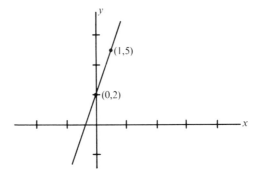

Figure 2-17

When $x = 0$, $y = 2$, and when $x = 1$, $y = 5$. The change in x is $1 - 0$ or 1 and the change in y is $5 - 2$ or 3. The change in y divided by the change in x is

$$\frac{5 - 2}{1 - 0} = \frac{3}{1} = 3.$$

The derivative of the function with respect to x, sometimes written $\dfrac{dy}{dx}$, is equal to 3. The derivative of a linear function is equal in value to the slope of the graph. Derivatives of different functions are found by other methods. However, the slope, the tangent of the angle with the x-axis, and the derivative all have the same value for a linear function.

2-24
Technical Example

The velocity of an automobile was noted during the first 500 feet traveled and a curve of distance versus time was plotted as shown in Figure 2-18.

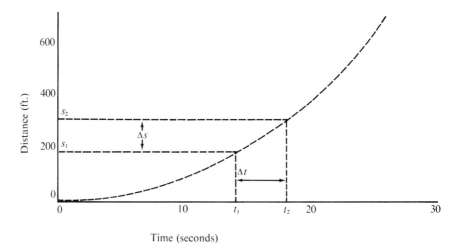

Time (seconds)

Figure 2-18

It was noted that the distance increases as the square of the time. This may be written as the equation $s = t^2$. Suppose you wish to determine the velocity of the vehicle after 16 seconds. Two means of solution can be used. We are interested in the velocity at a particular instant, or *instantaneous velocity*.

One means of solution is to draw a short line tangent to the curve at the point where t equals 16 seconds. From the ends of the tangent, project vertical lines to the time axis and horizontal lines to the distance axis. Identify the time corresponding to the vertical line on the right as t_2 and to the vertical line on the left as t_1. The distance corresponding to the upper horizontal line is s_2 and the lower horizontal line is s_1. Then let

$$t = t_2 - t_1$$

and

$$s = s_2 - s_1$$

The instantaneous velocity is then determined by the expression

$$v = \frac{s}{t}$$

From the curve we read the following:

$$t_2 = 18 \text{ sec}$$
$$t_1 = 14 \text{ sec}$$
$$s_2 = 320 \text{ ft}$$
$$s_1 = 194 \text{ ft}$$

Therefore:

$$t = t_2 - t_1 = 18 \text{ sec} - 14 \text{ sec} = 4.0 \text{ sec}$$
$$s = s_2 - s_1 = 320 \text{ ft} - 194 \text{ ft} = 126 \text{ ft}$$

$$v = \frac{s}{t}$$

$$= \frac{126 \text{ ft}}{4.0 \text{ sec}}$$

$$v = 31.5 \text{ ft/sec}$$

The second means of solution utilizes calculus to which you have just been introduced. The derivative of the equation of a line gives the slope of the line. The slope of a line is defined as the change in vertical distance divided by the change in horizontal distance, or $\frac{\Delta s}{\Delta t}$ in our problem. The equation for our curve is $s = t^2$. The derivative of this is found by

$$\frac{d(s)}{dt} = \frac{d(t^2)}{dt}$$

In other words, the derivative of distance s with respect to time t equals the derivative of time squared t^2 with respect to time t. Taking the derivative yields

$$\frac{ds}{dt} = 2t\frac{dt}{dt}$$

$$= 2t$$

Since velocity equals a change in distance divided by a change in time we can say

$$v = \frac{ds}{dt}$$

$$= 2t$$

or

$$v = 2t$$

Since we were interested in the velocity at 16 seconds we substitute this for t which gives

$$v = 2(16)$$
$$= 32 \text{ ft/sec}$$

This is very nearly the same as the result by the first method of solution. This is the exact velocity whereas the result of the first method of solution was a close approximation.

PROBLEM SET 2-24

Determine the slope and the y-intercept for the following equations, and draw the graph of each.

1. $y = 2x$

2. $y = 3x + 3$

3. $2x - y = 4$

4. $x + 3 = 3y$

5. $8x + 5 = 4y + 1$

6. $5(4x - 2) = 6y - 4x + 2$

7. $9x + 7y + 5 = 12 - 5x$

8. $3(x + y + 3) = 4(x - y - 3)$

9. $6(x - 3y) = 7x + 12$

10. $\dfrac{x + 6}{2} = y - 2$

11. $\dfrac{2(y - 5)}{8} = \dfrac{2x + 1}{4}$

12. $\dfrac{9}{x + 3} = \dfrac{3}{y - 2}$

13. $2x + 3y = 8x - 2y - 5$

14. $\dfrac{8x - 7y + 3}{5} = \dfrac{x + 3y + 3}{4}$

15. $\dfrac{4(2x + y + 1)}{6} = \dfrac{x + 3y + 3}{2}$

Determine the equation of the line having the following slope and y-intercept and draw the graph of each.

1. $m = 1, b = 2$ 2. $m = -2, b = 3$

3. $m = 4, b = -1$ 4. $m = \frac{1}{2}, b = 5$

5. $m = \frac{2}{3}, b = -2$ 6. $m = -\frac{3}{2}, b = 0$

7. $m = 3, b = -6$ 8. $m = 6, b = 1$

9. $m = -4, b = \frac{7}{2}$ 10. $m = 0, b = 3$

Determine the equation of a line passing through the given point with the given slope and draw the graph of each.

1. $(1,1), m = 1$ 2. $(2,3), m = 3$

3. $(2,0), m = 1$ 4. $(1,10), m = 7$

5. $(1,4), m = 3$ 6. $(-5,2), m = 2$

7. $(-1,4), m = -6$ 8. $(-3,-5), m = 1$

9. $(-6,\frac{1}{2}), m = -\frac{1}{2}$ 10. $(-5,-7), m = 0$

2-25
Perpendicular Lines

If two lines are perpendicular, their slopes are negative reciprocals of each other. Conversely, if the slopes of two lines are negative reciprocals of each other, the lines are *perpendicular*.

EXAMPLE 72   Are the lines $2x + y = 7$ and $2y = x + 3$ perpendicular?

Solution: Determine the slope of each.

$$2x + y = 7 \qquad\qquad 2y = x + 3$$
$$y = -2x + 7 \qquad\qquad y = \tfrac{1}{2}x + \tfrac{3}{2}$$
$$m = -2 \qquad\qquad m = \tfrac{1}{2}$$

Since -2 and $\frac{1}{2}$ are negative reciprocals of each other, the lines are perpendicular.

EXAMPLE 73   Are the lines $x + 37 = 15$ and $3x - y = 4$ perpendicular?

Solution: Determine the slope of each.

$$3x - y = 4 \qquad\qquad x + 3y = 15$$
$$-y = -3x + 4 \qquad\qquad 3y = -x + 15$$
$$y = 3x - 4 \qquad\qquad y = -\tfrac{1}{3}x + 5$$
$$m = 3 \qquad\qquad m = -\tfrac{1}{3}$$

Since 3 and $-\tfrac{1}{3}$ are negative reciprocals of each other, the lines are perpendicular.

2-26
Parallel Lines

If two lines have the same slope they are parallel lines.

EXAMPLE 74   Find the slope of $y = 3x - 1$ and $2y - 6x = 8$. Graph each equation.

Solution:

$$y = 3x - 1$$
$$m = 3$$

$$2y - 6x = 8$$
$$2y = 6x + 8$$
$$y = 3x + 4$$
$$m = 3$$

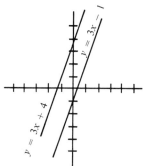

Figure 2-19

It can be seen by the graph that the two lines are parallel and that the slopes are equal.

EXAMPLE 75   Are $5x + 2y = 7$ and $x - y = 4$ parallel?

Solution: Find the slope of each.

$$5x + 2y = 7 \qquad\qquad x - y = 4$$
$$2y = -5x + 7 \qquad\qquad -y = -x + 4$$
$$y = -\tfrac{5}{2}x + \tfrac{7}{2} \qquad\qquad y = x - 4$$
$$m = -\tfrac{5}{2} \qquad\qquad m = 1$$

Since $-\tfrac{5}{2} \neq 1$ the lines are not parallel.

PROBLEM SET 2-27

Determine whether the following sets of equations represent parallel or perpendicular lines, or neither.

1. $x + y = 6$

   $x - y = 4$

2. $x - y = 8$

   $x + y = 2$

3. $5y + 12x = 3$

   $5x - 12y = 8$

4. $4x - 6y - 8 = 0$

   $6x - 9y + 12 = 0$

5. $2x + 6y = 4$

   $12x - 4y = -10$

6. $3x - y = 4$

   $2x - 6y = -4$

7. $3x + 2y = 10$

   $3y - 2x = 8$

8. $2x - y = 3$

   $2x - 3y = 9$

9. $4x - 3y - 10 = 0$

   $6y - 8x + 16 = 0$

10. $8x - 12y - 10 = 0$

    $18y - 12x + 12 = 0$

2-27
Solving Two Equations with Two Unknowns

A man has some chickens and some horses in his barnyard. He counted the heads of both and there were 26. The total number of feet was 70. How many of each were in the barnyard?

Solution: Let x = number of chickens

y = number of horses

Since the total number of chickens and horses is 26,

$$x + y = 26$$

If a chicken has two feet and a horse has four feet then $2x$ would equal the number of chickens feet and $4y$ would equal the number of horses feet. Therefore,

$$2x + 4y = 70$$

We now have the two equations:

$$x + y = 26$$

$$2x + 4y = 70$$

A battery of unknown voltage is connected to a circuit containing a 3 ohm resistor and an unknown resistor. The current through the circuit is 2 amperes and the voltage across the 3 ohm resistor is 6 volts. The battery is then connected to a circuit containing the same unknown resistor and an 8 ohm resistor. The current through this circuit is 1 ampere and the voltage across the unknown resistor is 2 volts. Determine the value of the voltage source and the unknown resistor.

Solution: Let Ex = unknown voltage

Rx = unknown resistor

I_1 = current of circuit A

I_2 = current in circuit B

R_1 = known resistor in circuit A

R_2 = known resistor in circuit B

$$Ex = I_1 Rx + I_1 R_1$$

$$Ex = I_2 Rx + I_2 R_2$$

An automobile traveling at 44 ft/sec accelerates at a constant rate for 10 seconds. A second automobile traveling at 60 ft/sec must accelerate at the same rate for only 8 seconds to reach the same final velocity as the first automobile. Determine the rate of acceleration and the final velocity.

Solution: Let v = final velocity of both autos

v_1 = initial velocity of auto A

v_2 = initial velocity of auto B

t_1 = time of acceleration for auto A

t_2 = time of acceleration for auto B

a = acceleration of both autos

Using the known data, write two equations in terms of the two unknowns, final velocity v and rate of acceleration a, in the form $v = v_0 + at$.

$$v = v_1 + at_1$$

$$v = v_2 + at_2$$

These are examples of problems which require two equations in two unknowns to find the solution. It is necessary to know how to solve two equations of this type.

Solving two equations in two unknowns is termed *solving simultaneous equations* or *solving equations simultaneously*. This means to solve the two equations at the same time, or to find the pair of numbers which is in the solution of both equations. It must be understood at this time that if we can find a solution, it will be a single *ordered pair*.

       Before we attempt to find the solution of the above problems, let us consider the following example.

EXAMPLE 76   Find the solution of $x + y = 4$ and $2x + y = 6$ graphically.

Solution: Graphing each we get the following:

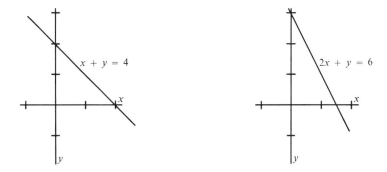

Figure 2-20

Now show the graph of both equations on one set of axes

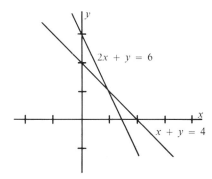

Figure 2-21

Notice that the two lines intersect in one and only one point. The coordinates of this point are the solution of the equations. How can we find the coordinates? One way is to estimate the values. It looks like the x-coordinate is 2 and the

y-coordinate is 2. We now check to see if our estimate (2,2) is the correct answer. Substitute (2,2) into both equations.

$$2x + y = 6 \qquad x + y = 4$$
$$2(2) + 2 = 6 \qquad 2 + 2 = 4$$
$$4 + 2 = 6 \qquad\quad 4 = 4$$
$$6 = 6$$

We see that (2,2) gave us a true statement in both equations, hence (2,2) is the solution of the equations $2x + y = 6$ and $x + y = 4$.

EXAMPLE 77   Find the solution of $x + 2y = 8$ and $x - y = 6$.

Solution: First, we will graph each equation on one set of axes.

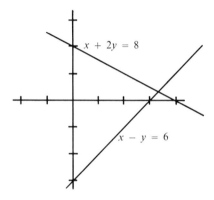

Figure 2-22

Next, we will estimate the coordinates of the intersection. It seems that the x-coordinate is approximately 7 and the y-coordinate is approximately 1. Now substitute (7,1) into both equations.

$$x + 2y = 8 \qquad x - y = 6$$
$$7 + 2(1) = 8 \qquad 7 - 1 = 6$$
$$7 + 2 = 8 \qquad\quad 6 = 6$$
$$9 = 8$$

We see that the point (7,1) satisfied $x - y = 6$, but did not satisfy $x + 2y = 8$. Therefore (7,1) *is not* the solution of the equations $x + 2y = 8$ and $x - y = 6$.

We must now make a different estimate of the intersection and substitute into both equations hoping to find a pair of numbers which will satisfy both equations.

Suppose we try point (6,1) in both equations.

$$x + 2y = 8 \qquad x - y = 6$$
$$6 + 2(1) = 8 \qquad 6 - 1 = 6$$
$$6 + 2 = 8 \qquad\quad 5 = 6$$
$$8 = 8$$

Again we see that one equation is satisfied and one is not satisfied. We must make still another estimate and hope both equations are satisfied.

It is obviously very difficult to solve two equations in two unknowns by graphing. Before we make an attempt to solve two equations simultaneously by another method, let us look at two other examples.

EXAMPLE 78   Find the solution of $x + y = 4$ and $2x + 2y = 8$.

Solution: First, we will graph both equations on the same axis.

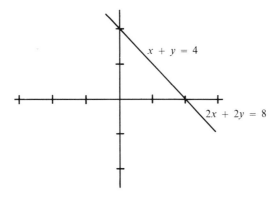

Figure 2-23

We see that both equations give us the same line. Do we have a solution? Yes, but not a particular solution as in the previous example. In this example any solution of $x + y = 4$ is also a solution of $2x + 2y = 8$.

EXAMPLE 79   Find the solution of $x + y = 3$ and $x + y = 7$.

Solution: First we will graph both equations on the same axis.

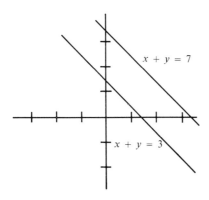

Figure 2-24

We see in this example that the two lines do not intersect. If this is the case, we do *not* have a solution.

We have seen examples of equations that intersect (Example 77), that are the same line (Example 78), and that are parallel (Example 79). We say that equations that intersect are *consistent*. Equations that are parallel are *inconsistent* and equations that are the same line are *dependent*. Normally, we will be concerned with equations which are consistent.

PROBLEM SET 2-28

Solve the following sets of equations graphically.

1. $2y = x + 1$
   $y = x + 4$

2. $2x - y = 7$
   $x + y = 2$

3. $t + 5r = 7$
   $t - 2r = 14$

4. $2u + 6v = 7$
   $4u + 8v = -2$

5. $a = b + 6$
   $a = -2b + 2$

6. $2y = x - 4$
   $y = x + 12$

7. $2e + r = 3$
   $3e + 4r = 7$

8. $x + y - 1 = 0$
   $6x + 2y - 3 = 0$

9. $4k - 2r = -6$
   $6k + 3r = -9$

10. $6x - 2y = 6$
    $2x + y = 2$

2-29

Solving Simultaneous Equations by Addition and Subtraction

If it is possible to add or subtract two linear equations and eliminate one of the unknowns, then we can solve for the unkown that has not been eliminated. Once one of the unknowns has been found, you can substitute into either of the two equations to find the other.

EXAMPLE 80   Solve for x and y in the following:

$$x + y = 10$$
$$x - y = 4$$

Solution: First, write the equations as follows:

$$x + y = 10$$
$$x - y = 4$$

Now add both sides together.

$$x + y = 10$$
$$x - y = 4$$
$$2x = 14$$

Solve $2x = 14$ for x.

$$2x = 14$$
$$x = 7$$

Substitute $x = 7$ into either of the equations and solve for y.

$$x + y = 10$$
$$7 + y = 10$$
$$y = 3$$

Now check the values $x = 7$ and $y = 3$ in the other equation, $x - y = 4$. If they satisfy $x - y = 4$, then they are the solution of both.

$$x - y = 4$$
$$7 - 3 = 4$$
$$4 = 4$$

They do satisfy $x - y = 4$, therefore $(7,3)$ is the solution of $x + y = 10$ and $x - y = 4$.

EXAMPLE 81   Solve for x and y in the following.

$$2x + y = 18$$
$$x - y = 6$$

Solution: Add both equations,

$$2x + y = 18$$
$$x - y = 6$$
$$3x = 24$$

Solve $3x = 24$ for x,

$$3x = 24$$
$$x = 8$$

now substitute $x = 8$ into one of the above equations,

$$2x + y = 18$$
$$2(8) + y = 18$$
$$16 + y = 18$$
$$y = 2$$

Substitute $x = 8$ and $y = 2$ into the equation $x - y = 6$ and check.

$$x - y = 6$$
$$8 - 2 = 6$$
$$6 = 6$$

Therefore, (8,2) is the solution.

EXAMPLE 82   Solve for x and y in the following.

$$4x + y = 10$$
$$x + y = 4$$

Solution: What happens if we add? We do not eliminate one of the unknowns. In this case we must subtract. Subtract the bottom equation from the top equation.

$$4x + \quad y = \quad 10$$
$$\underline{(-x) + (-y) = (-4)}$$
$$3x \qquad\quad = \quad 6$$

Solve for x.

$$3x = 6$$
$$x = 2$$

Substitute $x = 2$ into one of the above equations.

$$x + y = 4$$
$$2 + y = 4$$
$$y = 2$$

Check by substituting $x = 2$ and $y = 2$ into $4x + y = 10$.

$$4x + y = 10$$
$$4(2) + 2 = 10$$
$$8 + 2 = 10$$
$$10 = 10$$

Therefore, (2,2) is the solution.

EXAMPLE 83   Solve for x and y in the following.

$$3x - 2y = 12$$
$$x + y = 4$$

Solution: Before we can eliminate one of the unknowns by addition or subtraction, we must first make the coefficients of either x or y equal. Let us make the coefficients of y equal in both equations. To do this we must multiply both sides of $x + y = 4$ by 2.

$$2(x + y) = 2(4)$$
$$2x + 2y = 8$$

Now, if we add we will eliminate the y's.

$$3x - 2y = 12$$
$$2x + 2y = 8$$
$$\overline{5x \qquad = 20,}$$

solve for x $\qquad\qquad x = 4$

Substitute $x = 4$ into $3x - 2y = 12$.

$$3x - 2y = 12$$
$$3(4) - 2y = 12$$
$$12 - 2y = 12$$
$$-2y = 0$$
$$y = 0$$

Check: $\qquad\qquad x + y = 4$
$$4 + 0 = 4$$
$$4 = 4$$

Therefore, (4,0) is the solution.

EXAMPLE 84   Solve for x and y in the following.

$$2x + 3y = 15$$
$$3x + 2y = 20$$

Solution: In this problem it is necessary to change both equations before adding. Multiply $2x + 3y = 15$ by 3 and multiply $3x + 2y = 20$ by 2.

$$3(2x + 3y) = 3(15)$$
$$6x + 9y = 45$$
$$2(3x + 2y) = 2(20)$$
$$6x + 4y = 40$$

Subtract the bottom equation from the top:

$$6x + 9y = 45$$
$$(-6x) + (-4y) = -40$$
$$\overline{\qquad\qquad 5y = 5}$$
$$y = 1$$

Substitute $y = 1$ into $3x + 2y = 20$,

$$3x + 2y = 20$$
$$3x + 2(1) = 20$$
$$3x + 2 = 20$$
$$3x = 18$$
$$x = 6$$

Check the answer (6,1)

$$2x + 3y = 15$$
$$2(6) + 3(1) = 15$$
$$12 + 3 = 15$$
$$15 = 15$$

Therefore, (6,1) is the solution.

___

Solve the following simultaneous equations by addition and subtraction.

1. $2y + x = 10$
   $y + x = 8$

2. $2\alpha + \beta = 1$
   $5\alpha + 3\beta = 3$

3. $m - 2n = 5$
   $4m + n = 2$

4. $3e - 2i = 7$
   $5e + i = 3$

5. $4p - 3q = 1$
   $-2p + 7q = 5$

6. $2u + 6v = 5$
   $5u + 2v = 6$

7. $r - 3s = 6$
   $7r + s = -2$

8. $20x - 2y = 5$
   $4x + 7y = -2$

9. $3\theta + 5\phi = 7$
   $\theta - 2\phi = 4$

2-30
Solving Simultaneous Equations by Substitution

Another method of solving two equations in two unkowns is by substitution. This means to solve for one unknown in one equation, then substitute the value into the other equation.

___

EXAMPLE 85   Solve for x and y.

$$x + y = 7$$
$$x - y = 1$$

Solution: Suppose we solve $x + y = 7$ for x,

$$x + y = 7$$
$$x = 7 - y$$

Now substitute $7 - y$ for x in $x - y = 1$

$$x - y = 1$$
$$(7 - y) - y = 1$$
$$7 - y - y = 1$$
$$7 - 2y = 1$$
$$-2y = -6$$
$$y = 3$$

Using $x = 7 - y$, substitute $y = 3$ and find x.

$$x = 7 - y$$
$$x = 7 - 3$$
$$x = 4$$

Check your answer by substituting (4,3) into $x - y = 1$

$$x - y = 1$$
$$4 - 3 = 1$$
$$1 = 1$$

Therefore, (4,3) is the solution.

EXAMPLE 86 Solve for x and y.

$$2x + 5y = 24$$
$$x + 2y = 6$$

Solution: In $x + 2y = 6$, solve for x.

$$x + 2y = 6$$
$$x = 6 - 2y$$

Now substitute $(6 - 2y)$ for x in $2x + 5y = 24$ and solve for y.

$$2x + 5y = 24$$
$$2(6 - 2y) + 5y = 24$$
$$12 - 4y + 5y = 24$$
$$-4y + 5y = 12$$
$$y = 12$$

Use the equation $x = 6 - 2y$ and solve for x.

$$x = 6 - 2y$$
$$x = 6 - 2(12)$$
$$x = 6 - 24$$
$$x = -18$$

Check your results in $2x + 5y = 24$. If $(-18,12)$ satisfy $2x + 5y = 24$, then they are correct.

PROBLEM SET 2-30

Solve the following simultaneous equations by substitution.

1. $e + i = 5$
$5e + 3i = 17$

2. $3u + v = 7$
$2u - 5v = -1$

3. $a - b = 0$
   $4a - 2b = -6$

4. $m - 3n = 0$
   $2m - 3n = -6$

5. $8a - v - 29 = 0$
   $2a + v - 11 = 0$

6. $2x + y = 5$
   $4x + 2y = 6$

7. $2c + d + 2 = 0$
   $6c - 5d - 18 = 0$

8. $e - i = 52$
   $3e - 8i = 6$

9. $2t + 5s = 18$
   $3t + 4s = 7$

10. $6a + 9b - 4 = 0$
    $10a + 15b - 3 = 0$

2-31
Solving Word Problems in Two Unknowns

We are now ready to solve the problems stated in Section 2-27.

EXAMPLE 87 A man has some chickens and some horses in his barnyard. He counted the heads of both and there were 26. The total number of feet was 70. How many of each were in the barnyard?

Solution: $x =$ no. of chickens

$y =$ no. of horses

(1) $x + y = 26$ $x + y = 26$
(2) $2x + 4y = 70$ $x + 9 = 26$
(3) $2x + 2y = 52$ [multiply (1) by 2] $x = 17$
$\overline{}$
$2y = 18$ [subtract (3) from (2)]
$y = 9$

Therefore, there are 9 horses and 17 chickens.

EXAMPLE 88 Car A traveling at 44 ft/sec accelerates at a constant rate for 10 seconds. Car B traveling at 60 ft/sec must accelerate at the same rate for only 8 seconds to reach the same final velocity as the first automobile. Determine the rate of acceleration and the final velocity.

Solution: Let $v =$ final velocity of both autos

$v_1 =$ initial velocity of car A

$v_2 =$ initial velocity of car B

$t_1 =$ time of acceleration for auto A

$t_2 =$ time of acceleration for auto B

$a =$ acceleration of both autos

Using the given data, write two equations in terms of the two unknowns, final velocity v and rate of acceleration a, in the form $v = v_0 + at$.

$$\text{Car A} \quad v = v_1 + at_1$$
$$\text{Car B} \quad v = v_2 + at_2$$

Substitute known values into these equations.

$$\text{Car A} \quad v = 44 \text{ ft/sec} + a(10 \text{ sec})$$
$$\text{Car B} \quad v = 60 \text{ ft/sec} + a(8 \text{ sec})$$

Substituting 60 ft/sec $+$ a(8 sec) from the second equation into the first equation yields

$$60 \text{ ft/sec} + a(8 \text{ sec}) = 44 \text{ ft/sec} + a(10 \text{ sec})$$
$$60 \text{ ft/sec} - 44 \text{ ft/sec} = a(10 \text{ sec}) - a(8 \text{ sec})$$
$$16 \text{ ft/sec} = a(10 \text{ sec} - 8 \text{ sec})$$
$$16 \text{ ft/sec} = a(2 \text{ sec})$$
$$a = \frac{16 \text{ ft/sec}}{2 \text{ sec}}$$
$$= 8 \text{ ft/sec}^2$$

Substituting this value of a into either equation will permit one to determine the value of v.

$$v = 44 \text{ ft/sec} + (8 \text{ ft/sec}^2)(10 \text{ sec})$$
$$v = 44 \text{ ft/sec} + 80 \text{ ft/sec}$$
$$v = 124 \text{ ft/sec}$$

This is the final velocity of either vehicle.

EXAMPLE 89 A battery of unknown voltage is connected to a circuit containing a 3 ohm resistor and an unknown resistor. The current through the circuit is 2 amperes and the voltage across the 3 ohm resistor is 6 volts. The battery is then connected to a circuit containing the same unknown resistor and an 8 ohm resistor. The current through this circuit is 1 ampere and the voltage across the unknown resistor is 2 volts. Determine the value of the voltage source and the unknown resistor.

Solution: Let Ex = unknown voltage

Rx = unknown resistance

I_1 = current in circuit A

I_2 = current in circuit B

R_1 = known resistor in circuit A

R_2 = known resistor in circuit B

$$Ex = I_1 Rx + I_1 R_1$$
$$Ex = I_2 Rx + I_2 R_2$$

Substituting in known values.

$$Ex = 2Rx + 2(3)$$

$$Ex = 2Rx + 6$$

$$Ex = 1Rx + 1(8)$$

$$Ex = 1Rx + 8$$

Since $Ex = Rx + 8$, substitute this in place of Ex in the first equation:

$$Rx + 8 = 2Rx + 6$$

$$2Rx - Rx = 8 - 6$$

$$Rx = 2 \text{ ohms}$$

Substituting this for Rx in the first equation gives:

$$Ex = 2(2) + 6$$

$$= 10 \text{ volts}$$

PROBLEM SET 2-31

1. Two gears have a total of 96 teeth. The difference between the number of teeth on each gear is 48. Determine the number of teeth on each gear.

2. If 10 one-half watt and 20 one watt resistors cost $2.10 and 60 one-half watt and 40 one watt resistors cost $6.20, what is the cost of a single one-half watt and one watt resistor?

3. The atomic number of cesium is one less than four times the atomic number of silicon. The sum of the atomic number of cesium and silicon equals three times the atomic number of vanadium (atomic number of vanadium is 23). Determine the atomic number of cesium and silicon.

4. The beam shown in Figure 2-25 is balanced; therefore $xy = wz$. If $xy + wz = 90$, $y = 3z$, and y is 5 times greater than x, determine the value of w, x, y, and z.

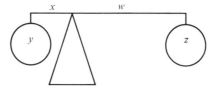

Figure 2-25

5. Referring to Figure 2-26, the following equations must be solved simultaneously to determine the value of the currents, I_1 and I_2.

$$E_1 = I_1 R_1 + I_1 R_2 + I_2 R_2$$
$$E_2 = I_2 R_3 + I_2 R_2 + I_1 R_2$$

Determine the value of I_1 and I_2.

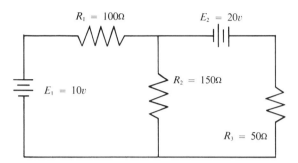

Figure 2-26

6. A set of 90 coins, consisting of nickels and dimes, is worth $5.00. Find the number of coins of each kind.

7. A set of 75 coins, consisting of nickels and quarters is worth $8.75. Find the number of coins of each kind.

8. How much of a solution which is 20% alcohol and how much of a 40% solution must be mixed together to make 60 gallons of a solution which will be 25% alcohol?

9. A chemist wants to form 40 oz. of an alloy consisting of 25% zinc, from an alloy which is 6% zinc and one which is 44% zinc. How much of each alloy should be used?

10. How many pounds of $.90 a pound coffee and how many pounds of $1.15 a pound coffee must be mixed for 75 pounds to sell for $1.00 a pound?

11. Find two complementary angles whose differences is 25°.

12. Find two supplementary angles whose difference is 25°.

13. A motor boat travels 96 miles upstream in 8 hrs. and makes the return trip in 6 hrs. Find the rate of the current and the rate of the boat in still water.

14. A plane travels 1800 miles in 5 hrs. The return trip requires 4.5 hrs. What is the speed of the wind and speed of the plane in still air?

15. A man makes a trip of 775 miles, part by car and the other part by plane. If the car averages 50 m.p.h. and the plane averages 200 m.p.h. and the total trip takes 5 hrs., how far does he travel in the car?

A rectangular array of numbers, arranged in rows and columns, enclosed in parentheses is called a *matrix*. Thus,

$$A = (3 \quad 2 \quad 7), \qquad B = \begin{pmatrix} 1 & 6 \\ 5 & 4 \end{pmatrix}, \qquad C = \begin{pmatrix} 3 & 5 & -2 \\ 8 & 9 & 20 \end{pmatrix},$$

$$D = \begin{pmatrix} 1 \\ 7 \end{pmatrix}, \qquad E = \begin{pmatrix} 6 & -2 & 0 \\ 4 & 5 & 1 \\ \frac{1}{2} & 9 & 4\frac{2}{3} \end{pmatrix}, \qquad F = \begin{pmatrix} 3 \\ -8 \\ -4 \end{pmatrix}$$

are examples of matrices.

A matrix is named with a capital letter and its size depends on the number of rows and the number of columns that make up the matrix. In the matrix $A = \begin{pmatrix} 1 & 0 & -3 \\ 4 & 2 & 7 \end{pmatrix}$, the first row consists of the numbers 1 0 -3. The numbers 4 2 7 make up the second row. The first, second, and third columns consist of $\frac{1}{4}, \frac{0}{2}, \frac{-3}{7}$, respectively. Since matrix A has two rows and three columns, it is a 2 by 3 matrix, usually written 2×3. Notice, that the number of rows is given first, and the number of columns is given second. Thus, the above matrix written

$$A = \begin{pmatrix} 1 & 0 & -3 \\ 4 & 2 & 7 \end{pmatrix}_{2 \times 3}$$

shows the correct placing of the numbers which indicate its size. If the matrix has the same number of rows and columns it is called a *square* matrix. Thus,

$$S = \begin{pmatrix} 2 & \frac{1}{2} \\ 0 & -1 \end{pmatrix}_{2 \times 2}, \quad T = \begin{pmatrix} -4 & 0 & 5 \\ 7 & 1 & 9 \\ 11 & 0 & -3 \end{pmatrix}_{3 \times 3} \quad \text{are examples of square matrices}$$

where S is a 2×2 matrix and T is a 3×3 matrix.

Each number of the matrix is called an *element*. The position of each element is determined by the row and column which contains the element. In T above, the element in the first row, first column is -4; the element in the first row, second column is 0; the element in the first row, third column is 5; the element in the second row, first column is 7; and so forth.

Many operations can be performed on matrices but our discussions will be limited to only one. This operation will be *finding the value of a determinant*.

A square matrix enclosed with vertical lines is called a *determinant*. Thus,

$$\begin{vmatrix} 2 & 7 \\ 8 & 5 \end{vmatrix}, \begin{vmatrix} 1 & 0 \\ 3 & 4 \end{vmatrix}, \begin{vmatrix} 1 & 5 & 1 \\ 2 & -9 & 4 \\ 7 & 6 & 6 \end{vmatrix} \quad \text{are examples of determinants. The following}$$

represents the general 2×2 (second order) determinant.

$$\begin{vmatrix} a_1 & b_1 \\ a_2 & b_2 \end{vmatrix}$$

The value of a 2 × 2 determinant is found as follows:

$$\begin{vmatrix} a_1 & b_1 \\ a_2 & b_2 \end{vmatrix} = a_1 b_2 - a_2 b_1.$$

Stated in words the value of a two by two determinant is the product of the element in the first row, first column and the element in the second row, second column minus the product of the element in the second row, first column and the element in first row, second column. Thus,

$$\begin{vmatrix} 2 & 4 \\ 3 & 1 \end{vmatrix} = (2 \times 1) - (3 \times 4)$$

$$= 2 - 12$$

$$= -10.$$

EXAMPLE 90 Evaluate the determinant $\begin{vmatrix} 3 & 8 \\ 7 & 5 \end{vmatrix}$

Solution:

$$\begin{vmatrix} 3 & 8 \\ 7 & 5 \end{vmatrix} = (3 \times 5) - (7 \times 8)$$

$$= 15 - 56$$

$$= -41$$

EXAMPLE 91 Evaluate the determinant $\begin{vmatrix} 0 & -5 \\ 7 & 2 \end{vmatrix}$

Solution:

$$\begin{vmatrix} 0 & -5 \\ 7 & 2 \end{vmatrix} = (0 \times 2) - 7 \times (-5)$$

$$= 0 - (-35)$$

$$= 0 + 35$$

$$= 35$$

PROBLEM SET 2-32

Solve the following determinants.

1. $\begin{vmatrix} 1 & 0 \\ 2 & -1 \end{vmatrix}$ 2. $\begin{vmatrix} 3 & 4 \\ 0 & 2 \end{vmatrix}$ 3. $\begin{vmatrix} 5 & -1 \\ 2 & 3 \end{vmatrix}$ 4. $\begin{vmatrix} -1 & 3 \\ \frac{1}{3} & 4 \end{vmatrix}$

5. $\begin{vmatrix} \frac{1}{2} & 5 \\ -1 & 10 \end{vmatrix}$ 6. $\begin{vmatrix} -8 & 9 \\ -3 & 2 \end{vmatrix}$ 7. $\begin{vmatrix} 5\frac{1}{2} & 3 \\ 6 & -2 \end{vmatrix}$ 8. $\begin{vmatrix} 0 & -3 \\ -2 & -4 \end{vmatrix}$

Solving a System of Two Linear Equations by Determinants

Determinants are useful in solving systems of linear equations. In order to see how determinants can be used in this manner, consider the following system of equations

$$a_1 x + b_1 y = c_1 \tag{1}$$
$$a_2 x + b_2 y = c_2 \tag{2}$$

Solve this system for x and y by the addition method. To solve for x we shall eliminate the y-terms. We must multiply equation (1) by b_2 and equation (2) by b_1.

$$a_1 b_2 x + b_1 b_2 y = b_2 c_1 \tag{3}$$
$$a_2 b_1 x + b_1 b_2 y = b_1 c_2 \tag{4}$$

Now subtract (4) from (3) and get

$$a_1 b_2 x - a_2 b_1 x = b_2 c_1 - b_1 c_2$$

factor the left member and solve for x,

$$(a_1 b_2 - a_2 b_1)x = b_2 c_1 - b_1 c_2$$
$$x = \frac{b_2 c_1 - b_1 c_2}{a_1 b_2 - a_2 b_1} \tag{5}$$

Now solve for y by eliminating the x-terms. Multiply equation (1) by a_2 and equation (2) by a_1.

$$a_1 a_2 x + a_2 b_1 y = a_2 c_1 \tag{6}$$
$$a_1 a_2 x + a_1 b_2 y = a_1 c_2 \tag{7}$$

Subtract (6) from (7) and get

$$a_1 b_2 y - a_2 b_1 y = a_1 c_2 - a_2 c_1$$

factor the left member and solve for y,

$$(a_1 b_2 - a_2 b_1)y = a_1 c_2 - a_2 c_1$$
$$y = \frac{a_1 c_2 - a_2 c_1}{a_1 b_2 - a_2 b_1} \tag{8}$$

Notice that the denominators in equations (5) and (8) are identical and are equal to the value of the determinant $\begin{vmatrix} a_1 & b_1 \\ a_2 & b_2 \end{vmatrix}$, because

$$\begin{vmatrix} a_1 & b_1 \\ a_2 & b_2 \end{vmatrix} = a_1 b_2 - a_2 b_1$$

A rearrangement in the numerator of equation (5) makes it equal to the value of the determinant $\begin{vmatrix} c_1 & b_1 \\ c_2 & b_2 \end{vmatrix}$, because

$$\begin{vmatrix} c_1 & b_1 \\ c_2 & b_2 \end{vmatrix} = c_1 b_2 - c_2 b_1$$

The numerator of equation (8) is equal to the value of the determinant $\begin{vmatrix} a_1 & c_1 \\ a_2 & c_2 \end{vmatrix}$,

because

$$\begin{vmatrix} a_1 & c_1 \\ a_2 & c_2 \end{vmatrix} = a_1 c_2 - a_2 c_1$$

Therefore equation (5) could be written as:

$$x = \frac{\begin{vmatrix} c_1 & b_1 \\ c_2 & b_2 \end{vmatrix}}{\begin{vmatrix} a_1 & b_1 \\ a_2 & b_2 \end{vmatrix}} \qquad (9)$$

and equation (8) as:

$$y = \frac{\begin{vmatrix} a_1 & c_1 \\ a_2 & c_2 \end{vmatrix}}{\begin{vmatrix} a_1 & b_1 \\ a_2 & b_2 \end{vmatrix}} \qquad (10)$$

Each of these determinants should be studied closely in order to determine which number in the system of equations corresponds to what element in each determinant. From the system of equations

$$a_1 x + b_1 y = c_1$$
$$a_2 x + b_2 y = c_2$$

let us call the determinant $\begin{vmatrix} a_1 & b_1 \\ a_2 & b_2 \end{vmatrix}$ the coefficient matrix. From the coefficient matrix, $\begin{vmatrix} c_1 & b_1 \\ c_2 & b_2 \end{vmatrix}$ can be formed by deleting the coefficients of the x-terms and replacing them with the constants. Likewise $\begin{vmatrix} a_1 & c_1 \\ a_2 & c_2 \end{vmatrix}$ can be formed from the coefficient matrix by deleting the coefficients of the y-terms and replacing them with the constants. For convenience let $|A| = \begin{vmatrix} a_1 & b_1 \\ a_2 & b_2 \end{vmatrix}$, $|A_x| = \begin{vmatrix} c_1 & b_1 \\ c_2 & b_2 \end{vmatrix}$, and $|A_y| = \begin{vmatrix} a_1 & c_1 \\ a_2 & c_2 \end{vmatrix}$; therefore, equation (5)

$$x = \frac{b_2 c_1 - b_1 c_2}{a_1 b_2 - a_2 b_1}$$

can be written as

$$x = \frac{|A_x|}{|A|}$$

and equation (8)

$$y = \frac{a_1 c_2 - a_2 c_1}{a_1 b_2 - a_2 b_1}$$

can be written as

$$y = \frac{|A_y|}{|A|}$$

The following examples will show how determinants can be used to solve a system of two linear equations in two unknowns.

EXAMPLE 92   Solve the following for x and y by determinants.

$$x + y = 5$$
$$2x - y = 1$$

Solution: Form the three determinants $|A|$, $|A_x|$, $|A_y|$.

$$|A| = \begin{vmatrix} 1 & 1 \\ 2 & -1 \end{vmatrix}, \qquad |A_x| = \begin{vmatrix} 5 & 1 \\ 1 & -1 \end{vmatrix}, \qquad |A_y| = \begin{vmatrix} 1 & 5 \\ 2 & 1 \end{vmatrix}$$

$$x = \frac{|A_x|}{|A|} \qquad\qquad\qquad y = \frac{|A_y|}{|A|}$$

$$= \frac{\begin{vmatrix} 5 & 1 \\ 1 & -1 \end{vmatrix}}{\begin{vmatrix} 1 & 1 \\ 2 & -1 \end{vmatrix}} \qquad\qquad = \frac{\begin{vmatrix} 1 & 5 \\ 2 & 1 \end{vmatrix}}{\begin{vmatrix} 1 & 1 \\ 2 & -1 \end{vmatrix}}$$

$$= \frac{[5 \times (-1)] - (1 \times 1)}{[1 \times (-1)] - (2 \times 1)} \qquad = \frac{(1 \times 1) - (2 \times 5)}{1 \times (-1) - (2 \times 1)}$$

$$= \frac{-5 - 1}{-1 - 2} \qquad\qquad\qquad = \frac{1 - 10}{-1 - 2}$$

$$= \frac{-6}{-3} \qquad\qquad\qquad\qquad = \frac{-9}{-3}$$

$$= 2 \qquad\qquad\qquad\qquad\quad = 3$$

EXAMPLE 93   Solve for x and y by determinants

$$2x + 5y = 24$$
$$x + 2y = 6$$

Solution: $|A| = \begin{vmatrix} 2 & 5 \\ 1 & 2 \end{vmatrix}, \ |Ax| = \begin{vmatrix} 24 & 5 \\ 6 & 2 \end{vmatrix}, \ |Ay| = \begin{vmatrix} 2 & 24 \\ 1 & 6 \end{vmatrix}$

$$x = \frac{|A_x|}{|A|} \qquad\qquad\qquad y = \frac{|A_y|}{|A|}$$

$$= \frac{\begin{vmatrix} 24 & 5 \\ 6 & 2 \end{vmatrix}}{\begin{vmatrix} 2 & 5 \\ 1 & 2 \end{vmatrix}} \qquad\qquad = \frac{\begin{vmatrix} 2 & 24 \\ 1 & 6 \end{vmatrix}}{\begin{vmatrix} 2 & 5 \\ 1 & 2 \end{vmatrix}}$$

$$= \frac{(24 \times 2) - (6 \times 5)}{(2 \times 2) - (1 \times 5)} \qquad = \frac{(2 \times 6) - (1 \times 24)}{(2 \times 2) - (1 \times 5)}$$

$$= \frac{48 - 30}{4 - 5} \qquad\qquad\qquad = \frac{12 - 24}{4 - 5}$$

$$= \frac{18}{-1} \qquad\qquad\qquad\qquad = \frac{-12}{-1}$$

$$= -18 \qquad\qquad\qquad\qquad = 12$$

Solve the following systems of equations by determinants.

1. $2x + 5y = \quad 13$

   $7x - 4y = -19$

3. $2m - n = -6$

   $2m + n = \quad 0$

5. $2\omega - 7\phi = 13$

   $5\omega - 4\phi = 19$

7. $\frac{1}{2}m - \frac{1}{5}n = \frac{21}{10}$

   $\frac{1}{6}m - \frac{1}{4}n = -\frac{1}{4}$

2. $20 - 5\alpha = \quad -7$

   $80 + 3\alpha = -11$

4. $r + s = 5$

   $2r + s = 6$

6. $2x - 3y = \quad 6$

   $4x - 2y = 12$

8. $\dfrac{x}{2} + \dfrac{y}{3} = 13$

   $\dfrac{x}{5} + \dfrac{y}{8} = 5$

2-34
Third-Order Determinants

A third-order (3×3) determinant such as

$$\begin{vmatrix} a_1 & b_1 & c_1 \\ a_2 & b_2 & c_2 \\ a_3 & b_3 & c_3 \end{vmatrix}$$

can be expanded to find its value as follows:

$$\begin{vmatrix} a_1 & b_1 & c_1 \\ a_2 & b_2 & c_2 \\ a_3 & b_3 & c_3 \end{vmatrix} = a_1 \begin{vmatrix} b_2 & c_2 \\ b_3 & c_3 \end{vmatrix} - a_2 \begin{vmatrix} b_1 & c_1 \\ b_3 & c_3 \end{vmatrix} + a_3 \begin{vmatrix} b_1 & c_1 \\ b_2 & c_2 \end{vmatrix}$$

$$= a_1(b_2c_3 - b_3c_2) - a_2(b_1c_3 - b_3c_1) + a_3(b_1c_2 - b_2c_1)$$

$$= a_1b_2c_3 - a_1b_3c_2 - a_2b_1c_3 + a_2b_3c_1 + a_3b_1c_2 - a_3b_2c_1$$

$$= a_1b_2c_3 + a_2b_3c_1 + a_3b_1c_2 - a_1b_3c_2 - a_2b_1c_3 - a_3b_2c_1 \quad (11)$$

The above method of solving this determinant is called "expanding by minors." A *minor* of an element is a determinant formed with the elements that do not appear in the same row or column as the element being considered.

For example in the determinant above the minor of a_1 is $\begin{vmatrix} b_2 & c_2 \\ b_3 & c_3 \end{vmatrix}$, the minor of b_1 is $\begin{vmatrix} a_2 & c_2 \\ a_3 & c_3 \end{vmatrix}$.

To expand a third-order determinant by minors, it is necessary that it be expanded along a row or column of elements. This means choosing a row or column and finding the algebraic sum of the products formed by each element in the indicated row or column with its minor. The sign which precedes each product of an element and its minor can be determined from the sum of the row and column in which the element appears. If the sum of the row and column in which the element being considered appears is an even number, the product will be preceded by a positive $(+)$ sign. If the sum is an odd number, the product will be preceded by a negative $(-)$ sign. For example, in the above expansion, the sign which precedes $a_1 \begin{vmatrix} b_2 & c_2 \\ b_3 & c_3 \end{vmatrix}$ is positive $(+)$ because a_1 is in the first row, first column and the sum of 1 and 1 is 2, an even number. Since a_2 is in the second row, first column the sign which precedes the product $a_2 \begin{vmatrix} b_1 & c_1 \\ b_3 & c_3 \end{vmatrix}$ is negative $(-)$ because the sum of 2 and 1, which is 3, is an odd number. The sign which precedes the product $a_3 \begin{vmatrix} b_1 & c_1 \\ b_2 & c_2 \end{vmatrix}$ is positive $(+)$ because a_3 is in the third row, first column and the sum of 3 and 1 is an even number.

Now expand $\begin{vmatrix} a_1 & b_1 & c_1 \\ a_2 & b_2 & c_2 \\ a_3 & b_3 & c_3 \end{vmatrix}$ along the third column by minors:

$$\begin{vmatrix} a_1 & b_1 & c_1 \\ a_2 & b_2 & c_2 \\ a_3 & b_3 & c_3 \end{vmatrix} = c_1 \begin{vmatrix} a_2 & b_2 \\ a_3 & b_3 \end{vmatrix} - c_2 \begin{vmatrix} a_1 & b_1 \\ a_3 & b_3 \end{vmatrix} + c_3 \begin{vmatrix} a_1 & b_1 \\ a_2 & b_2 \end{vmatrix}$$

At this time we point out that the signs preceding c_1, c_2, and c_3 were found in the following manner. c_1 is in the first row, third column, $1 + 3$ is even, hence the positive sign is used; c_2 is in the second row, third column, $2 + 3$ is odd, hence the negative sign is used; c_3 is in the third row, third column, $3 + 3$ is even, hence the positive sign is used.

$$\begin{vmatrix} a_1 & b_1 & c_1 \\ a_2 & b_2 & c_2 \\ a_3 & b_3 & c_3 \end{vmatrix} \begin{aligned} &= c_1(a_2b_3 - a_3b_2) - c_2(a_1b_3 - a_3b_1) + c_3(a_1b_2 - a_2b_1) \\ &= a_2b_3c_1 - a_3b_2c_1 - a_1b_3c_2 + a_3b_1c_2 + a_1b_2c_3 - a_2b_1c_3 \\ &= a_1b_2c_3 + a_2b_3c_1 + a_3b_1c_2 - a_1b_3c_2 - a_2b_1c_3 - a_3b_2c_1 \end{aligned}$$

$$(12)$$

Expand $\begin{vmatrix} a_1 & b_1 & c_1 \\ a_2 & b_2 & c_2 \\ a_3 & b_3 & c_3 \end{vmatrix}$ along the second row.

$$\begin{aligned} \begin{vmatrix} a_1 & b_1 & c_1 \\ a_2 & b_2 & c_2 \\ a_3 & b_3 & c_3 \end{vmatrix} &= -a_2 \begin{vmatrix} b_1 & c_1 \\ b_3 & c_3 \end{vmatrix} + b_2 \begin{vmatrix} a_1 & c_1 \\ a_3 & c_3 \end{vmatrix} - c_2 \begin{vmatrix} a_1 & b_1 \\ a_3 & b_3 \end{vmatrix} \\ &= -a_2(b_1c_3 - b_3c_1) + b_2(a_1c_3 - a_3c_1) - c_2(a_1b_3 - a_3b_1) \\ &= -a_2b_1c_3 + a_2b_3c_1 + a_1b_2c_3 - a_3b_2c_1 - a_1b_3c_2 + a_3b_1c_2 \\ &= a_1b_2c_3 + a_2b_3c_1 + a_3b_1c_2 - a_1b_3c_2 - a_2b_1c_3 - a_3b_2c_1 \quad (13) \end{aligned}$$

By now, it should be obvious from equations (11), (12), and (13), that it makes no difference as to which row or column is chosen for the expansion.

EXAMPLE 94   Find the value of $\begin{vmatrix} 1 & 5 & 2 \\ 4 & 7 & 3 \\ 2 & -3 & 6 \end{vmatrix}$.

Solution: Expanding along row 1,

$$\begin{vmatrix} 1 & 5 & 2 \\ 4 & 7 & 3 \\ 2 & -3 & 6 \end{vmatrix} = 1\begin{vmatrix} 7 & 3 \\ -3 & 6 \end{vmatrix} - 5\begin{vmatrix} 4 & 3 \\ 2 & 6 \end{vmatrix} + 2\begin{vmatrix} 4 & 7 \\ 2 & -3 \end{vmatrix}$$

$$= 1[42 - (-9)] - 5(24 - 6) + 2(-12 - 14)$$

$$= 51 - 90 - 52$$

$$= -91$$

EXAMPLE 95   Find the value of $\begin{vmatrix} 1 & -1 & 0 \\ 2 & 3 & -4 \\ 3 & 4 & -8 \end{vmatrix}$.

Solution: Expanding along column 2,

$$\begin{vmatrix} 1 & -1 & 0 \\ 2 & 3 & -4 \\ 3 & 4 & -8 \end{vmatrix} = -(-1)\begin{vmatrix} 2 & -4 \\ 3 & -8 \end{vmatrix} + 3\begin{vmatrix} 1 & 0 \\ 3 & -8 \end{vmatrix} - 4\begin{vmatrix} 1 & 0 \\ 2 & -4 \end{vmatrix}$$

$$= 1[-16 - (-12)] + 3(-8 - 0) - 4(-4 - 0)$$

$$= -4 - 24 + 16$$

$$= -12$$

PROBLEM SET 2-34

Solve the following determinants.

1. $\begin{vmatrix} 1 & 1 & 1 \\ 3 & -1 & 2 \\ 2 & 3 & -1 \end{vmatrix}$
   2. $\begin{vmatrix} 6 & 7 & 5 \\ 1 & 0 & 1 \\ 2 & 4 & 8 \end{vmatrix}$
   3. $\begin{vmatrix} 2 & 0 & 1 \\ 1 & 1 & 1 \\ -1 & -1 & -\frac{1}{2} \end{vmatrix}$

4. $\begin{vmatrix} 4 & 1 & 3 \\ 0 & -5 & 1 \\ 7 & 2 & -1 \end{vmatrix}$
   5. $\begin{vmatrix} 1 & 0 & 0 \\ 0 & 2 & 3 \\ 0 & 0 & -1 \end{vmatrix}$
   6. $\begin{vmatrix} 2 & -1 & \frac{1}{2} \\ \frac{1}{4} & 1 & 2 \\ \frac{1}{2} & 0 & \frac{1}{4} \end{vmatrix}$

2-35

Solving Systems of Three Equations in Three Unknowns with Determinants

The general equation in three unknowns is $ax + by + cz = d$. Consider three such equations and solve these for x, y, and z algebraically.

$$a_1x + b_1y + c_1z = d_1 \tag{14}$$

$$a_2x + b_2y + c_2z = d_2 \tag{15}$$

$$a_3x + b_3y + c_3z = d_3 \tag{16}$$

To solve this system for x, we first eliminate z and then y. To eliminate z, we pick two equations and eliminate z, then from another pair z is eliminated. In each case an equation results in x and y only. From the two resulting equations y can be eliminated. Hence

$$a_1x + b_1y + c_1z = d_1 \tag{14}$$

$$a_2x + b_2y + c_2z = d_2 \tag{15}$$

multiply (14) by c_2, (15) by c_1, and subtracting,

$$a_1c_2x + b_1c_2y + c_1c_2z = c_2d_1$$

$$a_2c_1x + b_2c_1y + c_1c_2z = c_1d_2$$

$$\overline{a_1c_2x - a_2c_1x + b_1c_2y - b_2c_1y = c_2d_1 - c_1d_2}$$

$$(a_1c_2 - a_2c_1)x + (b_1c_2 - b_2c_1)y = c_2d_1 - c_1d_2 \tag{17}$$

$$a_1x + b_1y + c_1z = d_1 \tag{14}$$

$$a_3x + b_3y + c_3z = d_3 \tag{16}$$

multiply (14) by c_3, (16) by c_1, and subtracting,

$$a_1c_3x + b_1c_3y + c_1c_3z = c_3d_1$$

$$a_3c_1x + b_3c_1y + c_1c_3z = c_1d_3$$

$$\overline{a_1c_3x - a_3c_1x + b_1c_3y - b_3c_1y = c_3d_1 - c_1d_3}$$

$$(a_1c_3 - a_3c_1)x + (b_1c_3 - b_3c_1)y = c_3d_1 - c_1d_3 \tag{18}$$

from equations (17) and (18) eliminate y

$$(a_1c_2 - a_2c_1)x + (b_1c_2 - b_2c_1)y = c_2d_1 - c_1d_2 \tag{17}$$

$$(a_1c_3 - a_3c_1)x + (b_1c_3 - b_3c_1)y = c_3d_1 - c_1d_3 \tag{18}$$

multiply (17) by $(b_1c_3 - b_3c_1)$, (18) by $(b_1c_2 - b_2c_1)$ and subtract,

$$(b_1c_3 - b_3c_1)(a_1c_2 - a_2c_1)x + (b_1c_3 - b_3c_1)(b_1c_2 - b_2c_1)y$$
$$= (b_1c_3 - b_3c_1)(c_2d_1 - c_1d_2) \tag{19}$$

$$(b_1c_2 - b_2c_1)(a_1c_3 - a_3c_1)x + (b_1c_2 - b_2c_1)(b_1c_3 - b_3c_1)y$$
$$= (b_1c_2 - b_2c_1)(c_3d_1 - c_1d_3) \tag{20}$$

$$(a_1b_1c_2c_3 - a_2b_1c_1c_3 - a_1b_3c_1c_2 + a_2b_3c_1c_1)x$$
$$+ (b_1b_1c_2c_3 - b_1b_2c_1c_3 - b_1b_3c_1c_2 + b_2b_3c_1c_1)y$$
$$= b_1c_2c_3d_1 - b_1c_1c_3d_2 - b_3c_1c_2d_1 + b_3c_1c_1d_2 \tag{19}$$

$$(a_1b_1c_2c_3 - a_3b_1c_1c_2 - a_1b_2c_1c_3 + a_3b_2c_1c_1)x$$
$$+ (b_1b_1c_2c_3 - b_1b_3c_1c_2 - b_1b_2c_1c_3 + b_2b_3c_1c_1)y$$
$$= b_1c_2c_3d_1 - b_1c_1c_2d_3 - b_2c_1c_3d_1 + b_2c_1c_1d_3 \tag{20}$$

subtracting (20) from (19) results in:

$$[(a_1b_1c_2c_3 - a_2b_1c_1c_3 - a_1b_3c_1c_2 + a_2b_3c_1c_1)$$
$$- (a_1b_1c_2c_3 - a_3b_1c_1c_2 - a_1b_2c_1c_3 + a_3b_2c_1c_1)]x$$
$$= (b_1c_2c_3d_1 - b_1c_1c_3d_2 - b_3c_1c_2d_1 + b_3c_1c_1d_2)$$
$$- (b_1c_2c_3d_1 - b_1c_1c_2d_3 - b_2c_1c_3d_1 + b_2c_1c_1d_3) \qquad (21)$$

$$(a_1b_1c_2c_3 - a_2b_1c_1c_3 - a_1b_3c_1c_2 + a_2b_3c_1c_1 - a_1b_1c_2c_3$$
$$+ a_3b_1c_1c_2 + a_1b_2c_1c_3 - a_3b_2c_1c_1)x$$
$$= b_1c_2c_3d_1 - b_1c_1c_3d_2 - b_3c_1c_2d_1 + b_3c_1c_1d_2 - b_1c_2c_3d_1$$
$$+ b_1c_1c_2d_3 + b_2c_1c_3d_1 - b_2c_1c_1d_3 \qquad (22)$$

$$(a_1b_2c_1c_3 + a_2b_3c_1c_1 + a_3b_1c_1c_2 - a_1b_3c_1c_2 - a_2b_1c_1c_3 - a_3b_2c_1c_1)x$$
$$= b_1c_1c_2d_3 + b_2c_1c_3d_1 + b_3c_1c_1d_2 - b_1c_1c_3d_2 - b_2c_1c_1d_3 - b_3c_1c_2d_1 \qquad (23)$$

$$c_1(a_1b_2c_3 + a_2b_3c_1 + a_3b_1c_2 - a_1b_3c_2 - a_2b_1c_3 - a_3b_2c_1)x$$
$$= c_1(b_1c_2d_3 + b_2c_3d_1 + b_3c_1d_2 - b_1c_3d_2 - b_2c_1d_3 - b_3c_2d_1) \qquad (24)$$

divide both sides by c_1, then solve for x,

$$x = \frac{b_1c_2d_3 + b_2c_3d_1 + b_3c_1d_2 - b_1c_3d_2 - b_2c_1d_3 - b_3c_2d_1}{a_1b_2c_3 + a_2b_3c_1 + a_3b_1c_2 - a_1b_3c_2 - a_2b_1c_3 - a_3b_2c_1} \qquad (25)$$

Equation (25) could be written as:

$$x = \frac{(b_2c_3d_1 - b_3c_2d_1) - (b_1c_3d_2 - b_3c_1d_2) + (b_1c_2d_3 - b_2c_1d_3)}{(a_1b_2c_3 - a_1b_3c_2) - (a_2b_1c_3 - a_2b_3c_1) + (a_3b_1c_2 - a_3b_2c_1)} \qquad (26)$$

$$x = \frac{d_1(b_2c_3 - b_3c_2) - d_2(b_1c_3 - b_3c_1) + d_3(b_1c_2 - b_2c_1)}{a_1(b_2c_3 - b_3c_2) - a_2(b_1c_3 - b_3c_1) + a_3(b_1c_2 - b_2c_1)} \qquad (27)$$

$$x = \frac{d_1 \begin{vmatrix} b_2 & c_2 \\ b_3 & c_3 \end{vmatrix} - d_2 \begin{vmatrix} b_1 & c_1 \\ b_3 & c_3 \end{vmatrix} + d_3 \begin{vmatrix} b_1 & c_1 \\ b_2 & c_2 \end{vmatrix}}{a_1 \begin{vmatrix} b_2 & c_2 \\ b_3 & c_3 \end{vmatrix} - a_2 \begin{vmatrix} b_1 & c_1 \\ b_3 & c_3 \end{vmatrix} + a_3 \begin{vmatrix} b_1 & c_1 \\ b_2 & c_2 \end{vmatrix}} \qquad (28)$$

$$x = \frac{\begin{vmatrix} d_1 & b_1 & c_1 \\ d_2 & b_2 & c_2 \\ d_3 & b_3 & c_3 \end{vmatrix}}{\begin{vmatrix} a_1 & b_1 & c_1 \\ a_2 & b_2 & c_2 \\ a_3 & b_3 & c_3 \end{vmatrix}} \qquad (29)$$

The denominator $\begin{vmatrix} a_1 & b_1 & c_1 \\ a_2 & b_2 & c_2 \\ a_3 & b_3 & c_3 \end{vmatrix}$ of (29) is the determinant formed with the

coefficients of x, y, and z in the system

$$a_1x + b_1y + c_1z = d_1$$
$$a_2x + b_2y + c_2z = d_2$$
$$a_3x + b_3y + c_3z = d_3$$

If the coefficients (a) are removed from this determinant and replaced with the constants (d) the resulting determinant

$$\begin{vmatrix} d_1 & b_1 & c_1 \\ d_2 & b_2 & c_2 \\ d_3 & b_3 & c_3 \end{vmatrix}$$

is equal to the numerator of equation (29).

It can be shown that solving for y and z in the above system results in the following:

$$y = \frac{\begin{vmatrix} a_1 & d_1 & c_1 \\ a_2 & d_2 & c_2 \\ a_3 & d_3 & c_3 \end{vmatrix}}{\begin{vmatrix} a_1 & b_1 & c_1 \\ a_2 & b_2 & c_2 \\ a_3 & b_3 & c_3 \end{vmatrix}} \tag{30}$$

$$z = \frac{\begin{vmatrix} a_1 & b_1 & d_1 \\ a_2 & b_2 & d_2 \\ a_3 & b_3 & d_3 \end{vmatrix}}{\begin{vmatrix} a_1 & b_1 & c_1 \\ a_2 & b_2 & c_2 \\ a_3 & b_3 & c_3 \end{vmatrix}} \tag{31}$$

Letting $|A| = \begin{vmatrix} a_1 & b_1 & c_1 \\ a_2 & b_2 & c_2 \\ a_3 & b_3 & c_3 \end{vmatrix}$, $|A_x| = \begin{vmatrix} d_1 & b_1 & c_1 \\ d_2 & b_2 & c_2 \\ d_3 & b_3 & c_3 \end{vmatrix}$, $|A_y| = \begin{vmatrix} a_1 & d_1 & c_1 \\ a_2 & d_2 & c_2 \\ a_3 & d_3 & c_3 \end{vmatrix}$, $A_z =$

$\begin{vmatrix} a_1 & b_1 & d_1 \\ a_2 & b_2 & d_2 \\ a_3 & b_3 & d_3 \end{vmatrix}$ equations (29), (30), and (31) can be written as:

$$x = \frac{|A_x|}{|A|} \tag{32}$$

$$y = \frac{|A_y|}{|A|} \tag{33}$$

$$z = \frac{|A_z|}{|A|} \tag{34}$$

EXAMPLE 96   Solve the following system of equations by determinants

$$3x + y - z = 14$$
$$x + 3y - z = 16$$
$$x + y - 3z = -10$$

Solution:
$$|A| = \begin{vmatrix} 3 & 1 & -1 \\ 1 & 3 & -1 \\ 1 & 1 & -3 \end{vmatrix}$$

$$|A_x| = \begin{vmatrix} 14 & 1 & -1 \\ 16 & 3 & -1 \\ -10 & 1 & -3 \end{vmatrix}$$

$$|A_y| = \begin{vmatrix} 3 & 14 & -1 \\ 1 & 16 & -1 \\ 1 & -10 & -3 \end{vmatrix}$$

$$|A_z| = \begin{vmatrix} 3 & 1 & 14 \\ 1 & 3 & 16 \\ 1 & 1 & -10 \end{vmatrix}$$

$$x = \frac{|A_x|}{|A|}$$

$$= \frac{\begin{vmatrix} 14 & 1 & -1 \\ 16 & 3 & -1 \\ -10 & 1 & -3 \end{vmatrix}}{\begin{vmatrix} 3 & 1 & -1 \\ 1 & 3 & -1 \\ 1 & 1 & -3 \end{vmatrix}}$$

$$= \frac{14\begin{vmatrix} 3 & -1 \\ 1 & -3 \end{vmatrix} - 1\begin{vmatrix} 16 & -1 \\ -10 & -3 \end{vmatrix} + (-1)\begin{vmatrix} 16 & 3 \\ -10 & 1 \end{vmatrix}}{3\begin{vmatrix} 3 & -1 \\ 1 & -3 \end{vmatrix} - 1\begin{vmatrix} 1 & -1 \\ 1 & -3 \end{vmatrix} + (-1)\begin{vmatrix} 1 & 3 \\ 1 & 1 \end{vmatrix}}$$

$$= \frac{14[(-9) - (-1)] - 1[(-48) - (10)] + (-1)[(16) - (-30)]}{3[(-9) - (-1)] - 1[(-3) - (-1)] + (-1)[(1) - (3)]}$$

$$= \frac{14(-8) - 1(-58) - 1(46)}{3(-8) - 1(-2) - 1(-2)}$$

$$= \frac{-112 + 58 - 46}{-24 + 2 + 2}$$

$$= \frac{-100}{-20}$$

$$= 5.$$

$$y = \frac{|A_y|}{|A|}$$

$$= \frac{\begin{vmatrix} 3 & 14 & -1 \\ 1 & 16 & -1 \\ 1 & -10 & -3 \end{vmatrix}}{\begin{vmatrix} 3 & 1 & -1 \\ 1 & 3 & -1 \\ 1 & 1 & -3 \end{vmatrix}}$$

$$= \frac{3\begin{vmatrix} 16 & -1 \\ -10 & -3 \end{vmatrix} - 1\begin{vmatrix} 14 & -1 \\ -10 & -3 \end{vmatrix} + 1\begin{vmatrix} 14 & -1 \\ 16 & -1 \end{vmatrix}}{3\begin{vmatrix} 3 & -1 \\ 1 & -3 \end{vmatrix} - 1\begin{vmatrix} 1 & -1 \\ 1 & -3 \end{vmatrix} + 1\begin{vmatrix} 1 & -1 \\ 3 & -1 \end{vmatrix}}$$

$$= \frac{3[(-48) - (10)] - 1[(-42) - (10)] + 1[(-14) - (-16)]}{3[(-9) - (-1)] - 1[(-3) - (-1)] + 1[(-1) - (-3)]}$$

$$= \frac{3(-58) - 1(-52) + 1(2)}{3(-8) - 1(-2) + 1(2)}$$

$$= \frac{-174 + 52 + 2}{-24 + 2 + 2}$$

$$= \frac{-120}{-20}$$

$$= 6.$$

$$z = \frac{|A_z|}{|A|}$$

$$= \frac{\begin{vmatrix} 3 & 1 & 14 \\ 1 & 3 & 16 \\ 1 & 1 & -10 \end{vmatrix}}{\begin{vmatrix} 3 & 1 & -1 \\ 1 & 3 & -1 \\ 1 & 1 & -3 \end{vmatrix}}$$

$$= \frac{-(1)\begin{vmatrix} 1 & 14 \\ 1 & -10 \end{vmatrix} + 3\begin{vmatrix} 3 & 14 \\ 1 & -10 \end{vmatrix} - 16\begin{vmatrix} 3 & 1 \\ 1 & 1 \end{vmatrix}}{(-1)\begin{vmatrix} 1 & 3 \\ 1 & 1 \end{vmatrix} - (-1)\begin{vmatrix} 3 & 1 \\ 1 & 1 \end{vmatrix} + (-3)\begin{vmatrix} 3 & 1 \\ 1 & 3 \end{vmatrix}}$$

$$= \frac{-1[(-10) - (14)] + 3[(-30) - (14)] - 16[(3) - (1)]}{-1[(1) - (3)] + 1[(3) - (1)] - 3[(9) - (1)]}$$

$$= \frac{-1(-24) + 3(-44) - 16(2)}{-1(-2) + 1(2) - 3(8)}$$

$$= \frac{+24 - 132 - 32}{2 + 2 - 24}$$

$$= \frac{-140}{-20}$$

$$= 7.$$

The answers $x = 5$, $y = 6$, $z = 7$, can be checked by substituting these values into each equation. The results must be a true statement in each of the three equations.

PROBLEM SET 2-35

Solve the following system of equations by determinants.

1. $x + y + z = 6$
   $2x + 3y - z = 5$
   $3x - y + 2z = 7$

2. $20 + \alpha + \beta = 8$
   $0 + \frac{1}{2}\alpha - \beta = -2$
   $40 - \alpha + 2\beta = 9$

3. $r + s + t = 3$
   $r - s + t = 2$
   $2r - s - t = 0$

4. $x + y = z$
   $y - z = -z$
   $2x + z = 7$

5. $\frac{1}{5}x + \frac{1}{5}y + z = 2$
   $4x + y - z = 10$
   $\frac{1}{2}x + \frac{1}{3}y - \frac{3}{2}z = \frac{7}{2}$

6. $.2x + .50y - .5z = 2$
   $4x - y - z = 0$
   $.50x + y - .5z = 3$

7. $2x + 3y + z = 4$
   $x + 5y - 2z = -1$
   $3x - 4y + 4z = -1$

8. $2r + 3s - t = -1$
   $3r + 4s + 2t = 14$
   $r - 6s - 5t = 4$

9. $5x + 2y + 4z = -5$
   $2x - 5y + 3z = 4$
   $7x + 8y - 2z = 13$

10. $x + 2y - z = 1$
    $3x + y - 4z = -13$
    $4x - y + 2z = -17$

2-36
Application of Determinants

Any system of linear equations may be solved by determinants. One advantage of using determinants, or matrices, is the ease with which a computer program may be written to solve the system of equations. The following example shows how a system of equations may be written for the series parallel electrical circuit depicted in Figure 2-27 and a solution obtained by the use of determinants.

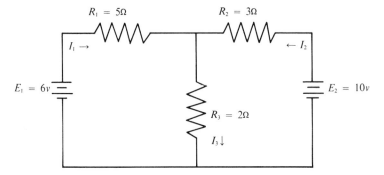

Figure 2-27

The unknowns in the circuit are the currents, I_1, I_2, and I_3. Having three unknowns we need a system of three equations. The following three equations may be written for the circuit.

$$I_1 R_1 + I_3 R_3 = E_1 \qquad I_1 R_1 + 0I_2 + I_3 R_3 = E_1$$
$$I_2 R_2 + I_3 R_3 = E_2 \quad \text{or} \quad 0I_1 + I_2 R_2 + I_3 R_3 = E_2$$
$$I_1 + I_2 - I_3 = 0 \qquad I_1 + I_2 - I_3 = 0$$

Substituting the known values of resistance and voltage, our equations become

$$5I_1 + 0I_2 + 2I_3 = 6$$
$$0I_1 + 3I_2 + 2I_3 = 10$$
$$1I_1 + 1I_2 - 1I_3 = 0$$

These equations may be put in determinant form as follows

$$|A| = \begin{vmatrix} 5 & 0 & 2 \\ 0 & 3 & 2 \\ 1 & 1 & -1 \end{vmatrix} \qquad |A_{I_1}| = \begin{vmatrix} 6 & 0 & 2 \\ 10 & 3 & 2 \\ 0 & 1 & -1 \end{vmatrix}$$

$$|A_{I_2}| = \begin{vmatrix} 5 & 6 & 2 \\ 0 & 10 & 2 \\ 1 & 0 & -1 \end{vmatrix} \qquad |A_{I_3}| = \begin{vmatrix} 5 & 0 & 6 \\ 0 & 3 & 10 \\ 1 & 1 & 0 \end{vmatrix}$$

Solving these determinants give the following results:

$$|A| = -31 \qquad |A_{I_1}| = -10 \qquad |A_{I_2}| = -58 \qquad |A_{I_3}| = -68$$

With these results, the value of the unknowns can be found as follows

$$I_1 = \frac{|A_{I_1}|}{|A|} = \frac{-10}{-31} = .319 \text{ amps}$$

$$I_2 = \frac{|A_{I_2}|}{|A|} = \frac{-58}{-31} = 1.88 \text{ amps}$$

$$I_3 = \frac{|A_{I_3}|}{|A|} = \frac{-68}{-31} = 2.19 \text{ amps}$$

3

Geometry

Certain ideas and properties from geometry are very important and useful in mathematics and technical areas. Only those which we feel are of most importance will be mentioned in this chapter. It is not our intent to give a rigorous account of geometry, therefore, very few proofs will be given and no proofs expected of the students. However, several exercises will be given so that the student who has not had a course in geometry will be able to solve the problems that he will encounter as he continues his studies in the technical areas.

A *point* is said to have position only, and can be represented as a dot ($\cdot$). A *line* is defined as an infinite number of points continuing without end in any direction. Normally, we think of a line as being straight. By straight we mean "in the same direction." Figure 3-1 shows most of the common representations of a line.

A line is usually referred to by the small letter or by two of its points. For example, in Figure 3-2, we could refer to the line as line m, line AB, line BD, or sometimes $\overleftrightarrow{CD}$.

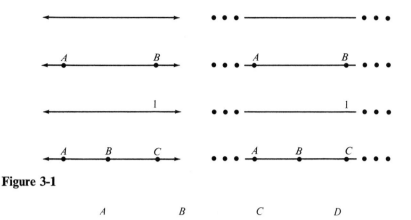

Figure 3-1

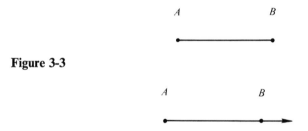

Figure 3-2

A *line segment* is defined as two points on a line and all points in between. For example, when we speak of line segment *BC* in Figure 3-3, we are referring to that part of the line from *B* to *C* and including *B* and *C*. Other ways of referring to line segment *BC* would be segment $\overline{BC}$ or *BC*.

A *ray* is a point on a line and all points to one side. Figure 3-4 shows the ray *AB*. Another way to identify ray *AB* is $\overrightarrow{AB}$.

Figure 3-3

Figure 3-4

We will use the idea of a flat surface extending in all directions without end as a *plane*. A plane can also be described as an infinite number of lines within a flat surface.

As our discussion of geometry and geometric ideas continues we assume everything to be in a plane unless otherwise stated.

3-3
Angles

An *angle* is defined as the rotation of a ray about its end point from one position (initial) to another position (terminal). We can also say that an angle is made by two rays, or half-lines, with a common endpoint called the vertex of the angle.

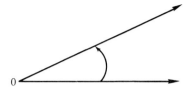

Figure 3-5

The measure of an angle is the amount of rotation from the initial side to the terminal side. Usually this measure is given in either degrees or radians. We will use degrees to define the different types of angles and discuss radian measure in a later chapter.

A complete rotation is said to be 360°; therefore, 1° is 1/360 of a rotation. Figure 3-6, shows a rotation of 360° and a rotation of 1°.

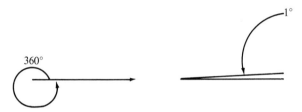

Figure 3-6

An *acute angle* is an angle whose measure is less than 90°. A *right angle* is an angle whose measure is 90°. An angle whose measure is more than 90° but less than 180° is an *obtuse angle*. An angle whose measure is 180° is a *straight angle*.

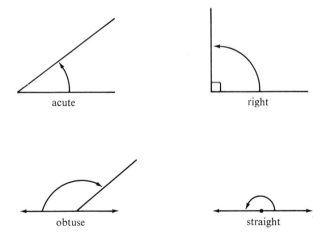

Figure 3-7

Two angles whose sum is 90° are *complementary angles*. Two angles that total 180° are *supplementary angles*.

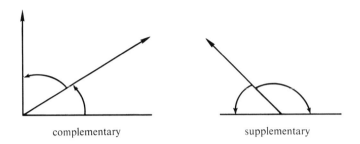

complementary   supplementary

Figure 3-8

Angles are labeled in several different ways. For example, in Figure 3-9, we could call this angle *ABC*, angle *CBA*, angle *B* or angle *b*.

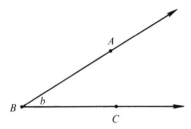

Figure 3-9

All are correct as long as it is understood which angle we are talking about. We use the symbol ($\angle$) to mean angle. Therefore we write *angle B* as $\angle B$. This symbol ($\angle$) will be used throughout the book to designate the word *angle*.

PROBLEM SET 3-1

Name each angle four different ways.

1.   2.

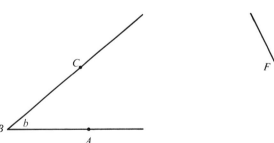

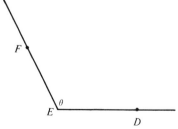

Identify each of the following.

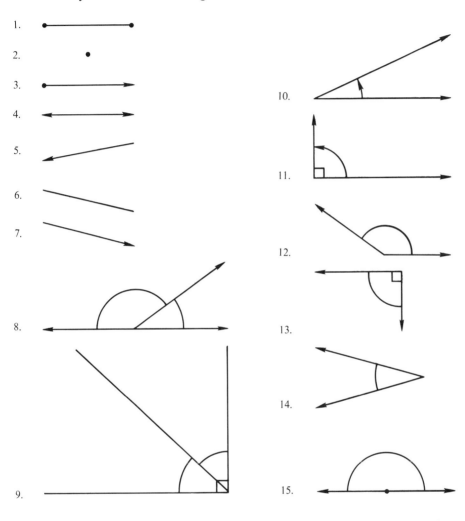

1.

2.

3.

4.

5.

6.

7.

8.

9.

10.

11.

12.

13.

14.

15.

3-4
Parallel and Perpendicular Lines

If two lines intersect at right angles they are said to be *perpendicular lines*. Two lines that are the same distance apart are called *parallel lines*.

A *transversal* is a line which intersects two or more parallel lines. In Figure 3-11, line *t* is a transversal.

From Figure 3-12, we see that eight angles are formed when two parallel lines are cut by a transversal. Angles *a* and *b* are called *adjacent angles*. Adjacent

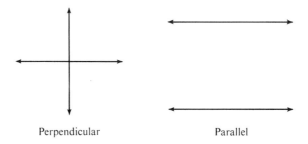

Perpendicular   Parallel

Figure 3-10

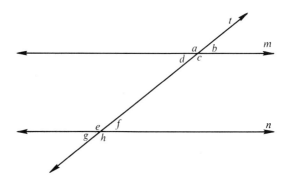

Figure 3-11

angles are two angles having a common side and a common vertex. The follow-
ing pairs of angles are called vertical angles, $\angle a$ and $\angle c$, $\angle b$ and $\angle d$, $\angle e$ and
$\angle g$, $\angle f$ and $\angle h$. *Vertical angles* are the opposite angles formed by two inter-
secting lines. Angles d and f are called *alternate interior angles* and are equal.
There is another pair of alternate interior angles. Can you name them? *Corresponding angles* are what we call $\angle a$ and $\angle e$. In Figure 3-12 there are four pairs
of corresponding angles. Sometimes we refer to $\angle b$ and $\angle h$ as *alternate
exterior angles*. What are the other alternate exterior angles? Which angles
do you think are supplementary? Which are corresponding?

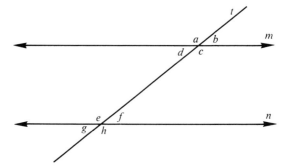

Figure 3-12

1. Draw two perpendicular lines.

2. Draw two parallel lines.

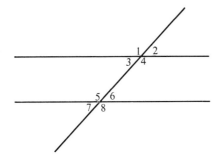

3. Use the drawing above and identify the following pairs of angles.

(a) $\angle 3$ and $\angle 6$ (b) $\angle 1$ and $\angle 2$

(c) $\angle 7$ and $\angle 6$ (d) $\angle 7$ and $\angle 3$

(e) $\angle 5$ and $\angle 8$ (f) $\angle 4$ and $\angle 5$

(g) $\angle 2$ and $\angle 6$ (h) $\angle 1$ and $\angle 8$

3-5
Polygons

A *polygon* is closed figure whose sides are line segments. If all the sides are equal, it is called a *regular* polygon. Polygons include triangles, quadrilaterals, pentagons, hexagons, and so forth. Our discussion will be limited to the triangle and quadrilateral.

3-6
Triangle

A *triangle* is a three-sided *polygon*. It forms three angles whose sum is 180°. There are several types of triangles and they are identified according to their angles or their sides.

An *acute triangle* is a triangle with three acute angles. A *right triangle* is a triangle with one right angle. An *obtuse triangle* is a triangle having one obtuse angle. In a right triangle and an obtuse triangle the other two angles are necessarily acute.

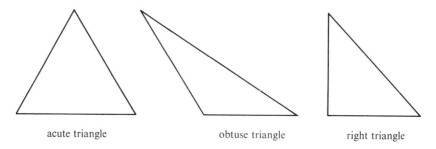

acute triangle   obtuse triangle   right triangle

Figure 3-13

If no two sides of a triangle are equal the triangle is *scalene*. A triangle with two equal sides is *isosceles*. In an isosceles triangle the angles opposite the equal sides are also equal. A triangle with three equal sides is *equilateral*. An equilateral triangle also has three equal angles.

As was previously stated, a triangle has three sides. These three sides form three angles and three *vertices*. Each vertex is usually labeled with a capital

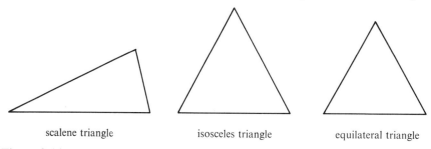

scalene triangle   isosceles triangle   equilateral triangle

Figure 3-14

letter. A triangle is named by its vertices. For example, Figure 3-15 is called triangle *ABC*. The order in which we write *ABC* makes no difference in naming the triangle. The symbol (△) may be used for the word triangle. Thus, instead of writing *triangle ABC*, we can write △*ABC*.

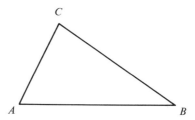

Figure 3-15

The side of a triangle can be named by the two end points or by a small letter which is the same as the capital letter used to name the angle opposite the side.

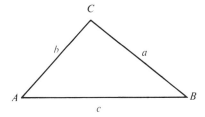

Figure 3-16

EXAMPLE 1   (a) Name the triangle in Figure 3-16 three different ways.

               (b) Name each side in two ways.

Solution: (a) 1. $\triangle ABC$ (b) 1. AC or b

                        2. $\triangle BCA$ 2. AB or c

                        3. $\triangle CBA$ 3. BC or a

PROBLEM SET 3-4

1. Identify the following triangles according to their angles.

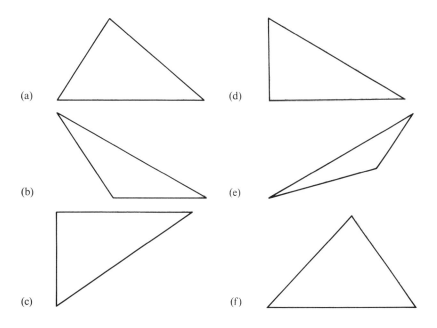

2. Identify the following triangles according to their sides.

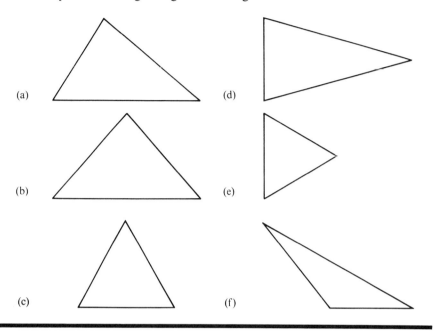

(a)

(b)

(c)

(d)

(e)

(f)

PROBLEM SET 3-5

1. Name the triangle at the right in four different ways.
2. Identify each side in two different ways.

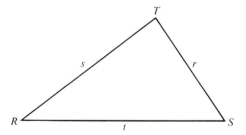

3-7
Medians, Angle Bisectors, and Altitudes

When a line is drawn from a vertex to the mid-point of the opposite side of a triangle it is called a *median*. An *angle bisector* is a line which bisects an angle of a triangle. An *altitude* of a triangle is a line drawn from a vertex perpendicular to the opposite side. An altitude is also called the *height* of the triangle. In *each* case, whether you are drawing medians, angle bisectors, or altitudes, all three lines will intersect at the same point.

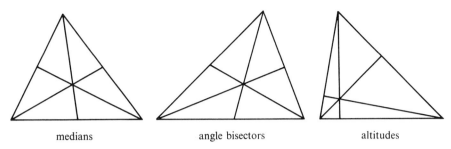

medians   angle bisectors   altitudes

Figure 3-17

3-8
Perimeter and Area

The perimeter of a triangle, as well as any polygon, is the distance around the triangle. This can be found by adding the three sides.

$$P = s_1 + s_2 + s_3$$

EXAMPLE 2   Find the perimeter of $\triangle ABC$

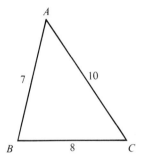

Figure 3-18

Solution: Let $s_1 = 7$ $P = s_1 + s_2 + s_3$
            $s_2 = 8$ $= 7 + 8 + 10$
            $s_3 = 10$ $= 25$

EXAMPLE 3   If two sides of a triangle are 10 and 17, find the third side if the perimeter is 38.

Solution: $P = s_1 + s_2 + s_3$
                            $38 = 10 + 17 + s_3$
                            $38 = 27 + s_3$
                      $38 - 27 = s_3$
                            $11 = s_3$

The *area* of a triangle is one-half the product of the base and height.

$$A = \tfrac{1}{2}bh$$

The height is equal to the altitude, and the base is the side to which the altitude is drawn.

EXAMPLE 4   Find the area of the following triangles.

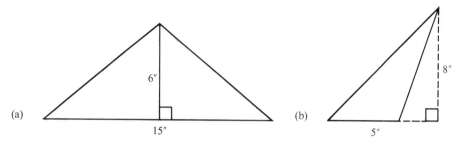

Figure 3-19

Solution: (a) $A = \tfrac{1}{2}bh$ (b) $A = \tfrac{1}{2}bh$

$= \tfrac{1}{2}(15'' \times 6'')$ $= \tfrac{1}{2}(5' \times 8')$

$= \tfrac{1}{2}(90 \text{ sq in.})$ $= \tfrac{1}{2}(40 \text{ sq ft})$

$= 45 \text{ sq in.}$ $= 20 \text{ sq ft}$

EXAMPLE 5   If the area of a triangle is 152 sq mi and the height is 20 mi, find the base.

Solution:

$$A = \tfrac{1}{2}bh$$

$$152 \text{ sq mi} = \tfrac{1}{2}(b \times 20 \text{ mi})$$

$$152 \text{ sq mi} = b \times 10 \text{ mi}$$

$$\frac{152 \text{ sq mi}}{10 \text{ mi}} = b$$

$$15.2 \text{ mi} = b$$

PROBLEM SET 3-6

1. If a triangle has sides of 6 in., 12 in., and 15 in., find the perimeter.
2. Find the perimeter of an equilateral triangle whose side is 11 ft.
3. Find the third side of a triangle if two of its sides are 4 in. and 9 in. and the perimeter is 20 in.
4. Find the perimeter of an isoceles triangle if one of the two equal sides is 25 in. and the other side is 7 in.
5. Find the two equal sides of an isosceles triangle if the third side is 16 in. and the perimeter is 3 ft.

6. The drawing is a dipole antenna which is cut to the proper length for 4 mega-hertz transmission. Determine the length of the antenna if the two guy wires form a vertical isosceles triangle with an area of 160 square feet.

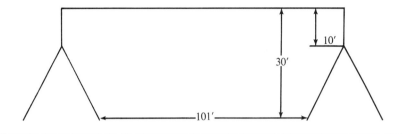

PROBLEM SET 3-7

Using the formula $A = \frac{1}{2}bh$ for the area of a triangle, find the missing letters.

	A	b	h			A	b	h
1.		10	12		2.		8 in.	6 in.
3.	91 sq in.	13 in.			4.	75 sq in.		15 in.
5.		11 ft	13 ft		6.	140 sq ft	10 ft	
7.	342 sq ft		20 ft		8.		2 ft	3 in.

3-9
Pythagorean Theorem

The Pythagorean Theorem refers to right triangles. Recall that a right triangle is a triangle with one right angle. The side opposite the right angle is called the *hypotenuse*. The other two sides are called legs.

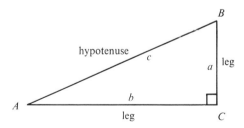

Figure 3-20

The *Pythagorean Theorem* states that the square of the hypotenuse is equal to the sum of the squares of the legs. Letting c equal the hypotenuse, a equal one leg and b equal the other leg, then

$$c^2 = a^2 + b^2$$

This formula should be committed to memory. The formula can be changed to several other forms such as:

$$c^2 = a^2 + b^2 \qquad a^2 = c^2 - b^2 \qquad b^2 = c^2 - a^2$$
$$c = \sqrt{a^2 + b^2} \qquad a = \sqrt{c^2 - b^2} \qquad b = \sqrt{c^2 - a^2}$$

No one of the above is more important than the other, therefore the student should be familiar with each of them.

EXAMPLE 6   Find the hypotenuse of a right triangle if the legs are 6 and 8.

Solution:
$$c^2 = a^2 + b^2$$
$$c^2 = \sqrt{6^2 + 8^2}$$
$$= \sqrt{36 + 64}$$
$$c^2 = \sqrt{100}$$
$$c = 10$$

EXAMPLE 7   Find the one leg of a right triangle if the hypotenuse is 18 and one leg is 7.

Solution:
$$a = \sqrt{c^2 - b^2}$$
$$a = \sqrt{18^2 - 7^2}$$
$$= \sqrt{324 - 49}$$
$$= \sqrt{275}$$
$$= 16.5$$

It is common practice to let c equal the hypotenuse and a and b equal the legs of a right triangle.

PROBLEM SET 3-8

Consider each of the following exercises to be referring to a right triangle and find the missing sides.

	c	a	b			c	a	b
1.		3	4		2.		5	12
3.	16	7			4.	15		10
5.		8	9		6.	25	24	
7.	20		16		8.	50	45	
9.		17	6		10.		14	10

3-10
Congruent and Similar Triangles

In two triangles like $\triangle ABC$ and $\triangle DEF$, $\angle A$ and $\angle D$, $\angle C$ and $\angle F$, and $\angle B$ and $\angle E$ are said to be *corresponding angles*. Sides AC and DF, AB and DE, and

BC and *EF* are called *corresponding sides*. If the corresponding sides and angles are equal, the two triangles are *congruent*. If only the corresponding angles are equal, then the triangles are *similar*. If two triangles are similar the ratios of the corresponding sides are equal.

In Figure 3-22, $\dfrac{AB}{DE} = \dfrac{BC}{EF} = \dfrac{AC}{DF}$. This triple equality states that the ratio of any two sides of the similar triangles equal the ratio of all other corresponding sides of the triangles. This very important idea concerning similar triangles will be used in solving many practical and technical problems.

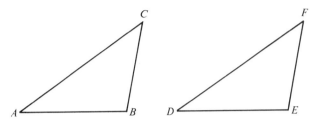

Figure 3-21

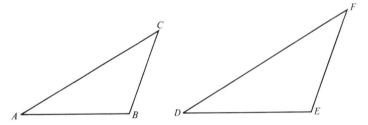

Figure 3-22

EXAMPLE 8   If △*ABC* is similar to △*DEF* and *AC* = 10, *AB* = 6, *DF* = 15, find *DE*.

Solution: It is best to draw the triangles in order to get an idea of what they look like.

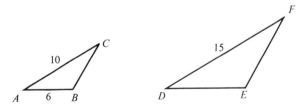

Figure 3-23

Since the triangles are similar, the ratios of their corresponding sides are equal.

$$\frac{AC}{DF} = \frac{AB}{DE}$$

Substituting the values of the sides we get,

$$\frac{10}{15} = \frac{6}{DE}$$

Solve for DE:

$$(DE)\left(\frac{10}{15}\right) = \left(\frac{6}{DE}\right)(DE)$$

$$\frac{10(DE)}{15} = 6$$

$$(15)\left(\frac{10DE}{15}\right) = (6)(15)$$

$$10DE = 90$$

$$\frac{10DE}{10} = \frac{90}{10}$$

$$DE = 9$$

EXAMPLE 9   *RST* is similar to *UVW*. Find the missing sides.

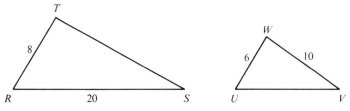

Figure 3-24

Solution: Our problem is to find *ST* and *UV*. Therefore,

$$\frac{ST}{VW} = \frac{RT}{UW} \qquad\qquad \frac{UV}{RS} = \frac{UW}{RT}$$

$$\frac{ST}{10} = \frac{8}{6} \qquad\qquad \frac{UV}{20} = \frac{6}{8}$$

$$6(ST) = 8(10) \qquad 8(UV) = 6(20)$$

$$6(ST) = 80 \qquad\quad 8(UV) = 120$$

$$ST = 13\tfrac{1}{3} \qquad\qquad UV = 15$$

EXAMPLE 10   If a boy 5 ft tall casts a shadow 4 ft long, find the height of a building whose shadow is 30 ft long.

Solution: This is another type of problem which can be solved by similar triangles.

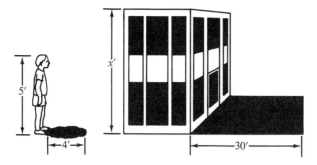

Figure 3-25

$$\frac{\text{Height of building}}{\text{Height of boy}} = \frac{\text{Shadow of building}}{\text{Shadow of boy}}$$

$$\frac{\text{Height of building}}{5'} = \frac{30'}{4'}$$

$$(4')(\text{Height of building}) = 5'(30')$$

$$(4')(\text{Height of building}) = 150 \text{ ft} \cdot \text{ft}$$

$$(\text{Height of building}) = 37\tfrac{1}{2}'$$

PROBLEM SET 3-9

Indicate which of the following pairs of triangles are congruent by a "yes" or "no" response.

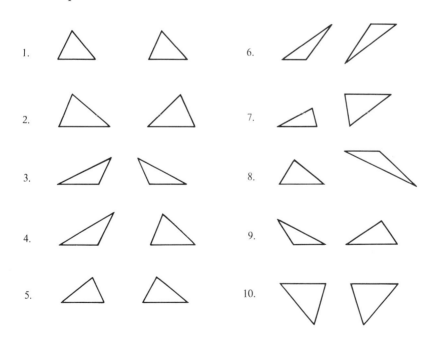

Consider the following pairs of triangles to be similar and find the missing sides.

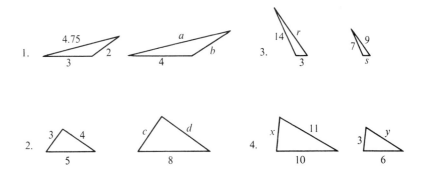

PROBLEM SET 3-11

Use the property of similar triangles to solve the following:

1. A telephone pole casts a shadow 25 ft long at the same time that a 4.0 ft post casts a shadow 2.5 ft long. How tall is the telephone pole?

2. A surveyor draws a triangular piece of land to scale. The lengths of the lines on the scale drawing are 5 in., 14 in., and 16 in. The shortest side of the plot of land is 800 ft. How long is each side of the plot of land?

3. The shadow cast by a tree is 45 ft when at the same time a yard stick casts a shadow of 15 in. How tall is the tree?

4. A certain building casts a shadow 75 ft long and at the same time a boy 6 ft tall casts a shadow 25 in. How tall is the building?

5. A boy sees a parachutist jump from an airplane. The boy sees the shadow of the airplane on the ground and estimates the shadow to be 1000 ft from the landing area. He knows his height is 5 ft and his shadow at the present time is 1.5 ft. How high was the airplane when the parachutist jumped?

3-11
Quadrilaterals

A *quadrilateral* is a four-sided polygon. The *sum of the angles* in any quadrilateral is equal to 360°. A *diagonal* is a line segment drawn from one vertex to the opposite vertex. Quadrilaterals, like triangles, are identified according to their sides and angles. Quadrilaterals are named by the letter at the vertices. In Figure 3-26, we could call this quadrilateral any of the following and be correct; quadrilateral *ABCD*, quadrilateral *ADCB*, or quadrilateral *BCDA*. It makes no difference where you begin but you must be consecutive.

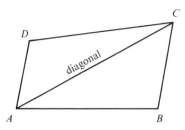

Figure 3-26

A *parallelogram* is a quadrilateral whose opposite sides are parallel and equal. The opposite angles are equal and any two consecutive angles are supplementary.

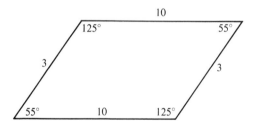

Figure 3-27

A *diagonal* divides a parallelogram into two congruent triangles as in Figure 3-28.

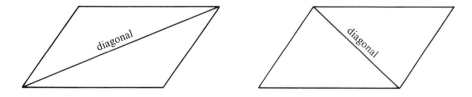

Figure 3-28

If both diagonals are drawn in the same parallelogram as in Figure 3-29, how many triangles are formed? How many pairs of triangles are congruent?

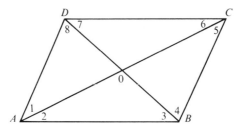

Figure 3-29

Do the diagonals bisect each other? What angles are equal?

The *area* of a parallelogram equals the product of the base and height where the base is any side and the height is the perpendicular distance between the base and the side parallel to the base.

$$A = bh$$

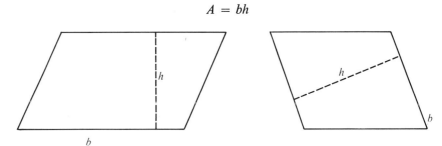

Figure 3-30

The following examples show how this formula is used.

EXAMPLE 11 Find the area of the following parallelograms.

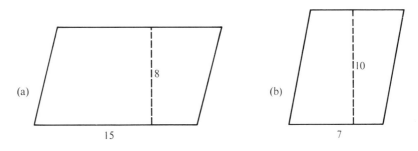

Figure 3-31

Solution: (a) $A = bh$ (b) $A = bh$

$A = 15(8)$ $A = 7(10)$

$= 120$ $= 70$

EXAMPLE 12 If the area of a parallelogram is 80 sq in. and the height is 5 in., find the base.

Solution:
$$A = bh$$
$$80 \text{ in.} \times \text{in.} = b(5 \text{ in.})$$
$$\frac{80 \text{ in.} \times \text{in.}}{5 \text{ in.}} = b$$
$$16 \text{ in.} = b$$

3-13
Rectangle

A *rectangle* is a parallelogram with one right angle. This necessarily makes each angle a right angle. The diagonals of a rectangle are equal.

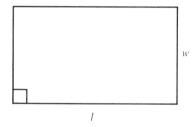

Figure 3-32

The height of a rectangle is equal to a side. Normally, the base is called the length and the height is called the width. Therefore, instead of writing $A = bh$ for the area of a rectangle we will write it as:

$$A = lw$$

The *perimeter* of a rectangle is the distance around the rectangle and can be given as:

$$P - l + w + l + w$$

or

$$P = 2l + 2w$$

or

$$P = 2(l + w)$$

EXAMPLE 13 Find the area and perimeter of a rectangle if the length is 25 ft and the width is 16 ft.

Solution:

$A = lw$	$P = 2(l + w)$
$A = 25 \text{ ft } (16 \text{ ft})$	$P = 2(25 \text{ ft} + 16 \text{ ft})$
$= 400 \text{ sq ft}$	$= 2(41 \text{ ft})$
	$= 82 \text{ ft}$

A *square* is a rectangle with all sides equal. The diagonals are equal and are perpendicular bisectors of each other.

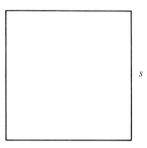

s

Figure 3-33

Since the sides are equal the area can be written

$$A = s^2$$

and the perimeter

$$P = 4s$$

EXAMPLE 14   Find the area and perimeter of a square whose side is 17 ft.

Solution: $A = s^2$ $P = 4s$
                        $= 17^2$ $= 4(17)$
                        $= 289$ sq ft $= 68$ ft

A *rhombus* is a parallelogram with equal sides. The diagonals are perpendicular bisectors of each other.

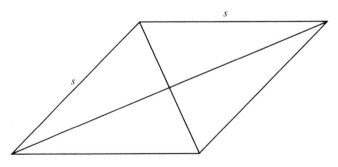

s

s

Figure 3-34

The perimeter is given

$$P = 4s$$

and the area is

$$A = bh$$

3-16
Trapezoid

A *trapezoid* is a quadrilateral with one pair of opposite sides parallel. If the two non-parallel sides are equal then it is an isosceles trapezoid. The two parallel sides are called bases.

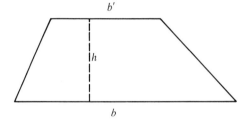

Figure 3-35

The area of a trapezoid is given as,

$$A = \tfrac{1}{2}h(b + b')$$

where b and b' represent the two parallel sides (bases).

EXAMPLE 15   Find the area of the following trapezoid.

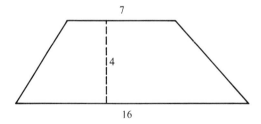

Figure 3-36

Solution:
$$A = \tfrac{1}{2}h(b + b')$$
$$= \tfrac{1}{2}(4)(16 + 7)$$
$$= 2(23)$$
$$= 46$$

PROBLEM SET 3-12

Solve the following problems concerning quadrilaterals.

1. If three angles of a quadrilateral are 55°, 70°, and 68°, how large is the fourth angle?

2. If the two diagonals of a quadrilateral are perpendicular bisectors of each other, what might the quadrilateral be?

3. Identify the possible quadrilaterals such that the diagonals bisect the angles.

4. If the diagonals of a quadrilateral are equal and bisect each other, identify the quadrilateral.

5. Draw a quadrilateral with only two sides parallel and only one right angle.

6. Draw a trapezoid with bases of 1 in. and 2 in., and altitude of $\frac{1}{2}$ in.

7. Draw a square with side $1\frac{1}{4}$ in.

8. Draw a parallelogram with sides of 2 in. and 1 in. How many such parallelograms can be drawn?

9. Draw a trapezoid, draw one diagonal. Are the two triangles formed congruent? Are the two triangles formed similar?

10. Draw a parallelogram and one diagonal. Which angles are equal?

PROBLEM SET 3-13

Solve each of the following:

1. Find the area of a parallelogram with a base of 14.7 in. and an altitude of 6.4 in.

2. Find the perimeter and area of a rectangle with length 19.1 ft and width 20.2 ft.

3. Find the perimeter and area of a square with a side 27.8 in.

4. Find the perimeter and the area of a rhombus with a side of 7.8 in. and height of 3 in.

5. Find the area of a trapezoid with bases of 17.2 in. and 1 ft and an altitude of 7 in.

6. Find the perimeter and area of a square with a side of 30.3 in.

7. Find the perimeter and area of a rectangle of length 14.9 ft and 21.3 ft width.

8. Find the area of a parallelogram with a base of 7.42 in. and a height of 6.43 in.

9. A baseball diamond is a square which is 90 ft between the bases. How far is it from home plate to second base?

10. If indoor-outdoor carpet cost 4.98 per sq yd. Find the cost of carpeting a square patio which is 12 ft on one side.

11. A yard is in the shape of a rectangle with width of 100 ft and depth of 150 ft. What will be the cost of fencing the yard if the fence is $1.95 per ft?

12. A rectanglar field is 75 yds wide. If the area contains 1 acre, what is the length? (1 acre = 4,840 sq yds.)

3-17
Circle

A *circle* is defined as the path of a point moving around and equal distance from a fixed point. We call the fixed point the center of the circle.

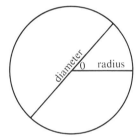

Figure 3-37

The distance from the center of the circle to any point on the circle is called a *radius*. A *diameter* is a line segment which passes through the center of the circle and whose endpoints are on the circle. Thus the diameter is equal in length to two radii.

$$d = 2r$$

The *circumference* of a circle is given as

$$C = \pi d$$

or

$$C = 2\pi r$$

and the area is given as

$$A = \pi r^2$$

where π is approximately 3.1416. Sometimes it is desirable to let $\pi = 3\frac{1}{7}$. Unless otherwise stated we will use 3.1416 for π. Referring back to the formula $C = \pi d$, one can readily see that this may be rewritten as $\pi = \dfrac{C}{d}$. It is worthy of note that $\pi = \dfrac{C}{d}$ for any circle.

EXAMPLE 16   Find the area and circumference of a circle whose radius is 16 in.

Solution: $A = \pi r^2$ $C = 2\pi r$

$A = (3.1416)(16^2)$ $C = 2(3.1416)(16)$

$= 3.1416(256)$ $= 100.5312$ in.

$= 804.2496$ sq in.

A *chord* is a line segment whose endpoints are on the circle. An arc is part of the circle. A *tangent* is a line that touches the circle at only one point. The tangent is also perpendicular to a radius at the point of intersection. In Figure 3-31, tangent *t* is perpendicular to radius *OP* at point *P*.

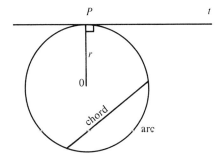

Figure 3-38

EXAMPLE 17   How long is the belt connecting the two pulleys of radius 5 in. and whose distance between centers is 36 in.?

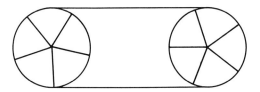

Figure 3-39

Solution:

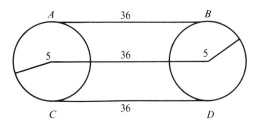

Figure 3-40

Since the distance between centers is 36", this means that the length AB is 36" and CD is 36".

We need to find each of the circles that the belt must go around. Since these halves are the same, it is equivalent to finding the circumference of one circle and adding AB and CD to this circumference. Therefore,

$$\text{Length} = \text{Circumference} + AB + CD$$
$$= \pi d + AB + CD$$
$$= 3.14(10) + 36 + 36$$
$$= 31.4 + 72$$
$$= 103.4 \text{ in.}$$

A *central angle* of a circle is an angle whose vertex is at the center of the circle. For example, in Figure 3-41, $\angle AOB$ is a central angle.

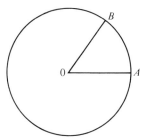

Figure 3-41

PROBLEM SET 3-14

1. Draw a circle, then draw and label each of the following: radius, diameter, chord, tangent, arc.
2. Draw two circles of the same size.
3. Draw a circle and then draw a central angle of 70°.
4. Draw two concentric circles (circles having the same center).
5. Draw a circle with a diameter AB. Using AB as one side construct a triangle with the third point C on the circle. What kind of angle is $\angle ACB$? Do this several times. Does $\angle ACB$ always seem to be a right angle?

PROBLEM SET 3-15

1. Find the area and circumference of a round table top 50 in. in diameter.
2. The minute hand of a clock is 6 in. long. How many feet does the tip move in 1 hr? in 12 hrs?
3. A tree in Sequoia National Park measures 30 ft in circumference. Find the cross-sectional area of the tree.

4. A piece of wire is 60 in. long. What is the area of a circle formed with the wire? What is the area of a square formed with the same wire?

5. Will two circles having the same radius contain the same amount of area as one circle whose radius is twice as large?

6. A circle is inscribed in a square that is 15 in. on each side. Compare the circumference of the circle to the perimeter of the square.

7. A circle with a diameter of 30 in. is to be divided into two circles of the same diameter, having the same total area. Find the diameters of the smaller circles.

8. Find the area of a circular mirror with a diameter of 42 in.

9. An automobile tire 26.6 in. in diameter makes 2 revolutions per second. How many feet will the car travel in 1 minute?

10. If a circle with a diameter of 5 in. is cut from a rectangular board that is 6 in. by 8 in., what is the area of the scrap material?

11. A 12 in. phonograph turntable is driven by a $\frac{1}{2}$ in. diameter spindle as shown in the drawing below. If the turntable turns at a rate of $33\frac{1}{3}$ revolutions per minute, how many revolutions does the spindle make in one minute?

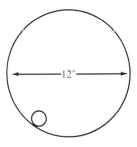

12. An electro-mechanical hoist may be made by connecting the output shaft of a variable-speed motor to a gear train as shown in the following drawing. If the motor has an angular velocity of 100 rpm, how far will the load move in one minute?

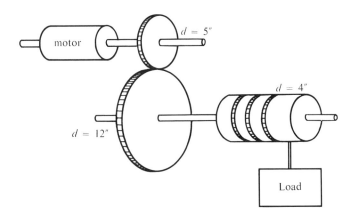

3-18
Geometric Solids

Our discussion of geometric figures thus far has concerned two-dimensional figures, usually referred to as plane figures. We would like to extend our thinking to three-dimensional figures. A geometric figure which is three-dimensional (having length, width, and depth) is called a *geometric solid*. Our discussion will be limited to prisms, pyramids, cylinders, cones, and spheres.

3-19
Prisms

A prism is a geometric solid formed by polygons (faces). Two of the polygons (faces) must be parallel and have the same size and shape. These two polygons are called the *bases*. The bases can be any kind of polygon. The remaining polygons are called *lateral faces* and must be parallelograms. The *name* of the prism is determined by its base. Thus, a prism whose base is a rectangle is called a *rectangular prism*. If the base is a triangle, it is a *triangular prism*. Figure 3-43 shows some different types of prisms. If the *lateral faces* are perpendicular to the bases it is a *right prism*.

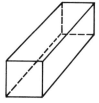

Figure 3-42

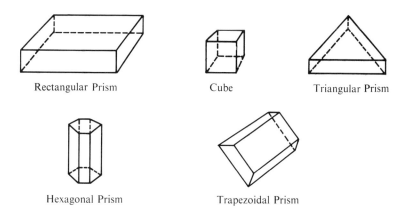

Rectangular Prism   Cube   Triangular Prism

Hexagonal Prism   Trapezoidal Prism

Figure 3-43

A prism has three measurements, the lateral area, total area, and volume. The *lateral area* is the sum of the area of the lateral faces. The symbol used for lateral area is *S*. The *total area* is the lateral area plus the sum of the areas of the two bases. The symbol used for the total area is *T*. The *volume* is the cubic measure of the inside of the prism. To find the *volume* of a prism multiply the area of the base by the *height* of the prism. The *height* of a prism is the perpendicular distance between the two bases.

EXAMPLE 18   Find the lateral area, total area, and volume of the rectangular prism in Figure 3-44.

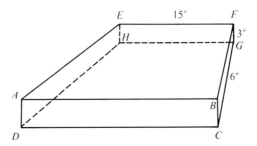

Figure 3-44

Solution: To find the lateral area we need to add the areas of rectangles *ABCD, BFGC, FEHG,* and *AEHD.* Hence,

$$S = \text{area } ABCD + \text{area } BFGC + \text{area } FEHG + \text{area } AEHD$$
$$= 15 \text{ in.} \times 3 \text{ in.} + 6 \text{ in.} \times 3 \text{ in.} + 15 \text{ in.} \times 3 \text{ in.} + 6 \text{ in.} \times 3 \text{ in.}$$
$$= 45 \text{ sq in.} + 18 \text{ sq in.} + 45 \text{ sq in.} + 18 \text{ sq in.}$$
$$= 126 \text{ sq in.}$$

The total area is equal to the lateral area plus the sum of the areas of the bases.

$$T = S + \text{area } ABFE + \text{area } DCGH$$
$$= 126 \text{ sq in.} + 15 \text{ in.} \times 6 \text{ in.} + 15 \text{ in.} \times 6 \text{ in.}$$
$$= 126 \text{ sq in.} + 90 \text{ sq in.} + 90 \text{ sq in.}$$
$$= 306 \text{ sq in.}$$

The volume is found by multiplying the area of the base by the height. Therefore.

$$V = \text{area } ABFE \times 3 \text{ in.}$$
$$= 15 \text{ in.} \times 6 \text{ in.} \times 3 \text{ in.}$$
$$= 270 \text{ cu in.}$$

EXAMPLE 19   Find *S*, *T*, and *V* of the triangular prism in Figure 3-45.

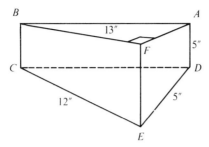

Figure 3-45

Solution:

$$S = \text{area } ABCD + \text{area } AFED + \text{area } BCEF$$
$$= (13 \text{ in.} \times 5 \text{ in.}) + (5 \text{ in.} \times 5 \text{ in.}) + (12 \text{ in.} \times 5 \text{ in.})$$
$$= 65 \text{ sq in.} + 25 \text{ sq in.} + 60 \text{ sq in.}$$
$$= 130 \text{ sq in.}$$

$$T = S + \text{area } ABF + \text{area } DCE$$
$$= 130 \text{ sq in.} + (\tfrac{1}{2} \times 12 \text{ in.} \times 5 \text{ in.}) + (\tfrac{1}{2} \times 12 \text{ in.} \times 5 \text{ in.})$$
$$= 130 \text{ sq in.} + 30 \text{ sq in.} + 30 \text{ sq in.}$$
$$= 190 \text{ sq in.}$$

$$V = \text{area } ABF \times 5 \text{ in.} = 150 \text{ in.}^3$$

PROBLEM SET 3-16

1. A piece of wood is 6 ft long, 10 in. wide and $\frac{3}{4}$ in. thick. Determine its total surface area in square inches.

2. A box is to be made from a piece of tin 18 inches long and 12 inches wide. If the tin is bent to make the sides and ends 3 inches high what is the volume of the box?

3. If one gallon of paint will cover 400 sq ft of wall space, how many gallons will be required to cover the walls of a room 28 ft × 18 ft if the ceiling is 8 ft high?

4. What is the volume of a building 160 meters high, 40 meters long, and 40 meters wide? What is the lateral surface area?

3-20
Cylinders

A geometric solid whose bases are parallel circles of the same size is a *cylinder*. A line connecting the centers of the bases is called the *axis* of the cylinder. The altitude of a cylinder is the perpendicular distance between the two bases. If the axis is perpendicular to the bases, it is called a *right cylinder*.

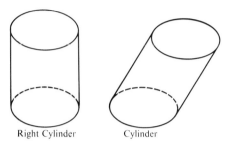

Right Cylinder   Cylinder

Figure 3-46

As with prisms, the cylinder has three measurements which we shall find: the lateral area, total area, and volume.

To find the lateral area is to find the area around the cylinder. The *total area* is equal to the lateral area plus the sum of the areas of the two bases. The *volume* is the number of cubic units inside the cylinder.

Suppose we take a cylinder and "open" it. Notice that the lateral surface becomes a rectangle whose length is equal to the circumference of the base and the width is equal to the height of the cylinder. The two bases are of course two circles

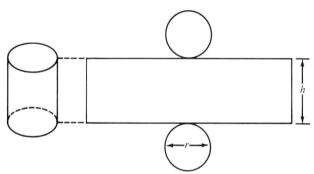

Figure 3-47

To find the lateral area (S), multiply the circumference of the base times the height. Hence,

$$S = Ch$$

or

$$S = \pi dh$$

since

$$C = \pi d$$

or

$$S = 2\pi rh$$

since

$$C = 2\pi r$$

Since the total area (T) is equal to (S) plus the sum of the area of the bases it can be written as:

$$T = S + \text{(area of bottom} + \text{area of top)}$$
$$= S + (\pi r^2 + \pi r^2)$$
$$= S + 2\pi r^2$$

or

$$T = 2\pi rh + 2\pi r^2$$

since

$$S = 2\pi rh$$

or

$$T = 2\pi r(h + r)$$

The volume (V) is equal to the area of the base times the height. Hence,

$$V = \text{area of base} \times \text{height}$$
$$= \pi r^2 \times h$$

EXAMPLE 20 Find S, T, V for the cylinder in Figure 3-48.

Solution:
$$S = 2\pi rh$$
$$= 2(3.14)(6 \text{ in.})(12 \text{ in.})$$
$$= 452.16 \text{ sq in.}$$

$$T = 2\pi r(h + r)$$
$$= 2(3.14)(6 \text{ in.})(18 \text{ in.})$$
$$= 678.24 \text{ sq in.}$$

$$V = \pi r^2 h$$
$$= (3.14)(6 \text{ in.})^2(12 \text{ in.})$$
$$= (3.14)(36 \text{ sq in.})(12 \text{ in.})$$
$$= 1356.48 \text{ cu in.}$$

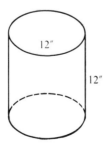

Figure 3-48

1. A rectangular sheet of paper containing 110 sq in. area is rolled to form a cylinder with a diameter of 3.5 in. Find the height, volume, lateral surface area, and total surface of the cylinder.

2. A cubic foot of copper is made into a wire $\frac{1}{8}$ in. in diameter. How long is the wire?

3. A cylindrical container has a radius of 6 in. and is 12 in. high. Another container with a diameter of 10 in. has the same volume as the first can. What is the height of the second can?

4. A sidewalk is to be poured around a fish pond as shown in Figure 3-49 If the sidewalk is to be 3 in. uniform thickness, how many cubic feet of concrete will be required?

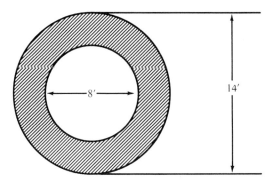

Figure 3-49

5. The rainwater which falls on a flat roof 20 ft by 40 ft drains into a cylindrical tank which is 10 ft in diameter. How much rainfall would fill the tank to a depth of 8 ft?

3-21
Pyramids and Cones

Pyramids and cones are similar in the sense that in each case the lateral surface extends upward to a point.

The main difference is the base of each. A cone has a circular base whereas a pyramid has some type of polygon for its base. Figure 3-51 shows three types of pyramids.

The lateral area, or surface area, of a pyramid, is equal to the sum of the areas of the faces. Each face of a pyramid is a triangle. The total area of a

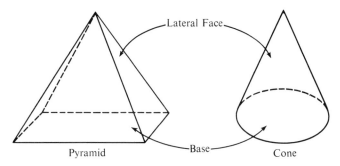

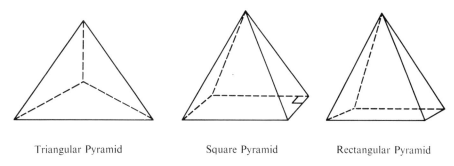

Figure 3-50

| Triangular Pyramid | Square Pyramid | Rectangular Pyramid |

Figure 3-51

pyramid is equal to the lateral area plus the area of the base. Generally, the *volume* is the measurement of most concern and is found as follows:

$$V = \tfrac{1}{3}Bh$$

where B stands for the area of the base, and altitude h, is the perpendicular distance from the point where the faces meet called the *apex* of the pyramid. In Figure 3-52, E is the *apex*. The slant height l is the altitude of a lateral face. Thus, in Figure 3-52, EF is the slant altitude.

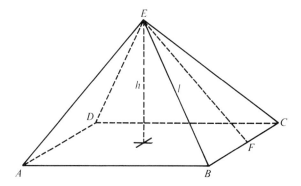

Figure 3-52

EXAMPLE 21   Find the volume of the pyramid in Figure 3-53.

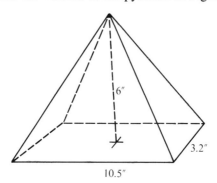

Figure 3-53

Solution: Since $V = \frac{1}{3}Bh$, we must find the area of the base which is a rectangle, then multiply by the height.

$$V = \frac{1}{3}Bh$$
$$= \frac{1}{3}(L)(W)(h)$$
$$= \frac{1}{3}(10.5 \text{ in.})(3.2 \text{ in.})(6 \text{ in.}$$
$$= \frac{1}{3}(201.60 \text{ cu in.})$$
$$= 67.20 \text{ cu in.}$$

PROBLEM SET 3-18

1. A pyramid has a slant height of 12 in. and each side forms an equilateral triangle 8 in. on a side. Find the lateral area of the pyramid if the base is square.

2. A pyramid has a lateral surface area of 56 sq in. Its sides are an equilateral triangle 4 in. on a side. Find the slant height of the pyramid if the base is square.

3. A pyramid has an altitude of 12 in. and a square base 10 in. on a side. Determine the cross-section area 4 in. above the base.

4. The Great Pyramid of Egypt is 762 sq ft at the base and 484 ft high. Determine its volume in cubic feet.

5. The spire on a church is a pyramid with a hexagonal base 5 ft on each side and an altitude of 75 ft. If a quart of paint will cover 100 sq ft, how many quarts will be required to complete the job?

The formulas $A = \pi r l$ and $V = \frac{1}{3}Bh$ are used to find the surface area and volume of a cone, respectively. In the formula $A = \pi r l$, r is the radius of the circular base, and l is the slant height. In Figure 3-54, AC or BC is equal to the slant height. In $V = \frac{1}{3}Bh$, B is the area of the base and h is the altitude.

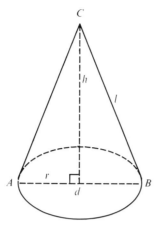

Figure 3-54

Usually, the altitude and either the radius or diameter of the base are the only dimensions of a cone that will be given. The slant height is very easily found by use of the Pythagorean Theorem.

EXAMPLE 22   Find the volume and surface area of the cone in Figure 3-55.

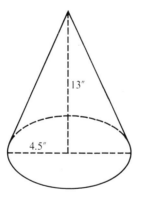

Figure 3-55

Solution: $V = \frac{1}{3}Bh$

$= \frac{1}{3}(\pi r^2)h$

$= \frac{1}{3}(3.14 \times 4.5 \text{ in.} \times 4.5 \text{ in.})13 \text{ in.}$

$= \frac{1}{3}(63.585 \text{ sq in.})13 \text{ in.}$

$= \frac{1}{3}(826.605 \text{ cu in.})$

$= 275.535 \text{ cu in.}$

Before we can find the surface area we must find the slant height.

$$l = \sqrt{(4.5)^2 + 13^2}$$
$$= \sqrt{20.25 + 169}$$
$$= \sqrt{189.25}$$
$$= 13.75 \text{ in.}$$

$$A = \pi r l$$
$$= 3.14 \times 4.5 \text{ in.} \times 13.75 \text{ in.}$$
$$= 3.14 \times 61.875 \text{ sq in.}$$
$$= 194.2875 \text{ sq in.}$$

EXAMPLE 23   Find the volume of the cone in Figure 3-56.

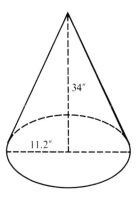

Figure 3-56

Solution:

$$V = \tfrac{1}{3}Bh$$
$$= \tfrac{1}{3}(\pi r^2)h$$
$$= \tfrac{1}{3}(3.14 \times (11.2)^2)(34)$$
$$= \tfrac{1}{3}(393.88)(34)$$
$$= \tfrac{1}{3}(13391.92)$$
$$= 4463.97 \text{ cu units}$$

PROBLEM SET 3-19

1. Determine the lateral surface area, total surface area, and volume of a cone with a diameter of 14 in. and an altitude of 20 in.

2. A cone (excluding the base) with an altitude of 6 in. and a radius of 4 in. is cut from an 8.4 in. × 11 in. sheet of paper. How much paper is wasted?

3. A cone has a base area of 254 sq in. and a volume of 1016 cu in. Determine its slant height.

4. A solid cone is made from a 1 ft cube of wood. The altitude of the cone is 12 in. and the radius of 5 in. How much wood was wasted?

5. A cone with an altitude of 14 in. and a radius of 6 in. is cut 7 in. from its vertex. The result is a cone with an altitude of 7 in. and a frustum with an altitude of 7 in. as shown in Figure 3-57. Determine the ratio of their volumes.

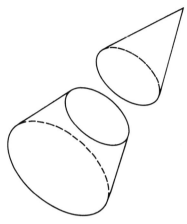

Figure 3-57

3-22

Sphere

A geometric solid such that the distance from a given point, called the *center*, to any point on the surface is always the same is called a *sphere*. The formula for the volume of a sphere is given as $V = \frac{4}{3}\pi r^3$. The formula for the surface area is $A = 4\pi r^2$.

EXAMPLE 24   Find the volume and surface area of the sphere in Figure 3-58.

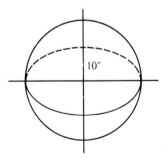

Figure 3-58

Solution:
$$V = \frac{4}{3}\pi r^3$$
$$= \frac{4}{3}(3.14)(10 \text{ in.})^3$$
$$= \frac{4}{3}(3.14)(1000 \text{ cu in.})$$
$$= \frac{4}{3}(3140 \text{ cu in.})$$
$$= 4186.66\frac{2}{3} \text{ cu in.}$$

$$A = 4\pi r^2$$
$$= 4(3.14)(10 \text{ in.})^2$$
$$= 4(3.14)(100 \text{ sq in.})$$
$$= 4(314 \text{ sq in.})$$
$$= 1256 \text{ sq in.}$$

EXAMPLE 25 How much material will it take to make a globe, representing the world, 24 in. in diameter?

Figure 3-59

Solution:

$$A = 4\pi r^2$$
$$= 4(3.14)(12 \text{ in.})^2$$
$$= 4(3.14)(144 \text{ sq in.})$$
$$= 1808.64 \text{ sq in.}$$

EXAMPLE 26 A town would like to install a spherical water storage tank which would hold 100,000 gallons of water. The city engineer says the diameter should be 25 ft. Is he correct? (1 cu ft = 7.48 gal).

Solution: Find the volume and convert the cu ft to gals.

$$V = \tfrac{4}{3}\pi r^3$$
$$= \tfrac{4}{3}(3.14)(12.5 \text{ ft})^3$$
$$= \tfrac{4}{3}(3.14)(1953.125 \text{ cu ft})$$
$$= \tfrac{4}{3}(6122.81250 \text{ cu ft})$$
$$= 8163.75 \text{ cu ft}$$
$$8163.75 \text{ cu ft} = (8163.75 \times 7.48) \text{ gal}$$
$$= 61064.75$$

Since the tank will hold only 61064.75 gal, the answer to the question is no.

PROBLEM SET 3-20

1. Find the surface area and volume of a sphere 10 in. in diameter.
2. A sphere 5 in. in diameter is to be cut from a 5 in. cube. Determine the amount of material which is wasted.

3. The mean radius of the earth is 3960 miles and of the moon 1080 miles. Find the ratio of their surface areas and volumes.

4. A ball of clay 3 in. in diameter is flattened into a circular disk $\frac{1}{8}$ in. thick. What is the diameter of the disk?

5. Spherical lead buckshot $\frac{1}{8}$ in. in diameter is melted and poured into a hemisphere mold 2.5 in. in diameter. How many must be melted to fill the hemisphere?

3-23
Useful Formulas from Geometry

The following figures and formulas from geometry are listed here for convenience.

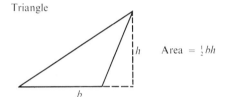

Triangle   Area $= \frac{1}{2}bh$

Figure 3-60

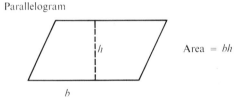

Parallelogram   Area $= bh$

Figure 3-61

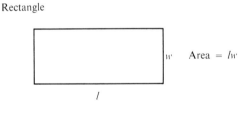

Rectangle   Area $= lw$

Figure 3-62

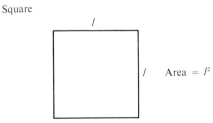

Square   Area $= l^2$

Figure 3-63

Rhombus

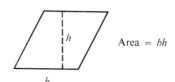

Area = bh

Figure 3-64

Trapezoid

Area = $\frac{1}{2}(b + b')$

Figure 3-65

Circle

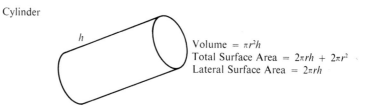

Area = πr^2
Circumference = $2\pi r$

Figure 3-66

Cylinder

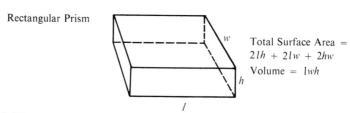

Volume = $\pi r^2 h$
Total Surface Area = $2\pi rh + 2\pi r^2$
Lateral Surface Area = $2\pi rh$

Figure 3-67

Rectangular Prism

Total Surface Area =
$2lh + 2lw + 2hw$
Volume = lwh

Figure 3-68

Sphere

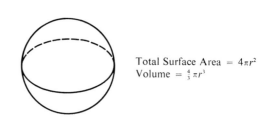

Total Surface Area = $4\pi r^2$
Volume = $\frac{4}{3}\pi r^3$

Figure 3-69

Cone

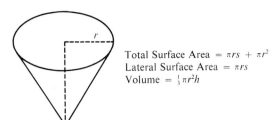

Total Surface Area $= \pi rs + \pi r^2$
Lateral Surface Area $= \pi rs$
Volume $= \frac{1}{3}\pi r^2 h$

Figure 3-70

PROBLEM SET 3-21

Determine the volume and total surface area of the following figures.

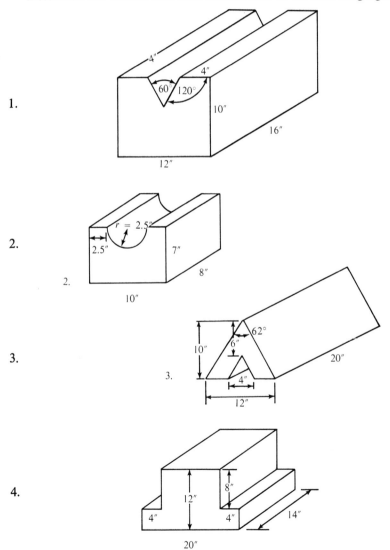

1.

2.

3.

4.

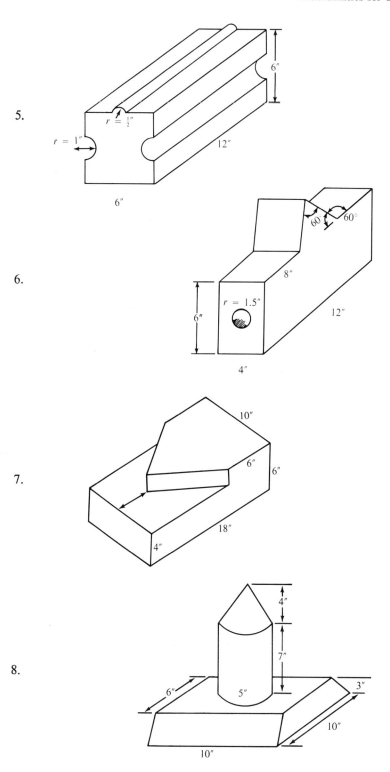

5.

6.

7.

8.

4

Ratio, Proportion and Variation

4-1
Ratio

The ratio of two quantities is a comparison between them. The most common way of writing the ratio of two numbers is to write them as a fraction. The numerator refers to the first number and the denominator refers to the second. Thus $\frac{2}{3}$ is the ratio of 2 to 3. A ratio is usually given in lowest terms.

A more complete definition of ratio is a comparison of two quantities by division. It is not necessary that both quantities have the same unit of measure. If the units of measure are equal, the ratio becomes a number. If the units of measure are not equal, the units must be included when expressing the ratio.

EXAMPLE 1   Write the ratios of the following.

(a) 1 to 4

(b) 2 in. to 6 in.

(c) 2 ft to 4 in.

(d) 100 mi to 4 hr

Solution:

(a) $\dfrac{1}{4}$

(b) $\dfrac{2 \text{ in.}}{6 \text{ in.}} = \dfrac{1}{3}$

133

(c) $\dfrac{2 \text{ ft}}{4 \text{ in.}} = \dfrac{24 \text{ in.}}{4 \text{ in.}} = \dfrac{6}{1} = 6$

(d) $\dfrac{100 \text{ mi}}{4 \text{ hr}} = 25 \text{ mi/hr}$

A ratio can be written with a colon. Therefore, a solution to the above example would be:

Solution: (a) $1 : 4$

                      (b) $2 \text{ in.} : 6 \text{ in.} = 1 : 3$

                      (c) $2 \text{ ft} : 4 \text{ in.} = 24 \text{ in.} : 4 \text{ in.} = 6 : 1 = 6$

                      (d) $100 \text{ mi} : 4 \text{ hr} = 25 \text{ mi/hr}$

PROBLEM SET 4-1

Write the ratio for the following.

1. 3 in. to 9 in.	2. 3.75 lbs to 15 lbs
3. 64 kg to 4 kg	4. 30 in. to 10 ft
5. 1 hr to 1440 sec	6. 2 mi to 1760 ft
7. 8 lbs to 8 oz	8. 30 yds to 72 in.
9. 250 mi to .5 hr	10. 3600 sec to 1 day

4-2
Proportion

A proportion is a statement of equality between the two equal ratios. Suppose we have the two ratios:

$$2 \text{ mi} : 3 \text{ mi} \quad\quad \text{and} \quad\quad 4 \text{ in.} : 6 \text{ in.}$$

Both ratios are equal to $\frac{2}{3}$, therefore, we can write a statement of equality between the two ratios.

$$2 : 3 = 4 : 6$$

When you have a proportion such as,

$$a : b = c : d$$

this should be read "a is to b as c is to d."

EXAMPLE 2   Read the following.

$$\text{(a) } 1 : 2 = 3 : 6$$

$$\text{(b) } 4 : 7 = 8 : 14$$

Solution: (a) 1 is to 2 as 3 is to 6

(b) 4 is to 7 as 8 is to 14

In a proportion such as $a : b = c : d$, a and d are called the *extremes*. The numbers b and c are called the *means*. The product of the means (bc) is equal to the product of the extremes (ad). Thus, if

$$a : b = c : d$$

then

$$bc = ad$$

This very important principle of proportions enable us to find any missing term of a proportion.

EXAMPLE 3   In the proportion $7 : X = 2 : 10$, find X.

Solution: $7 : X = 2 : 10$

The product of the means is $2X$, and the product of the extremes is 70, therefore

$$2X = 70$$
$$X = 35$$

EXAMPLE 4   In the proportion $X : 4 = 3 : 8$, find X.

Solution: $X : 4 = 3 : 8$

The product of the means is 12, and the product of the extremes is $8X$, therefore

$$8X = 12$$
$$X = 1\tfrac{1}{2}$$

PROBLEM SET 4-2

Solve for the unknown in the following.

1. $8 : x = 64 : 16$ 2. $35 : 5 = 49 : y$

3. $250 : 10 = 5 : e$ 4. $v : 210 = 14 : .35$

5. $2^8 : 2^3 = 2^5 : t$ 6. $625 : a = 25 : 5^2$

7. $(49)^{\frac{1}{2}} : 350 = .7 : u$ 8. $\dfrac{4.5}{81} : b = 60 : 3$

9. $.02 : 40 = 200 : z$ 10. $\frac{64}{36} : r = 30 : .1$

When two or more quantities are involved in such a way that as one quantity varies the other quantity will also change, we call this *variation*. The three types of variation are *direct*, *inverse*, and *joint*.

When two quantities increase or decrease at the same rate, *direct variation* occurs. A very common example of this is the distance traveled by an automobile at a constant rate of speed. For example, if an automobile travels at a constant rate of 50 m.p.h. for 1 hour, the distance traveled will be 50 miles. If the time is 2 hours the distance will be 100 miles; for 3 hours the distance will be 150 miles. From this example we see that the distance varies directly as the time varies. Since we know that distance d is equal to the product of the rate r and time t, we can write the equation

$$d = rt.$$

Applying this equation to the example above we have,

$$d = 50t$$

If we let t vary from 1 to 5 we have,

d	50	100	150	200	250
t	1	2	3	4	5

Taking any two pairs of these numbers we can form a proportion. Thus,

$$\frac{50}{1} = \frac{100}{2}$$

If two quantities are always in proportion, we say that one is proportional to the other, or that one varies directly as the other. This is usually written as $y = kx$ where k is called the constant of proportionality or the constant of variation.

In the example $d = 50t$, the number 50 is the constant of variation.

In each of the following,

$$y = 2x \qquad y = \tfrac{2}{3}x \qquad y = 25x$$

we say that y varies directly as x. In the first example we see that the constant of variation k equals 2. In the second example, k equals $\tfrac{2}{3}$ and k equals 25 in the third example.

EXAMPLE 5 Express the following as an equation of direct variation.

(a) The cost of a certain number of items varies directly as the price of each.

(b) The distance traveled by an automobile varies directly as the time.

(c) The amount of work done varies directly as the number of hours worked.

Solution: (a) $C = kp$

(b) $d = kt$

(c) $w = kh$

4-5

Inverse Variation

Sometimes two quantities are related in such a way that one decreases as the other increases. If this is the case, then we say that one quantity varies inversely as the other quantity. The equation for inverse variation is

$$y = \frac{k}{x}.$$

This equation states that *y varies inversely as x.*

Suppose we must drive a distance of 200 miles. In this case the distance 200 is constant. The time to drive 200 miles depends upon the speed of the auto. The time and speed are related in such a way that as one increases the other decreases. Suppose we write this example as the equation

$$t = \frac{200}{r}$$

where *t* equals time and *r* equals rate.

Letting *r* equal 25, 40, 50 we get:

r	25	40	50
t	8	5	4

We can see that as the speed increases the time decreases.

EXAMPLE 6 Express the following as an equation of inverse variation.

(a) Boyle's laws states that *the pressure of gas under constant temperature varies inversely as the volume.*

(b) The rate of a certain trip varies inversely as the time.

(c) The length of a rectangle of constant area varies inversely as the width.

Solution: (a) $P = \dfrac{k}{V}$

(b) $r = \dfrac{k}{t}$

(c) $l = \dfrac{k}{w}$

Sometimes a quantity varies as the product of two or more quantities. We call this type of variation *joint variation*. For example, the area of a rectangle which varies as its length and width would be written as

$$A = k(lw)$$

In general if y varies jointly as x and z, we write the equation

$$y = k(xz)$$

where k is the constant of variation.

EXAMPLE 7   Express the following as an equation of joint variation.

(a) The volume of a rectangle prism varies jointly as the length, width, and height.

(b) The cost of a tank of gasoline varies jointly as the amount per gallon and the number of gallons.

(c) The cost of painting a wall varies jointly as the length and width of the wall and the cost per sq ft of the paint.

Solution: (a) $V = k(lwh)$

(b) $C = k(an)$

(c) $C = k(lwc)$

It is possible to have a combination of variations. For example, Newton's law of gravitation, which states that *the gravitational attraction between two bodies varies directly as the product of their masses and inversely as the square of the distance between their centers of gravity*, could be expressed as

$$G = \left(\frac{M \cdot m}{d^2}\right) k$$

where G equals gravitational attraction, M and m equal the two masses, and d equals the distance between their center of gravity.

PROBLEM SET 4-3

Express the following statements as equations.

1. d varies directly as v.

2. i varies inversely as e.

3. a varies inversely as the square of t.

4. KE varies jointly as one-half m and the square of v.

5. w varies directly as W and inversely as the product of l and A.

PROBLEM SET 4-4

1. If y varies directly as x and $y = 30$ when $x = 6$, find y when $x = 15$.
2. If e varies directly as r and $e = 75$ when $r = 3$, find e when $r = 39$.
3. If F varies jointly as m and a, and $F = 120$ when $m = .6$ and $a = 100$, find F when $m = 3$ and $a = 80$.
4. If P varies directly as the square of e and inversely as r, and $P = 25$ when $e = 10$ and $r = 4$, find P when $e = 15$ and $r = 5$.
5. If the square of v varies jointly as F and d and inversely as m, and $v = 10$ when $F = 60$, $d = 10$, and $m = 12$, determine v when $F = 64$, $d = 24$, and $m = 12$.

PROBLEM SET 4-5

Solve the following.

1. The ratio of the cruising speed of an airplane to its maximum speed is $7 : 8$. If the cruising speed is 490 m.p.h., what is the maximum speed?

2. The formula, $\dfrac{d}{D} = \dfrac{W}{w}$, expresses an inverse relationship between the diameter and the number of revolutions of two meshed gears. If gear A is 2 in. in diameter, gear B is 5 in. in diameter, and gear A is turning at 500 r.p.m., how many revolutions per minute does gear B make.

3. If gear A is 3 in. in diameter and turns at 900 r.p.m., how fast does gear B turn if its diameter is 1.75 in?

4. In a cement mix, the ratio of cement, sand, and stone is $1 : 2 : 4$. How many cubic feet of each are needed for a wall of volume 840 cu ft?

5. Determine the width of the river in the figure below.

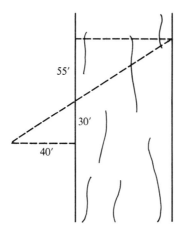

Figure 4-1

Suppose y varies directly as x and $y = 18$ when $x = 6$, find y when $x = 2$. In this problem we first need to find the constant of variation. First, we write the general equation

$$y = kx$$

Substituting 18 for y and 6 for x and solving for k, we get

$$18 = k(6)$$
$$3 = k$$

Hence, we see that the constant of variation is 3. Substituting this in the general equations, we have

$$y = 3x$$

Substitute 2 for x and solve

$$y = 3(2)$$
$$= 6$$

EXAMPLE 8   If y varies inversely as x and is equal to 12 when $x = 4$, find the value of y when $x = 10$.

Solution: The general equation would be

$$y = \frac{k}{x}$$

Substitute 12 for y and 4 for x and solve for k.

$$12 = \frac{k}{4}$$

$$48 = k$$

Now substitute 48 for k in the general equation.

$$y = \frac{48}{x}$$

Next substitute 10 for x and solve for y

$$y = \frac{48}{10}$$

$$y = 4.8$$

EXAMPLE 9   The resistance to the flow of electricity in a wire varies directly as the length and inversely as the square of the diameter of the wire. If the resistance is .26 ohm when the diameter is .02 in. for a wire 10 ft long, what is the resistance of a wire of the same material 15 ft long and .01 in. in diameter?

Solution: $R = k \dfrac{L}{d^2}$

Substitute .26 for R, .02 for d, and 10 for L and solve for k.

$$.26 = k \frac{10}{(.02)^2}$$

$$.26 = k \frac{10}{.0004}$$

$$.000104 = 10k$$

$$.0000104 = k$$

Substitute for k in the original equation and solve for R.

$$R = k \frac{L}{d^2}$$

$$= (.0000104) \frac{15}{(.01)^2}$$

$$= (.0000104) \frac{15}{.0001}$$

$$= \frac{.0001560}{.0001}$$

$$= 1.56 \text{ ohms}$$

PROBLEM SET 4-6

Solve the following.

1. The equation $F = ks$, known as Hooke's law, states that a spring stretches proportionally to the force applied to it. If $F = 20$ lbs and $s = .4$ in., determine the value of k and the units associated with it. If F is increased to 70 lbs, how much will the spring be stretched?

2. The equation, $KE = kmv^2$, expresses the kinetic energy of a body of mass m and velocity v. If $KE = 25 \frac{kg \cdot m^2}{sec^2}$, $m = 2$ kg, and $v = 5m/sec$, determine k. If v is increased to $8m/sec$, determine KE.

3. The period of a simple pendulum is given by the equation $T = k\sqrt{\frac{L}{g}}$. If the length of the pendulum, L, is .25 meter, the acceleration due to gravity, g, equals 9.8 meter/sec^2, and the period, T, equals 1 sec, determine the value of k. If L is increased to 1 meter, determine the period T.

4. The resonant frequency of a circuit containing inductance and capacitance is given by the equation $f = \frac{1}{k\sqrt{LC}}$. If f has a value of 500, L a value of 10×10^{-3}, and C a value of 10×10^{-6}, determine the value of k. If C is changed to 1×10^{-9} determine the value of f.

5. The temperature in degrees Celsius can be determined by the equation $C = k(F - 32)$ where F is in Fahrenheit degrees. If $C = 30$ and $F = 86$ determine the value of k. If F is increased to 122, determine the value of C.

4-8

Application of Ratio and Proportion to Simple Machines

In physics or mechanics you learn that the most complex machines are made from various combinations of simple machines such as levers, pulleys, gears, or the hydraulic press. A mathematical analysis of any of these is based on the principle of ratio and proportion.

4-9

Levers

A lever is a rigid bar pivoted at a point called the fulcrum. It is most often used to provide mechanical advantage. Using a lever, a heavy object can be lifted with little force by applying the force a considerable distance from the fulcrum. The distance from the fulcrum to the point where the force is applied, or the point where the load is applied, is called the moment arm, l. If we designate the applied force as f_1, the input moment arm as l_1, the load as f_2, and the output moment arm as l_2, we can express mathematically the following inverse proportional relationship between these parameters

$$\frac{f_1}{f_2} = \frac{l_2}{l_1}$$

This may be written as $f_1 l_1 = f_2 l_2$. Figure 4-2 shows a drawing of a lever.

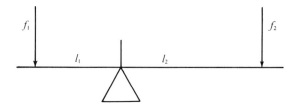

Figure 4-2

EXAMPLE 10   What is the force required to move the drum in Figure 4-3?

Solution:

$$f_2 \times l_2 = f_1 \times l_1$$
$$300 \text{ lbs} \times 1 \text{ ft} = f_1 \times 3 \text{ ft}$$
$$300 \text{ ft lbs} = f_1 \times 3 \text{ ft}$$
$$\frac{300 \text{ ft lbs}}{3 \text{ ft}} = f_1$$
$$100 \text{ lbs} = f_1$$

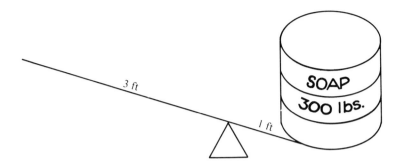

Figure 4-3

EXAMPLE 11   A chemist uses the analytical balance in Figure 4-4 to weigh out a certain amount of calcium sulfate. If the scales are balanced as drawn, the weight weighs 10 ounces and the pan weighs 10 ounces, how much does the calcium weigh?

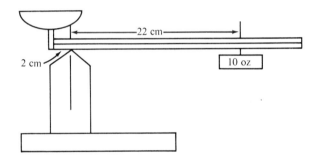

Figure 4-4

Solution: The weight times the moment arm to the left of the fulcrum equals the weight times the moment arm to the right of the fulcrum.

$$f_1 l_1 = f_2 l_2$$

$$f_1 = \frac{f_2 l_2}{l_1}$$

$$f_1 = \frac{(10 \text{ oz})(22 \text{ cm})}{2 \text{ cm}} = 110 \text{ oz}$$

The pan holding the calcium sulfate weighs 10 oz, therefore this must be subtracted which means the calcium sulfate weighs 100 oz.

PROBLEM SET 4-7

1. Determine the weight of F_1 in the drawing below to balance the lever.

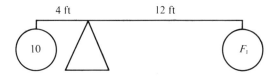

2. If an 80 lb boy sits 6 ft from the fulcrum on a seesaw how far must a 64 lb boy sit to balance the seesaw?

3. On an analytical balance one pan is 10 cm from the fulcrum. In this pan is a 200 gram weight and an unknown weight. The second pan is 15 cm from the fulcrum. In this pan are two 50 gram, three 20 gram, two 10 gram, and a 100 gram weight. How many grams does the unknown weigh if the system is balanced?

4. If $f_t l_t = f_1 l_1 + f_2 l_2 + f_3 l_3$, how much does F_x weigh in the following drawing?

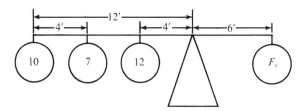

4-10

Pulleys and Gears

When two pulleys are connected together by a belt of some type they form the following mathematical relationship: $\omega \times d = \omega_1 \times d_1$ where ω represents the revolutions per unit of time and d represents the diameters of the pulleys.

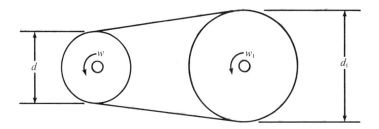

Figure 4-5

Suppose you have a motor driving a grinding wheel as in Figure 4-5. If the motor turns a pulley with a 4 in. diameter at 1600 r.p.m., how fast will the grinder be turning if the pulley on the grinder has a 3 in. diameter?

Using the formula

$$\omega \times d = \omega_1 \times d_1$$

where

$$d = 3 \text{ in.} \qquad d_1 = 4 \text{ in.}$$

and

$$\omega_1 = 1600 \text{ r.p.m.}$$

we get

$$3 \text{ in.} \times \omega = 4 \text{ in.} \times 1600 \text{ r.p.m.}$$
$$3\omega = 6400 \text{ r.p.m.}$$
$$\omega = 2133\tfrac{1}{3} \text{ r.p.m.}$$

When two gears are connected they have the mathematical relationship:

$$\frac{d}{d_1} = \frac{N}{N_1}$$

where N equals the number of teeth and d equals the diameter of the gears.

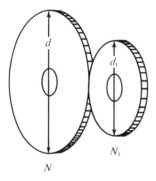

Figure 4-6

EXAMPLE 12   Determine the diameter of the large gear in Figure 4.7.

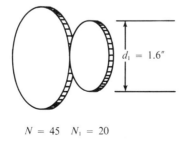

$N = 45 \quad N_1 = 20$

Figure 4-7

Solution: Let $N_1 = 20$

$$d_1 = 1.6 \text{ in.}$$

$$N = 45$$

then

$$\frac{d}{d_1} = \frac{N}{N_1}$$

$$\frac{d}{1.6} = \frac{45}{20}$$

$$d = 1.6 \left(\frac{45}{20}\right)$$

$$= \frac{72.0}{20}$$

$$= 3.6 \text{ in.}$$

PROBLEM SET 4-8

1. A motor with an angular velocity of 1750 r.p.m. and a 2 in. pulley drives a 10 in. pulley connected to the squirrel cage of an air conditioner. What is the angular velocity of the squirrel cage?

2. The rewind mechanism of a tape recorder is a belt-driven pulley arrangement. If the motor turns at a rate of 435 r.p.m. and has a .5 in. diameter, what is angular velocity of the tape reel if it is connected to a 2 in. diameter pulley?

3. A drill press is driven by a 1740 r.p.m. motor and two 3-step pulleys with diameters of 2 in., 3 in., and 4 in. Determine the three possible speeds of the drill.

4. Determine the angular velocity of gear 4 in the drawing.

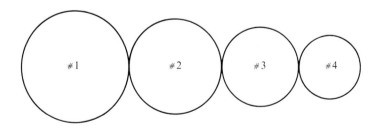

5. Determine the angular velocity of gear 7 in the drawing if gear 1 turns at 1750 r.p.m.

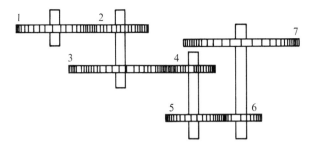

4-11

The Hydraulic Press

The final application of ratio and proportion which we will consider is the hydraulic press. A drawing of the hydraulic press, a simple machine based on Pascal's principle, is shown in Figure 4-8.

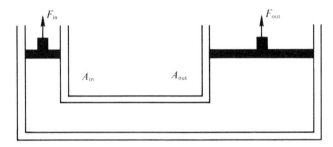

Figure 4-8

The input force, F_{in}, causes a pressure to be exerted on the fluid in the input piston. This pressure is transmitted by the incompressible fluid to the output piston. According to Pascal's principle the same pressure is applied to all points on the face of the output piston. Since force equals pressure times area ($F = pA$), the greater area of the output piston means a greater force at the output. Applying the principle of ratio and proportion, we can say that the ratio of the force equals the ratio of the areas of the pistons. As an equation,

$$\frac{F_{out}}{F_{in}} = \frac{A_{out}}{A_{in}}$$

If the output piston has a greater area, the output force will be greater according to our equation. Some industrial applications of the hydraulic press are hydraulic jacks, such as car lifts, brake systems, control systems in airplanes, presses, and barber chairs.

EXAMPLE 13   An automobile weighing 3400 lbs is to be raised on a hydraulic lift. The diameter of the output piston is 20 in. and the input piston has a 3 in. diameter. What input force is needed to lift the car?

Solution: The area of the pistons is

$$A_{in} = \pi r^2 = (3.14)(1.5)^2 = 7 \text{ in.}^2$$

$$A_{out} = \pi r^2 = (3.14)(10)^2 = 314 \text{ in.}^2$$

$$\frac{F_{out}}{F_{in}} = \frac{A_{out}}{A_{in}}$$

$$(F_{in})(A_{out}) = (F_{out})(A_{in})$$

$$F_{in} = F_{out}\left(\frac{A_{in}}{A_{out}}\right)$$

$$F_{in} = 3400 \text{ lbs}\left(\frac{7}{314}\right)$$

$$F_{in} = 75.8 \text{ lbs}$$

Knowing the distance either piston moved, we could determine how far the other piston moved by the same technique. The ratio of the forces also equals the ratio of the distance the pistons move. However, in this case, the relationship is an inverse ratio given by

$$\frac{F_{out}}{F_{in}} - \frac{l_{in}}{l_{out}}$$

Suppose the output piston moves 1 ft. How far must the input piston move?

$$l_{in} = \frac{F_{out}}{F_{in}}(l_{out})$$

$$l_{in} = \frac{3400}{75.8}(1 \text{ ft})$$

$$l_{in} = 44.8 \text{ ft}$$

Obviously this would have to be accomplished by a pump mechanism since an input piston 44 ft long is hardly practical.

PROBLEM SET 4-9

1. If the pump piston of a hydraulic press is $\frac{3}{4}$ in. in diameter and the output piston is 8 in. in diameter, what input force is required to lift a 500 lb load?

2. A large industrial press has an output piston 4 ft in diameter and an input piston 4 in. in diameter. How far does the output piston move when the input piston moves 1 ft?

3. What must the value of w_2 be if neither piston moves in the following drawing?

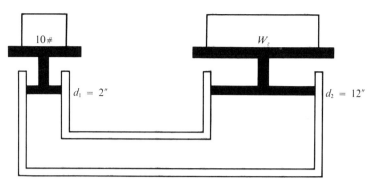

4. A lever is connected to the input piston of a hydraulic jack. The input piston is 3 in. in diameter and the output piston is 15 in. in diameter. The fulcrum on the lever is 2 in. from the input piston and 2 ft from the input end of the lever. How much force is required on the lever to raise an 800 lb load on the hydraulic press?

5

Trigonometry

Trigonometry is that branch of mathematics which deals primarily with angles and triangles. Problems such as finding the height of a building, the distance from one point to another, certain electrical problems, and physics and applied mechanics problems can be solved by triangles. Of the different types of triangles, the right triangle is used most often in trigonometry. Since a triangle has three sides and three angles, let us begin our study of trigonometry with the angle.

5-2
Angle

An angle is made by two rays which have a common endpoint. The common endpoint of the two rays is called the vertex.

A trigonometric angle will be defined as a ray in one position (*initial*) moving to a new position (*terminal*).

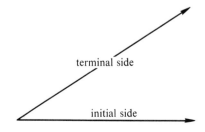

Figure 5-1

The amount of rotation from the initial side to the terminal side is the measure of the angle.

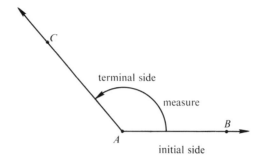

Figure 5-2

In Figure 5-2, $\overrightarrow{AB}$ is the initial side, $\overrightarrow{AC}$ is the terminal side, and A is the vertex.

An angle can be named several ways. Any way is correct as long as it is clear as to which angle we are talking about.

EXAMPLE 1

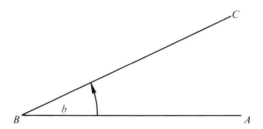

Figure 5-3

This angle could be called $\angle ABC$, $\angle CBA$, $\angle B$, or $\angle b$. In using three letters, such as $\angle ABC$ to name an angle, be sure the letter of the vertex is in the middle.

EXAMPLE 2

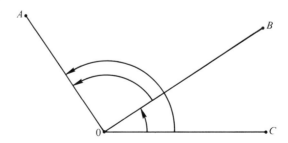

Figure 5-4

Which angle is ∠0? Do you see that it is not clear just which angle we are speaking of? This is why we must be careful how we name an angle.

Identify the following angles in three ways.

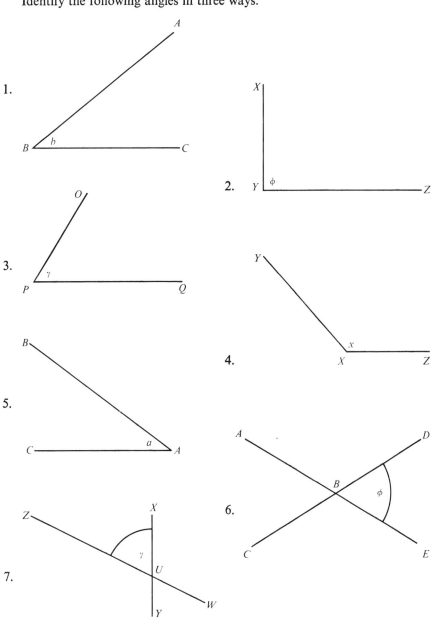

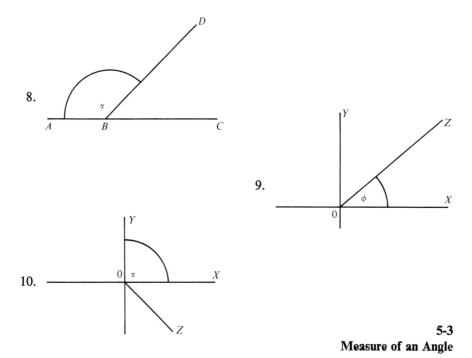

8.

9.

10.

5-3
Measure of an Angle

The measure of an angle can be given in two ways, in *degrees* or in *radians*. A degree is $\frac{1}{360}$ of a complete rotation or circle.

$1°$

Figure 5-5

A *radian* is an angle which if placed at the center of a circle will make an arc equal in length to the radius of the circle. How many radians are in a circle?

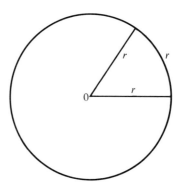

Figure 5-6

If $C = 2\pi r$ (2π times the length of the radius) and if a radian cuts an arc equal in length to the radius, then:

$$C = 2\pi \text{ radians.}$$

If $C = 2\pi$ radians, and C is also equal to $360°$, then:

$$360° = 2\pi \text{ radians}$$
$$180° = \pi \text{ radians} \quad \text{(dividing by 2)}$$
$$\frac{180°}{\pi} = 1 \text{ radian} \quad \text{(dividing by } \pi \text{)}$$
$$57.3° \cong 1 \text{ radian}$$
$$1° = .01745 \text{ radian}$$
$$1° = \frac{\pi}{180} \text{ radian}$$

EXAMPLE 3   Change radians to degrees.

(1) $2\pi = 2\pi$ radians $= 360$

(2) $\pi = \pi$ radians $= 180°$

(3) $\dfrac{\pi}{2} = \dfrac{\pi}{2}$ radians $= \dfrac{180°}{2} = 90°$

(4) $\dfrac{3\pi}{2} = \dfrac{3\pi}{2}$ radians $= \dfrac{3 \times 180°}{2} = 270°$

EXAMPLE 4   Change degrees to radians.

(1) $30° = 30\,\dfrac{(\pi)}{180} = \dfrac{\pi}{6}$

(2) $40° = 40\,\dfrac{(\pi)}{180} = \dfrac{2\pi}{9}$

(3) $120° = 120\,\dfrac{(\pi)}{180} = \dfrac{2\pi}{3}$

(4) $330° = 330\,\dfrac{(\pi)}{180} = \dfrac{11\pi}{6}$

PROBLEM SET 5-2

Express the following in radian measure.

1. $360°$ 　　　 2. $57.3°$ 　　　 3. $1°$ 　　　 4. $45°$ 　　　 5. $90°$

6. 120°   7. 30°   8. 225°   9. 540°   10. 75°

11. 22°   12. 139°   13. 310°   14. 65°   15. 5°

Express the following angular measurements in degrees.

1. 1 radian   2. $\frac{2}{3}$ radian

3. $\frac{1}{\pi}$ radian   4. 3 radians

5. .4 radian   6. $\frac{7}{8}$ radian

7. .08 radian   8. .75 radian

9. 4.5 radians   10. 12.56 radians

5-4
Direction of Angles

An angle is *positive* if its terminal side moves (rotates) in a counterclockwise direction. An angle is *negative* if its terminal side moves (rotates) in a clockwise direction. A curved arrow is used to indicate the direction.

EXAMPLE 5   Draw the following angles.

30°   135°   −45°   270°   −400°

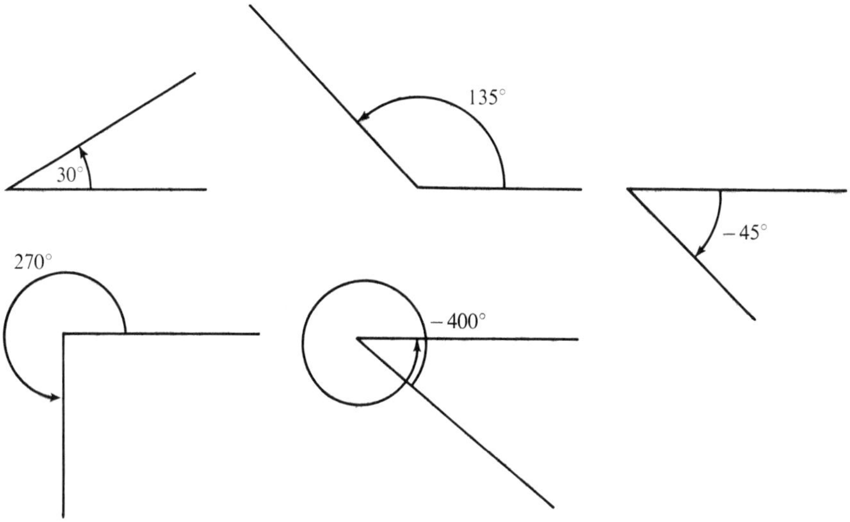

Figure 5-7

Draw the following angles.

1. 45°   2. 75°   3. −30°   4. 150°   5. −75°

6. 240°   7. −180°   8. 400°   9. −375°   10. −225°

5-5
Angles in Standard Position

An angle is in standard position if its vertex is at the origin and its initial side is on the x-axis. The following graphs show angles in standard position.

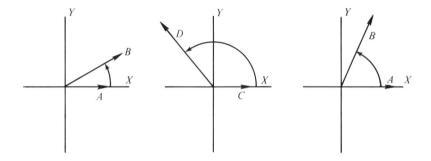

Figure 5-8

5-6
Coterminal Angles

Coterminal angles are angles in standard position whose terminal sides coincide. For example, 30° and 390° are coterminal angles.

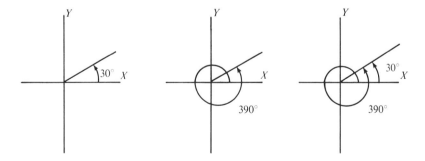

Figure 5-9

PROBLEM SET 5-4

Determine the coterminal angle for the following angles.

1. 405°   2. 540°   3. 725°   4. 375°   5. 580°

6. 611°   7. 900°   8. 361°   9. 702°   10. 1050°

5-7
The Right Triangle

The Pythagorean Theorem states that *the square of the hypotenuse of a right triangle equals the sum of the squares of the legs*. If we have a triangle like the one in Figure 5-10, we can state the Pythagorean Theorem as:

$$c^2 = a^2 + b^2$$

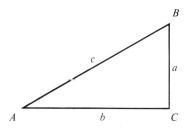

Figure 5-10

EXAMPLE 6   Find c, if $a = 8$ and $b = 6$.

$$c^2 = a^2 + b^2$$
$$c^2 = 8^2 + 6^2$$
$$c^2 = 64 + 36$$
$$c^2 = 100$$
$$c = 10$$

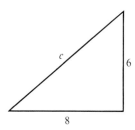

Figure 5-11

EXAMPLE 7 Find a, if $c = 13$, $b = 12$.

$$c^2 = a^2 + b^2$$
$$13^2 = a^2 + 12^2$$
$$169 = a^2 + 144$$
$$25 = a^2$$
$$5 = a$$

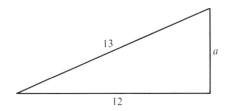

Figure 5-12

EXAMPLE 8 Suppose we plot the point $P(3,4)$ on a rectangular coordinate system. What would be the distance of OP?

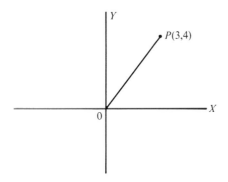

Figure 5-13

If we draw a line from P, perpendicular to the x-axis, it will form a right triangle. OP becomes the hypotenuse of a right triangle and its length can be found by the Pythagorean Theorem if we can determine the length of the two sides, OR and RP.

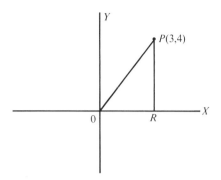

Figure 5-14

To find the lengths of OR and RP we need to know the coordinates of O and R. Since O is the origin (0,0) are its coordinates. Do you see that R is the same distance from the y-axis as P, and since it is on the x-axis its coordinates would be (3,0).

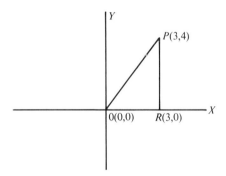

Figure 5-15

Now that we have the coordinates of each point, the length of the horizontal side and the vertical side can be found. In Chapter 2 we found horizontal distance by subtracting the x coordinates and vertical distance by subtracting the y coordinates.

$$OR = 3 - 0 \qquad RP = 4 - 0$$

$$= 3 \qquad\qquad = 4$$

$$(OP)^2 = (OR)^2 + (RP)^2$$

$$(OP)^2 = 3^2 + 4^2$$

$$(OP)^2 = 9 + 16$$

$$(OP)^2 = 25$$

$$OP = 5$$

EXAMPLE 9   Given: $R(-2,1)$, $S(4,1)$, and $T(4,5)$, find RS, ST, and RT.

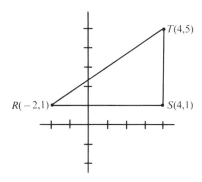

Figure 5-16

Solution:

$$RS = 4 - (-2) \qquad ST = 5 - 1$$
$$RS = 4 + 2 \qquad ST = 4$$
$$RS = 6$$

$$(RT)^2 = (RS)^2 + (ST)^2$$
$$(RT)^2 = 6^2 + 4^2$$
$$(RT)^2 = 36 + 16$$
$$(RT)^2 = 52$$
$$RT = \sqrt{52}$$

From the above examples we can make a general formula for finding the distance of any line segment.

Suppose we wish to find the length of PQ where $P(x_1,y_1)$ and $Q(x_2,y_2)$.

First we will draw the right triangle PQR. The coordinates of R will be (x_2,y_2).

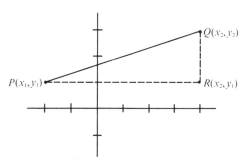

Figure 5-17

Now we find PQ by the Pythagorean Theorem.

$$(PQ)^2 = (PR)^2 + (RQ)^2$$
$$(PQ)^2 = (x_2 - x_1)^2 + (y_2 - y_1)^2$$
$$PQ = \sqrt{(x_2 - x_1)^2 + (y_2 - y_1)^2}$$

$\sqrt{(x_2 - x_1)^2 + (y_2 - y_1)^2}$ is the distance formula and can be used to find the distance of a line segment if the coordinates of the endpoints are known.

EXAMPLE 10   Given: $A(-2,6)$, $B(8,-1)$ find AB.

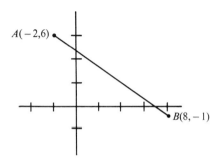

$A(-2,6)$

$B(8,-1)$

Figure 5-18

Solution:

$$AB = \sqrt{[8 - (-2)]^2 + [(-1) - 6]^2}$$
$$AB = \sqrt{(8 + 2)^2 + (-1 - 6)^2}$$
$$AB = \sqrt{10^2 + 7^2}$$
$$AB = \sqrt{100 + 49}$$
$$AB = \sqrt{149}$$

PROBLEM SET 5-5

Using Pythagorean's theorem, determine the length of the hypotenuse for each of the following.

1. $x = 3$, $y = 4$ 2. $x = 60$, $y = 50$

3. $x = 1$, $y = 6$ 4. $x = 28$, $y = 21$

5. $A(3,4)$, $B(0,0)$ 6. $A(-3,5)$, $B(4,-2)$

7. $x = 579$, $y = 2410$ 8. $x = 30.6$, $y = 78.3$

9. $A(-5,4)$, $B(-1,-3)$ 10. $x = .207$, $y = .421$

11. $x = 3650$, $y = 3810$ 12. $A(-6,-1)$, $B(5,-3)$

Because a right triangle is so important to trigonometry, let us again define the right triangle and its sides.

A *right triangle* is a triangle with one right angle. The other two angles will be acute angles and complementary.

The side opposite the right angle is called the *hypotenuse*.

The other two sides are called the *opposite* or *adjacent* sides, according to which angle you are talking about.

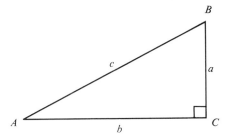

Figure 5-19

In Figure 5-19, AB is the hypotenuse. If you are referring to $\angle A$, then BC would be the opposite side and AC would be the adjacent side. If you are referring to $\angle B$, then AC is the opposite side and BC is the adjacent side.

In a right triangle it is possible to form six different ratios of the sides. For example, in Figure 5-20 we have the following ratios:

$$\frac{BC}{AB} \quad \frac{AC}{AB} \quad \frac{BC}{AC} \quad \frac{AB}{BC} \quad \frac{AB}{AC} \quad \frac{AC}{BC}$$

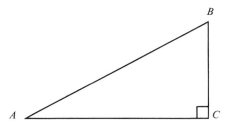

Figure 5-20

These ratios when used in reference to one of the acute angles give us six very important functions or ratios used in trigonometry. The following is a list of the six trigonometric functions with respect to $\angle A$.

$$\text{sine } A = \frac{\text{side opposite } \angle A}{\text{hypotenuse}}$$

$$\text{cosine } A = \frac{\text{side adjacent } \angle A}{\text{hypotenuse}}$$

$$\text{tangent } A = \frac{\text{side opposite } \angle A}{\text{side adjacent } \angle A}$$

$$\text{cotangent } A = \frac{\text{side adjacent } \angle A}{\text{side opposite } \angle A}$$

$$\text{secant } A = \frac{\text{hypotenuse}}{\text{side adjacent } \angle A}$$

$$\text{cosecant } A = \frac{\text{hypotenuse}}{\text{side opposite } \angle A}$$

Notice that the following are *reciprocals* of each other:

(a) sine and cosecant

(b) cosine and secant

(c) tangent and cotangent

Using sin, cos, tan, cot, sec, and csc as abbreviations for sine, cosine, tangent, cotangent, secant, and cosecant, respectively, and θ as a general angle we can write the following:

$$\sin \theta = \frac{\text{opposite}}{\text{hypotenuse}} \qquad \cot \theta = \frac{\text{adjacent}}{\text{opposite}}$$

$$\cos \theta = \frac{\text{adjacent}}{\text{hypotenuse}} \qquad \sec \theta = \frac{\text{hypotenuse}}{\text{adjacent}}$$

$$\tan \theta = \frac{\text{opposite}}{\text{adjacent}} \qquad \csc \theta = \frac{\text{hypotenuse}}{\text{opposite}}$$

These six functions are so important that they must be memorized. However, you can see that the csc, sec, cot can be derived from the sin, cos, and tan, respectively.

EXAMPLE 11   Using the above functions and the triangle in Figure 5-21, we have the following.

Solution: $\sin A = 6/10$ $\sin B = 8/10$

$\cos A = 8/10$ $\cos B = 6/10$

$\tan A = 6/8$ $\tan B = 8/6$

$$\cot A = 8/6 \qquad \cot B = 6/8$$
$$\sec A = 10/8 \qquad \sec B = 10/6$$
$$\csc A = 10/6 \qquad \csc B = 10/8$$

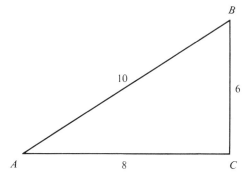

Figure 5-21

5-9
Cofunctions

Example 11 shows another important relation. Notice that the sin A is equal to the cos B. What other functions are equal? Do you see that tan A = cot B? sec A = csc B? These are called *cofunctions*. The cofunctions of complementary angles are equal.

(1) If sin $30° = \frac{1}{2}$, then cos $60° = \frac{1}{2}$

(2) If tan $45° = 1$, then cot $45° = 1$

(3) If sec $30° = \dfrac{2}{\sqrt{3}}$, then csc $60° = \dfrac{2}{\sqrt{3}}$

5-10
Trigonometric Functions for Angles in Standard Position

Suppose we are given angle θ in standard position. Let $P(x,y)$ be any point on the terminal side. OP can be thought of as a radius vector of length r. Since $OR = x$ and $RP = y$, we can write the trigonometric functions of θ as follows:

$$\sin \theta = \frac{y}{r} \qquad \cot \theta = \frac{x}{y}$$

$$\cos \theta = \frac{x}{r} \qquad \sec \theta = \frac{r}{x}$$

$$\tan \theta = \frac{y}{x} \qquad \csc \theta = \frac{r}{y}$$

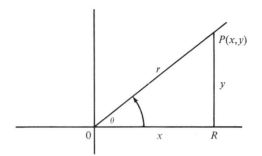

Figure 5-22

If the value of x and y are known we can find r by the Pythagorean theorem.

EXAMPLE 12   Find the six trigonometric functions of θ if the terminal side passes through the point (4,3).

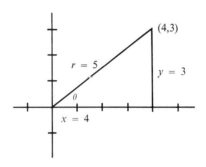

Figure 5-23

Since the terminal side passes through the point (4,3) by the Pythagorean theorem $r = 5$, we have the following:

$$\sin \theta = \tfrac{3}{5} \qquad \cot \theta = \tfrac{4}{3}$$
$$\cos \theta = \tfrac{4}{5} \qquad \sec \theta = \tfrac{5}{4}$$
$$\tan \theta = \tfrac{3}{4} \qquad \csc \theta = \tfrac{5}{3}$$

EXAMPLE 13   Find the six trigonometric functions of θ if the terminal side passes through the point (3, 7).

Solution: By the Pythagorean theorem $r = \sqrt{58}$, therefore

$$\sin \theta = \frac{7}{\sqrt{58}} \qquad \cot \theta = \frac{3}{7}$$

$$\cos \theta = \frac{3}{\sqrt{58}} \qquad \sec \theta = \frac{\sqrt{58}}{3}$$

$$\tan \theta = \frac{7}{3} \qquad \csc \theta = \frac{\sqrt{58}}{7}$$

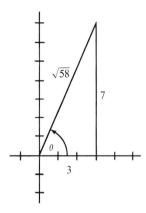

Figure 5-24

PROBLEM SET 5-6

If the given point falls on the terminal side of an angle in the 1st quadrant determine the value of the six trigonometric functions.

1. (4,3)   2. (3,4)   3. (8,8)   4. (2,1)   5. (1,5)
6. (4,5)   7. (7,3)   8. (5,2)   9. (3,1)   10. (5,7)

5-11
Special Angles

There are certain angles in which we can find the trigonometric functions.

EXAMPLE 14   Find the trigonometric functions of a 45° angle.

Solution: If one of the acute angles is 45°, then the other acute angle must be 45°. Since both acute angles are 45°, this makes the triangle isosceles.

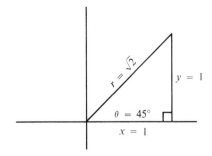

Figure 5-25

In an isosceles triangle the sides opposite the equal angles are equal. Therefore, $x = y$. Let us agree that $x = y = 1$. Then

$$r = \sqrt{x^2 + y^2} \qquad \sin 45° = \frac{1}{\sqrt{2}} = \frac{\sqrt{2}}{2} \qquad \cot 45° = \frac{1}{1} = 1$$

$$= \sqrt{1^2 + 1^2}$$

$$\cos 45° = \frac{1}{\sqrt{2}} = \frac{\sqrt{2}}{2} \qquad \sec 45° = \frac{\sqrt{2}}{1} = \sqrt{2}$$

$$= \sqrt{1 + 1}$$

$$= \sqrt{2} \qquad \tan 45° = \frac{1}{1} = 1 \qquad \csc 45° = \frac{\sqrt{2}}{1} = \sqrt{2}$$

EXAMPLE 15   Find the trigonometric functions of a 30° angle.

Solution: If we draw an equilateral triangle as in Figure 5-26, so that the x-axis bisects the angle at the origin, we have a 30° angle. Since all sides of the triangle are equal, we will let them be 2 units in length. If the x-axis bisects the angle it also bisects the side opposite. Therefore, the side opposite the 30° angle is 1.

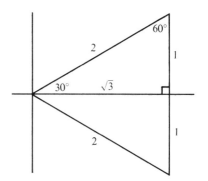

Figure 5-26

Now if $x = 1$ and $r = 2$ then,

$$y = \sqrt{r^2 - x^2} \qquad \sin 30° = \frac{1}{2} \qquad \cot 30° = \frac{\sqrt{3}}{1} = \sqrt{3}$$

$$= \sqrt{2^2 - 1^2}$$

$$\cos 30° = \frac{\sqrt{3}}{2} \qquad \sec 30° = \frac{2}{\sqrt{3}} = \frac{2\sqrt{3}}{3}$$

$$= \sqrt{4 - 1}$$

$$= \sqrt{3} \qquad \tan 30° = \frac{1}{\sqrt{3}} = \frac{\sqrt{3}}{3} \qquad \csc 30° = \frac{2}{1} = 2$$

EXAMPLE 16   Use the drawing in Example 14 and find the functions of 60°.

Solution: $\sin 60° = \dfrac{\sqrt{3}}{2}$ $\qquad$ $\cot 60° = \dfrac{1}{\sqrt{3}} = \dfrac{\sqrt{3}}{3}$

$\cos 60° = \dfrac{1}{2}$ $\qquad$ $\sec 60° = \dfrac{2}{1} = 2$

$\tan 60° = \dfrac{\sqrt{3}}{1} = \sqrt{3}$ $\quad$ $\csc 60° = \dfrac{2}{\sqrt{3}} = \dfrac{2\sqrt{3}}{3}$

In finding functions of angles we can express the ratio in decimal form.

$$\sin 30° = \frac{1}{2} = .500$$

$$\cos 30° = \frac{\sqrt{3}}{2} = .866$$

$$\tan 60° = \sqrt{3} = 1.732$$

Always remember that a function of an angle is simply the ratio of two sides, whether the ratio is expressed as a fraction or as a decimal.

5-12
Functions of Angles Between 0° and 90°

Table 3 in Appendix B shows the decimal representations of the functions for the sine, cosine, tangent, and cotangent functions of all angles from 0° through 90°.

You will notice that the tables have headings at both top and bottom and angle measures on the left and right sides. In looking for the function of any angle from 0° to 45°, find the angle on the left side of the chart and use the headings at the top. If the angle is between 45° and 90°, find the angle on the right side of the chart and use the headings at the bottom of the page.

EXAMPLE 17   Find sin 25°.

Solution: Look on the left side until you find 25°. Now find the sin column at the top. To the right of 25° under the sin column is .4226. Therefore,

$$\sin 25° = .4226$$

EXAMPLE 18   Find tan 85°.

Solution: Look on the right side until you find 85°. Now looking at the bottom find the tan column. To the left of 85° in the tan column (reading from the bottom) you will find 11.43. Therefore,

$$\tan 85° = 11.43$$

EXAMPLE 19   Find θ if sin θ = .3420.

Solution: This time you are given the value of the sin of some angle and are asked to find the size of the angle. To find θ look under the sin column until you come to .3420. Remember if you are reading down, find the angle at the left, if you are reading up, find the angle on the right. In this particular problem you are reading down. At the left you find 20°00'. Therefore,

$$\sin \theta = .3420$$
$$\theta = 20°$$

PROBLEM SET 5-7

Determine sin, cos, cot, and tan of θ if θ equals the following.

1. 45°	2. 60°	3. 80°
4. 7°	5. 23°	6. 67°
7. 48°	8. 54°	9. 88°
10. 1°	11. 16°	12. 72°
13. 5°	14. 75°	15. 13°
16. 30°	17. 62°	18. 90°
19. 0°	20. 35°	21. 65°

PROBLEM SET 5-8

Using the table of trigonometric functions determine the angle ϕ for the following.

1. sin ϕ = .5000	2. cos ϕ = .8090	3. tan ϕ = 1.0000
4. sin ϕ = .7771	5. cos ϕ = .9848	6. tan ϕ = 4.011
7. sin ϕ = 1.0000	8. cos ϕ = .3090	9. tan ϕ = .0524
10. sin ϕ = .8660	11. cos ϕ = .8660	12. tan ϕ = 9.514
13. sin ϕ = .7071	14. cos ϕ = .0175	15. tan ϕ = .4040

5-13
Interpolation

Sometimes we cannot find the value of the function of an angle directly from the tables. For example, you can find sin 13° 10' but not find sin 13° 14'. To find a function of an angle like 13° 14', you must make use of the angles given in the table that are immediately larger and smaller than the angle you are working with. This process is called interpolation.

EXAMPLE 20 Find sin 13° 14′.

Solution: Sin 13° 10′ and sin 13° 20′ can be found in the tables. The sin 13° 14′ is somewhere between these two angles and is found as follows:

$$10'\begin{bmatrix}\begin{array}{l}\text{— sin } 13°\ 20' = .2306\text{—}\\ 4'\begin{bmatrix}\text{sin } 13°\ 14' = \\ \text{sin } 13°\ 10' = .2278\end{bmatrix}\end{array}\end{bmatrix}28$$

We see that the difference between 13° 14′ and 13° 10′ is 4′, and the difference between 13° 20′ is 10′. Therefore sin 13° 14′ is $\frac{4}{10}$ the way between sin 13° 10′ and 13° 20′. The difference between .2306 and .2278 is 28, therefore sin 13° 14′ is ($\frac{4}{10}$ × 28) between .2278 and .2306. Therefore:

$$\frac{4}{10} \times 28 = .4 \times 28$$
$$= 11.2$$
$$= 11 \text{ (rounded off to nearest whole number)}$$

Add 11 to 2278.

$$\begin{array}{r}2278\\ 11\\ \hline 2289\end{array}$$

Therefore:

$$\text{sin } 13°\ 14' = .2289$$

EXAMPLE 21 Find tan 21° 27′.

Solution:

$$10\begin{bmatrix}\begin{array}{l}\text{—tan } 21°\ 30' = .3665\text{—}\\ 7\begin{bmatrix}\text{tan } 21°\ 27' = \\ \text{tan } 21°\ 20' = .3638\end{bmatrix}\end{array}\end{bmatrix}27$$

The difference between 21° 30′ and 21° 20′ is 10.
The difference between 21° 27′ and 21° 20′ is 7.
The difference between .3665 and .3638 is 27.
Therefore:

$$\frac{7}{10} \times 27 = .7 \times 27$$
$$= 18.9$$
$$= 19 \text{ (rounded to nearest whole number)}$$

Add 19 to 3638.

$$\begin{array}{r}3638\\ 19\\ \hline 3657\end{array}$$

Therefore:

$$\tan 21° 27' = .3657$$

EXAMPLE 22   Find sin 53° 22'.

Solution:

$$10 \left[\begin{array}{l} \sin 53° 30' = .8039 \\ 2 \left[\begin{array}{l} \sin 53° 22' = \\ \sin 53° 20' = .8021 \end{array} \right] \end{array} \right] 18$$

$$\tfrac{2}{10} \times 18 = .2 \times 18$$
$$= 3.6$$
$$= 4 \text{ (rounded)}$$
$$8021 + 4 = 8025$$

Therefore:

$$\sin 53° 22' = .8025$$

EXAMPLE 23   Find θ if sin θ = .8146.

Solution: This time we are asked to find the angle whose sine is .8146. Looking in Table 3, we find sin 54° 30' = .8141 and sin 54° 40' = .8158, therefore:

$$\begin{array}{l} \sin 54° 40' = .8158 \\ \sin \theta \quad\;\; = .8146 \\ \sin 54° 30' = .8141 \end{array} \left. \begin{array}{l} \\ 5 \\ \end{array} \right] 17$$

We see that the difference between .8146 and .8141 is 5, and the difference between .8158 and .8141 is 17. Therefore, .8146 is $\tfrac{5}{17}$ of the way between 54° 30' and 54° 40'. The difference between 54° 30' and 54° 40' is 10. Therefore:

$$\tfrac{5}{17} \times 10 = \tfrac{50}{17}$$
$$= 2.9$$
$$= 3$$

Add 3' to 54° 30'.

$$\begin{array}{r} 54° 30' \\ +3' \\ \hline 54° 33' \end{array}$$

Therefore:

$$\text{if } \sin \theta = .8146$$
$$\theta = 54° 33'$$

EXAMPLE 24   Find θ if tan θ = .9030.

Solution:

$$\begin{array}{l} \tan 42° 10' = .9057 \\ \tan \theta \quad\;\; = .9030 \\ \tan 42° 00' = .9004 \end{array} \left. \begin{array}{l} \\ 26 \\ \end{array} \right] 53$$

$$10 \times \tfrac{26}{53} = \tfrac{260}{53} = 4.9 \text{ or } 5$$

Therefore:

$$\text{if } \tan \theta = .9030$$
$$\theta = 42° \, 05'$$

For the following angles determine the value of the specified functions by interpolation.

1. sin 10° 25'   2. cos 24° 35'

3. tan 34° 05'   4. cot 39° 15'

5. sin 44° 44'   6. cos 51° 43'

7. tan 62° 36'   8. cot 71° 21'

9. sin 76° 37'   10. cos 81° 02'

Determine the angle θ for the following.

1. $\sin \theta = .2504$ 2. $\tan \theta = .6619$ 3. $\cot \theta = 1.1850$

4. $\cos \theta = .6248$ 5. $\tan \theta = .4954$ 6. $\cos \theta = .2581$

7. $\sin \theta = .3711$ 8. $\cot \theta = 2.0400$ 9. $\sin \theta = .6152$

10. $\cos \theta = .7127$ 11. $\sin \theta = .3955$ 12. $\tan \theta = 1.8130$

13. $\cot \theta = .6049$ 14. $\tan \theta = .5403$ 15. $\cos \theta = .3732$

5-14
Solutions of Right Triangles

To solve a right triangle means to find all angles and all sides.

EXAMPLE 25 Solve the right triangle ABC.

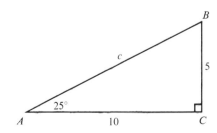

Figure 5-27

Solution: $\angle A = 25°$ $\angle C = 90°$

$a = 5$ $\angle B = ?$ $c = \sqrt{a^2 + b^2}$

$b = 10$ $c = ?$ $c = \sqrt{5^2 + 10^2}$

$\angle B = 90° - \angle A$ $= \sqrt{25 + 100}$

$= 90° - 25°$ $= \sqrt{125}$

$= 65°$ $= 5\sqrt{5}$

Suppose you are asked to solve a problem like the following.

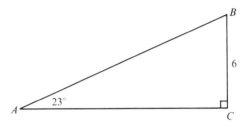

Figure 5-28

Angle B could be found by subtracting $\angle A$ from 90° but sides b and c could not be found readily with the Pythagorean theorem. In this problem we must make use of some of the trigonometric functions.

EXAMPLE 26   Solve the right triangle ABC.

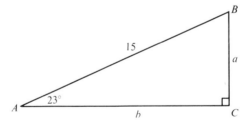

Figure 5-29

Solution: $\angle A = 23°$ $\angle B = ?$

$\angle C = 90°$ $a = ?$

$c = 15$ $b = ?$

If we form the ratio $\dfrac{a}{15}$, this gives us the ratio for the sin A in this triangle.

Therefore:

$$\sin A = \frac{a}{15}$$

Substituting 23° for A we have

$$\sin 23° = \frac{a}{15}$$

$$15(\sin 23°) = a$$

In Table 3, we find $\sin 23° = .3907$, therefore:

$$15(.3907) = a$$

$$5.8605 = a$$

$$\angle B = 90° - 23° = 67°$$

If we form the ratio $\dfrac{b}{15}$, this gives us the ratio for the cos A in this triangle.

Therefore:

$$\cos A = \frac{b}{15}$$

Substituting 23° for A, we have

$$\cos 23° = \frac{b}{15}$$

$$15(\cos 23°) = b$$

In Table 3, we find $\cos 23° = .9205$, therefore:

$$15(.9205) = b$$

$$13.8075 = b$$

EXAMPLE 27 Solve triangle ABC.

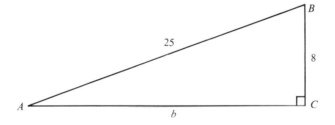

Figure 5-30

Solution: $\angle C = 90°$ $b = \sqrt{c^2 - a^2}$ $\sin A = \frac{8}{26}$

$\quad\quad\quad\quad a = 8$ $\quad\quad = \sqrt{25^2 - 8^2}$ $\sin A = .320$

$\quad\quad\quad\quad c = 25$ $\quad\quad = \sqrt{625 - 64}$ $\angle A = 18° \, 40'$

$\quad\quad\quad\quad\quad\quad\quad\quad\quad\quad = \sqrt{561}$

$\quad\quad\quad \angle A = ?$ $\quad\quad = 23.68$ $\cos B = \frac{8}{25}$

$\quad\quad\quad \angle B = ?$ $\quad\quad\quad\quad\quad\quad\quad\quad \cos B = .320$

$\quad\quad\quad\quad b = ?$ $\quad\quad\quad\quad\quad\quad\quad\quad \angle B = 71° \, 20'$

$\angle B$ could have been found by subtracting 18° 40′ from 90°.

In solving right triangles it is best to find as many unknown parts as possible from the given parts. Also, your answers should be given to the same degree of accuracy as the given measurements. Hence, in Example 26, $a = 5.8605$ should be

$$a = 5.9$$

and $b = 13.8075$ should be

$$b = 14$$

In Example 27, $b = 23.68$ should be

$$b = 24$$

Normally, when the measures of the sides are given to two-figure accuracy the angles found are given to the nearest 10 minutes. Therefore: $A = 18° 40'$ would be correct, and $B = 71° 20'$ would be correct.

PROBLEM SET 5-11

Determine the length of the hypotenuse where the following are the lengths of the sides of a right triangle.

1. $x = 3, y = 4$
2. $x = 10, y = 10$
3. $x = 28, y = 21$
4. $x = 40, y = 8$
5. $x = 75, y = 200$
6. $x = 550, y = 1225$
7. $x = .250, y = .400$
8. $x = \frac{3}{4}, y = \frac{1}{5}$
9. $x = 150, y = 20$
10. $x = 34,250, y = 29,725$

11. In walking to school a boy walks 500 yards north, then turns east and walks 800 yards. What is the distance from the boy's home to the school?

12. Two trains leave a station at the same time. Train A travels south at 50 m.p.h. Train B travels west at 60 m.p.h. How far apart are the trains after 3 hours?

13. A surveyor wishes to determine the distance across a lake at a particular point. He sights a point across the lake from his position, then he moves 200 ft and sights on the same point across the lake. He then determines that the angle between his original sighting and his second sighting is 20°. How wide is the lake at the point of interest?

14. The Pythagorean theorem is useful in determining the impedance in an electrical circuit where resistance is plotted along the x-axis, reactance along the y-axis, and the hypotenuse of the right triangle is the impedance. If an electrical circuit contains 30 ohms resistance and 40 ohms reactance, what is the impedance of the circuit?

15. A fly crosses a room diagonally from one corner at floor level to the opposite corner at ceiling level. If the diagonal makes an angle of 20° with the floor,

30° with the longer wall, and the fly traveled 20 ft, determine the dimensions
of the room.

Solve for X in the following problems.

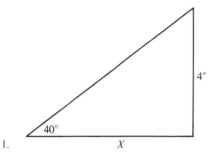

1.

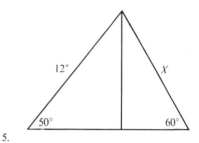

5.

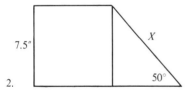

2.

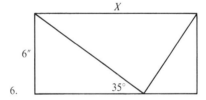

6.

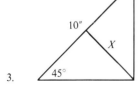

3.

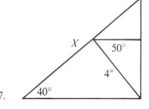

7.

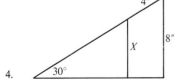

4.

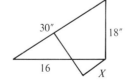

8.

Solve for the unknowns in the following right triangles.

	side a	side b	side c	angle A	angle B	angle C
1.	8.35			62.7°		
2.		1,290		41.9°		
3.			45.3		20.3°	
4.			265	22.4°		
5.	60		100			
6.	12	8				
7.		52	60			
8.	40			75°		
9.		5,300			47.9°	
10.	.237				16.8°	

5-15
Angles of Elevation and Depression

The angle made by the line of sight of an observer and the horizontal is called
the *angle of elevation* if it is measured upward and the *angle of depression* if it is
measured downward.

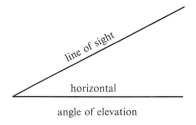

angle of elevation

Figure 5-31

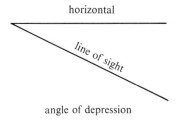

angle of depression

Figure 5-32

EXAMPLE 28   Find the angle of elevation of the sun if a building 100 ft tall casts a shadow 60 ft long.

Solution: To find θ, the angle of elevation, use the tangent ratio.

$$\tan \theta = \tfrac{100}{60}$$
$$\tan \theta = 1.666$$
$$\theta = 59° \ 02'$$
$$\theta = 59°$$

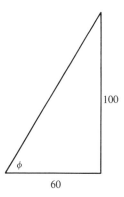

Figure 5-33

EXAMPLE 29   Suppose you are standing on a cliff which is 75 ft high and observe a boat in the water. If the angle of depression is 35°, how far is the boat from the base of the cliff?

Solution: If the angle of depression is 35°, then $\theta = 55°$. Therefore:

$$\tan 55° = \frac{x}{75}$$
$$75(\tan 55°) = x$$
$$75(1.428) = x$$
$$107.1 = x$$
$$107 = x \text{ (rounded to two-figure accuracy)}$$

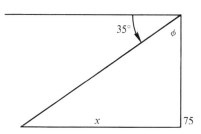

Figure 5-34

The *bearing of a line* is its direction from a point. Normally, this is given in either a north or south direction with the number of degrees to the east or west.

EXAMPLE 30 A bearing north 25° west would be written as N25°W and drawn as in Figure 5-35.

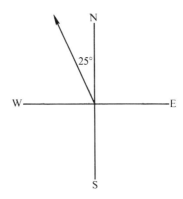

Figure 5-35

EXAMPLE 31 Draw the following headings: (1) S30°E, (2) N75°E, (3) S45°W.

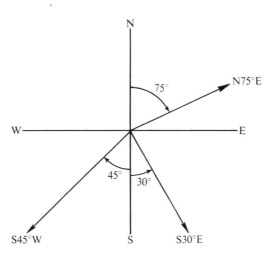

Figure 5-36

Lines of bearings can also be thought of as *vectors*. *Vectors* have magnitude and direction. The *magnitude* of a vector is the length of the vector and may represent inches, feet, miles, pounds, and so on. The *direction* can be given as the bearing of a line or as an angle in standard position.

Figure 5-37 shows a vector of magnitude *a*. This vector could mean any one of the following.

(a) *a* at θ

(b) $a \angle \theta$

(c) *a* at N α E

(d) an airplane flying N α E at "*a*" m.p.h.

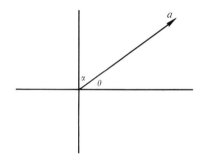

Figure 5-37

EXAMPLE 32 Draw the following as vectors.

(a) 5 at 25°

(b) $10 \angle 120°$

(c) 7 at S15°E

Solution:

(a)   (b)   (c)

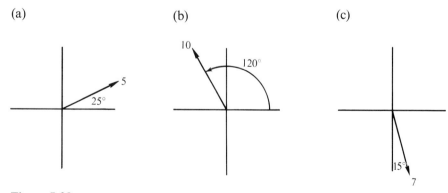

Figure 5-38

EXAMPLE 33   Determine the direction of the following vectors as (a) the bearing of a line, (b) an angle in standard position.

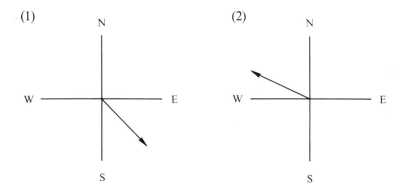

Figure 5-39

Solution:

$$(1)\ a.\ S60°E \qquad (2)\ a.\ N70°W$$
$$b.\quad 30° \qquad\quad b.\ 160°$$

Vectors can be used to help solve problems in physics and other fields. For example, if an airplane is flying due north at the rate of 100 m.p.h., what is the actual distance it will have traveled in one hour if the wind is blowing from the west at 25 m.p.h.?

Letting V_a and V_w represent the velocity of the airplane and wind respectively, we have Figure 5-40.

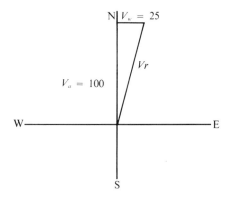

Figure 5-40

If we draw vector V_r, which is the *resultant*, we get a representation of where the plane would be relative to the point of departure. The *resultant* tells us the distance and the direction traveled.

The *resultant* is a vector connecting the tail of the first vector to the head of the last vector, after each vector has been connected head to tail.

EXAMPLE 34   Draw the resultant of the following vectors.

     (a)   (b)   (c)

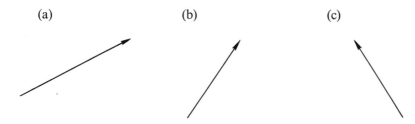

Figure 5-41

Solution:

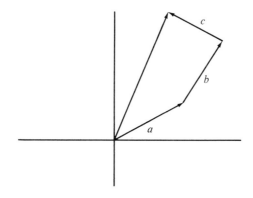

Figure 5-42

If the magnitude and direction of two or more vectors are given the resultant can be found. Consider the example of the airplane which was traveling at 100 m.p.h. in a northerly direction with the wind blowing from the west at 25 m.p.h. The resultant can be found as follows.

Since the wind is blowing from the west we have a right triangle whose legs are 100 and 25. We are now concerned with finding the hypotenuse which is the resultant. This also gives the distance traveled in one hour. Solving for d we have:

$$d^2 = 100^2 + 25^2$$
$$d^2 = 10000 + 625$$
$$d^2 = 10625$$
$$d = 103 \text{ miles}$$

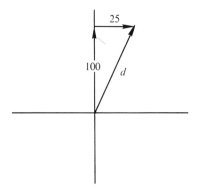

Figure 5-43

EXAMPLE 35 A barrel which weighs 275 lbs rests on a ramp which makes a 15° angle with the horizontal. What force must be exerted to keep the barrel from rolling down the ramp? What is the force of the barrel against the plane of the ramp?

Solution:

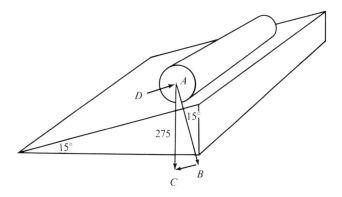

Figure 5-44

AC is the weight of the barrel and is drawn perpendicular to the horizontal. AD is the force required to keep the barrel from rolling down the ramp and is drawn parallel to the ramp. AB is force against the ramp and is drawn perpendicular to the ramp. AD, which is equal to BC, can be found as follows:

$$\sin 15° = \frac{BC}{275}$$

$$275(\sin 15°) = BC = AD$$

$$275(.2588) = AD$$

$$71.2 \text{ lbs} = AD$$

AB can be found as follows:

$$\cos 15° = \frac{AB}{275}$$

$$275(\cos 15°) = AB$$

$$275(.9659) = AB$$

$$265.6 \text{ lbs} = AB$$

PROBLEM SET 5-14

1. Two forces act on an object at right angles to one another. If force *A* is 100 lbs and force *B* is 50 lbs determine the magnitude of the resultant force.

2. Two forces act on an object. A force of 256 lbs pulls horizontally and a force of 164 lbs pulls upward vertically. Determine the magnitude of the resultant force.

3. The resultant of two forces acting at right angles to each other is 25 lbs. If one of the forces is 20 lbs determine the other force.

4. An automobile travels 80 miles in a straight line from town *A* to town *B* then makes a 90 degree turn and travels 40 miles in a straight line to town *C*. If a new highway is built connecting town *A* and *C* what will be its minimum length.

5. One end of a 12 ft ladder is placed against a building. The other end of the ladder rests on the ground 3 ft from the building. How high from the ground is the point where the ladder touches the building?

6. A boy standing on a cliff throws a rock over the cliff. The rock lands 360 ft from the base of the cliff. The straight line distance from the boy to the point of impact of the stone is 540 ft. How high is the cliff?

7. A television transmitting tower is 500 ft high. A guy wire is attached 80 ft from its top to a point 250 ft from the base of the tower. How long is the guy wire?

8. One end of a 10 ft board rests on the bed of a truck and the other on the ground 9.5 ft from the truck. How high is the bed of the truck?

9. A ship leaves port and travels 900 miles due west. It then turns north and travels 480 miles. What is the displacement of the ship from port?

10. Town *B* is 80 miles northwest of town *A*. Town *C* is 65 miles northwest of town *B*. How far from town *A* is town *C*?

11. A man pulls a crate weighing 200 lbs up a ramp to the bed of a truck. The bed of the truck is 4 ft high and one end of the ramp rests on the ground 10 ft from the end of the truck. What force must the man apply parallel to the crate to keep it from sliding?

12. How much must block *A* weigh for neither block to move in the drawing below?

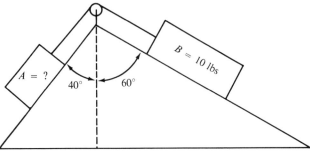

13. What should the spring balance read in the following drawing?

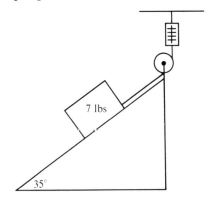

14. A boy is pulling a 20 lb wagon filled with 60 lbs of sand. If the wagon tongue makes a 25° angle with the ground, what minimum force must the boy apply to move the wagon if a 10 lb frictional force acts on the wagon?

15. What must the force *F* equal to balance the bar in the following drawing?

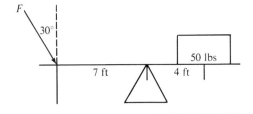

PROBLEM SET 5-15

Given the following vectors, graphically determine the resultant vector in the following problems.

$$A = 15 \text{ at } 30°$$
$$B = 5 \text{ at } 105°$$
$$C = 7W20°S$$
$$D = 10\angle 270°$$

1. $A + B + C$
2. $A - C + D$
3. $B + C - D$
4. $A - B + D$
5. $C + D - B$
6. $A + B + C + D$
7. $A - B + C - D$
8. $-A + B + C - D$
9. $A + B + C - D$
10. $A + B - C - D$

5-17
Reference Angles

Our discussion of trigonometric functions thus far has been concerned with angles between 0^0 and 90^0 (angles in the first quadrant).

What about angles larger than 90°, or negative angles? Is it possible to find trigonometric functions of these kinds of angles?

We have defined the sine of an angle in standard position to be:

$$\sin A = \frac{y}{r}$$

where y is the ordinate of a point on the terminal side and r is the length from the origin to the point (radius vector).

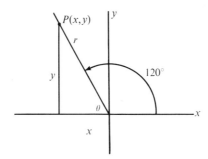

Figure 5-45

Let us examine an angle of 120°. Suppose the terminal side passes through the point $P(x,y)$, and the distance from 0 to P is r. What is the sin 120°? By definition it is the ordinate of the point P divided by the length of $0P$, therefore:

$$\sin 120° = \frac{y}{r}$$

In the same figure what is sin θ?

$$\sin \theta = \frac{y}{r}$$

How large is θ? 60°? Then could we say from this drawing that

$$\sin 60° = \frac{y}{r} ?$$

Does it seem that the sin 120° = sin 60°.

It is true that the sin 120° = sin 60°. 60° is the reference angle for 120°. Instead of having tables of functions for all sizes of angles, we use reference angles instead.

To find the reference angle of any angle θ:

(1) Draw θ in standard position.

(2) Choose any point P on the terminal side of θ, and draw a perpendicular line from P to the x-axis.

(3) The angle formed by the terminal side and the x-axis will be the *reference angle*.

EXAMPLE 36   What is the reference angle for 150°

Solution: (1) Draw an angle of 150° in standard position.

(2) Draw a perpendicular from any point P to the x-axis.

(3) θ will be the reference angle.

How large is θ?

$$\theta = 180° - 150°$$

$$= 30°$$

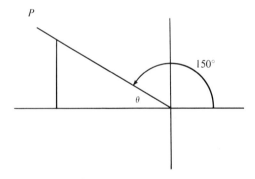

Figure 5-46

EXAMPLE 37   What is the reference angle for 230°?

Solution: How large is θ?

$$\theta = 230° - 180°$$

$$= 50°$$

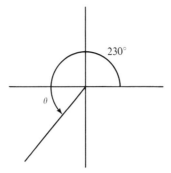

Figure 5-47

To find the functions of an angle that involves the use of a reference angle, it is necessary to be aware of the signs of the x, y, and r values. The radius vector r is considered to be positive in all quadrants but x and y are positive in some quadrants and negative in other quadrants. Figure 5-48 shows the signs of x and y in the four quadrants. In quadrant I the signs of x and y are positive. In the second quadrant x is negative and y is positive. In the third quadrant both x and y are negative. In the fourth quadrant x is positive and y is negative.

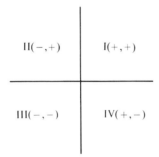

Figure 5-48

EXAMPLE 38 Find the value of the sin 120°, cos 120°, and tan 120°.

Solution: First we must find the reference angle. We see that 120° is in the second quadrant. Therefore θ, the reference angle, is 60°.

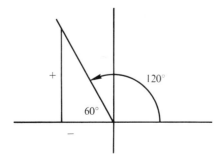

Figure 5-49

(1) y is positive in the second quadrant, therefore sin 120° = sin 60°.

$$\sin 120° = .8660$$

(2) x is negative in the second quadrant, therefore cos 120° = −cos 60°.

$$\cos 120° = -.5000$$

(3) Since y is positive and x is negative the tan 120° = −tan 60°.

$$\tan 120° = -1.732$$

By now, it should be understood that to find the functions of an angle greater than 90°, you must first find the reference angle. After the reference angle has been found, you must determine whether the function is positive or negative, look up the value of the reference angle, and attach the proper sign.

EXAMPLE 39   Find the tan 305°.

Solution: (1) θ the reference angle is

$$\theta = 360° - 305°$$

$$= 55°$$

              (2) The sign of the tangent in the fourth quadrant is negative. Therefore: tan 305° = −tan 55°.

              (3) tan 305° = −1.428.

PROBLEM SET 5-16

Determine the reference angle for the following angles.

1. 135°	2. 150°	3. 210°	4. 300°	5. 45°
6. 170°	7. −40°	8. 225°	9. 390°	10. 180°
11. −300°	12. 80°	13. 105°	14. −100°	15. 480°

PROBLEM SET 5-17

Determine which quadrant the terminal side of ϕ lies in for the following conditions.

	sin ϕ	cos ϕ	tan ϕ
1.	+	+	
2.	+		−
3.		+	−
4.	−		+
5.	−	−	
6.		+	+
7.	−	+	
8.	+		+
9.		−	+
10.		−	−

5-18
Horizontal and Vertical Components

The *horizontal component* of a vector is the x coordinate of the end point of the vector. The *vertical component* of a vector is the y coordinate of the end point of the vector.

EXAMPLE 40   Find the horizontal and vertical components of 5 at 40°.

Solution:

$$\cos 40° = \frac{x}{5}$$

$$5 \cos 40° = x, \text{ horizontal component}$$

$$\sin 40° = \frac{y}{5}$$

$$5 \sin 40° = y, \text{ vertical component}$$

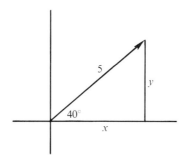

Figure 5-50

EXAMPLE 41   Find the horizontal and vertical components of 10 at N25°W.

Solution: $\cos 115° = \dfrac{x}{10}$

$$10 \cos 115° = x$$

or

$$-10 \cos 65° = x$$

Why?

$$\sin 115° = \dfrac{y}{10}$$

$$10 \sin 115° = y$$

or

$$10 \sin 65° = y$$

Why?

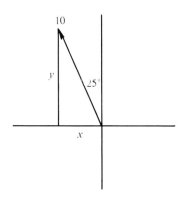

Figure 5-51

PROBLEM SET 5-18

Find the horizontal and vertical components for the following vectors.

1. 5 at 36.9°   2. 35 at 53.1°

3. 420 at 81.2°   4. 19.2 at 40°

5. .108 at 10.9°   6. 60 at 75°

7. 100 at 30°   8. 2.8 at 7.5°

9. 5540 at 45°   10. 12.4 at 14°

11. Two forces acting perpendicular to one another have a resultant force of 300 lbs acting at 20° above the horizontal. Determine the forces.

12. How far east and how far north is town *B* from town *A* in the following drawing?

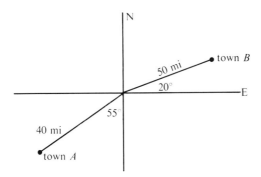

5-19
Arc Length

Radian measure has many uses in mathematics and technological fields. We will discuss two of these methods which are used extensively in physics and applied mechanics.

We know that a central angle whose measure is one radian subtends an arc equal in length to the radius. Therefore, if the central angle was equal to two radians, the length of the subtended arc would equal two times the radius. If the angle was three-fourths radian, the length of the arc would be 3/4 times the radius. In other words letting s equal the length of the subtended arc, θ equal the angle in radians, and r equal the radius of the circle, we can say:

$$s = \theta \cdot r$$

EXAMPLE 42   Find the length of an arc of a circle if $\theta = \dfrac{\pi}{3}$, and $r = 6$ in.

Solution: Let $\pi = 3.14$

$$s = \theta \cdot r$$

$$= \frac{\pi}{3} \cdot 6$$

$$= \frac{3.14}{3} \cdot 6$$

$$= 3.14 \cdot 2$$

$$= 6.28 \text{ in.}$$

EXAMPLE 43   A pendulum 3 ft long makes an arc of 9°. What is the length of the arc made by the pendulum?

Solution: Since θ must be in radians we need to convert $9°$ to radians.

$$9° = 9 \times \frac{\pi}{180} \text{ radians} \qquad s = \theta \cdot r$$

$$= \frac{\pi}{20} \text{ radians} \qquad\qquad = \frac{\pi}{20} \cdot 3$$

$$\qquad\qquad\qquad\qquad = .47 \text{ ft (approx.)}$$

$$\qquad\qquad\qquad\qquad = 5.64 \text{ in.}$$

5-20
Linear and Angular Velocity

Suppose a point P moves around a circle with a constant rate of speed. The actual distance traveled by the point P is equal to the length of the arc traversed. If we divide the length of the arc s by some unit of time t, we can find the *linear velocity* v, of point P. Thus:

$$v = \frac{s}{t}$$

Letting O be the center of the circle and OP the radius we see also that as P moves around the circle, OP swings through θ angular units. If we divide the angular measure θ by some unit of time t, we can find the angular velocity ω of the radius OP. Thus:

$$\omega = \frac{\theta}{t}$$

If we divide the equation $s = \theta \cdot r$ by t, we get:

$$s = \theta \cdot r$$

$$\frac{s}{t} = \frac{\theta}{t} \cdot r$$

Substitute v for $\frac{s}{t}$ and ω for $\frac{\theta}{t}$.

$$v = \omega r$$

This says that the linear velocity of a point on the circumference of a circle is equal to the angular velocity of the point times the radius.

The angular velocity must be expressed in radians per unit time. Angular velocity is often expressed as r.p.m. and can be converted to radians by remembering that 1 r.p.m. $= 2\pi$ radians per min.

EXAMPLE 44 A pulley 10 in. in diameter makes 500 r.p.m. Find (1) the angular velocity in radians per second and (2) speed per second of the belt that drives the pulley.

Solution: (1) We must change 500 r.p.m. to radians per minute and then to radians per second.

$$\omega = 500 \text{ r.p.m.} = 500(2\pi) \text{ radians per min}$$
$$= 1000\pi \text{ radians per min}$$
$$= \frac{1000\pi}{60} \text{ radians per sec}$$
$$= 50\,\frac{\pi}{3} \text{ radians per sec (angular velocity)}$$

(2) The speed of the belt is the same as the speed of a point on the pulley.

$$v = \omega r$$
$$= 50\,\frac{\pi}{3}\,(5)$$
$$= 250\,\frac{\pi}{3} \text{ in. per sec}$$
$$= 263 \text{ (approx.) in. per sec}$$

PROBLEM SET 5-19

1. Find the arc length subtended on the circumference of a gear of radius 2 in. by a central angle of 60°.

2. Determine the linear velocity of a point on the circumference of a $33\frac{1}{3}$ r.p.m., 12 in. diameter record. Express the results in inches per second.

3. A pendulum 1 meter long oscillates through an angle which subtends an arc length of .12 meters. Determine the angle and express the results in degrees.

4. The linear velocity on the circumference of an automobile tire 30 in. in diameter is 88 ft per second. Determine the angular velocity and express the results in degrees per minute.

5. A circular saw blade 10 in. in diameter rotates at 1750 r.p.m. At what linear velocity in inches per second do the teeth strike a piece of wood?

6. Two gears are meshed together. Gear A is 5.5 in. in diameter and gear B is 1.5 in. in diameter. If gear A rotates through 90° determine the distance a point on the circumference of gear B will travel.

7. The diameter of the moon's orbital path about the earth is approximately 440,000 miles. If it takes 28 days for the moon to complete one orbit determine its angular velocity in radians per second.

8. A see-saw rotates through a 40° angle. If the end of the see-saw travels 5.8 ft., how long is it?

In previous sections the trigonometric functions of an angle have been discussed. We shall now consider relations which can be derived directly from these functions or a combination of them. These are very important to your success in more advanced mathematics, physics and other courses related to engineering. Therefore, they should be memorized as quickly as possible.

Given any angle θ the following relations are true.

$$(1) \ \ \sin \theta = \frac{1}{\csc \theta}$$

$$(2) \ \ \cos \theta = \frac{1}{\sec \theta}$$

$$(3) \ \ \tan \theta = \frac{1}{\cot \theta}$$

$$(4) \ \ \tan \theta = \frac{\sin \theta}{\cos \theta}$$

$$(5) \ \ \cot \theta = \frac{\cos \theta}{\sin \theta}$$

$$(6) \ \ \sin^2 \theta + \cos^2 \theta = 1$$

$$(7) \ \ 1 + \tan^2 \theta = \sec^2 \theta$$

$$(8) \ \ 1 + \cot^2 \theta = \csc^2 \theta$$

We shall show a proof for (1), (4), and (6) and leave the proofs of the others for the student.

Proof: (1)　　　　　　　$\sin \theta = \dfrac{1}{\csc \theta}$

$$\sin \theta = \frac{1}{\csc \theta}$$

$$= \frac{1}{\dfrac{r}{y}}, \quad \csc \theta = \frac{r}{y}$$

$$= \frac{y}{r}$$

$$= \sin \theta$$

Proof: (4)
$$\tan\theta = \frac{\sin\theta}{\cos\theta}$$

$$\tan\theta = \frac{\sin\theta}{\cos\theta}$$

$$= \frac{\frac{y}{r}}{\frac{x}{r}}$$

$$= \frac{y}{x}$$

$$= \tan\theta$$

Proof: (6)
$$\sin^2\theta + \cos^2\theta = 1$$

In order to prove (6), we shall make use of the relation $x^2 + y^2 = r^2$.

$$x^2 + y^2 = r^2$$

Dividing both sides by r^2, results in

$$\frac{x^2}{r^2} + \frac{y^2}{r^2} = \frac{r^2}{r^2}$$

$$\left(\frac{x}{r}\right)^2 + \left(\frac{y}{r}\right)^2 = 1$$

Since $\cos\theta = \frac{x}{r}$ and $\sin\theta = \frac{y}{r}$, substituting results in

$$(\cos\theta)^2 + (\sin\theta)^2 = 1$$

$$\cos^2\theta + \sin^2\theta = 1$$

$$\sin^2\theta + \cos^2\theta = 1$$

To prove (7), divide $x^2 + y^2 = r^2$ by x^2.

To prove (8), divide $x^2 + y^2 = r^2$ by y^2.

The student should be able to recognize these relations in other forms. For example:

(1) $\sin\theta = \dfrac{1}{\csc\theta}$ can be written as

$$\sin\theta\,\csc\theta = 1, \quad \text{or}$$

$$\csc\theta = \frac{1}{\sin\theta}$$

(2) $\cos \theta = \dfrac{1}{\sec \theta}$ can be written as

$$\cos \theta \sec \theta = 1, \quad \text{or}$$

$$\sec \theta = \frac{1}{\cos \theta}$$

(3) $\tan \theta = \dfrac{1}{\cot \theta}$ can be written as

$$\tan \theta \cot \theta = 1, \quad \text{or}$$

$$\cot \theta = \frac{1}{\tan \theta}$$

(4) $\tan \theta = \dfrac{\sin \theta}{\cos \theta}$ can be written as

$$\cos \theta \tan \theta = \sin \theta, \quad \text{or}$$

$$\cos \theta = \frac{\sin \theta}{\tan \theta}$$

(5) $\cot \theta = \dfrac{\cos \theta}{\sin \theta}$ can be written as

$$\sin \theta \cot \theta = \cos \theta, \quad \text{or}$$

$$\sin \theta - \frac{\cos \theta}{\cot \theta}$$

(6) $\sin^2 \theta + \cos^2 \theta = 1$ can be written as

$$\sin \theta = \pm \sqrt{1 - \cos^2 \theta}, \quad \text{or}$$
$$\cos \theta = \pm \sqrt{1 - \sin^2 \theta}$$

(7) $1 + \tan^2 \theta = \sec^2 \theta$ can be written as

$$\tan \theta = \pm \sqrt{\sec^2 \theta - 1}, \quad \text{or}$$
$$\sec \theta = \pm \sqrt{1 + \tan^2 \theta}$$

(8) $1 + \cot^2 \theta = \csc^2 \theta$ can be written as

$$\cot \theta = \pm \sqrt{\csc^2 \theta - 1}, \quad \text{or}$$
$$\csc^2 \theta = \pm \sqrt{1 + \cot^2 \theta}$$

In (6), (7), and (8) the positive or negative sign would be determined by the quadrant in which θ lies. For example, if θ were in quadrant I, you would choose the positive sign for all three, but if θ were in quadrant II, you would choose the positive sign for $\sqrt{1 - \cos^2 \theta}$ and the negative sign for $\sqrt{1 - \sin^2 \theta}$ in (6).

The eight relations can be thought of as identities and can be used to prove other identities. There is no set rule for proving identities but it is best to leave one side unaltered, thus transforming one side of the equation into the other side.

EXAMPLE 45 Prove $\dfrac{\sin\theta}{\tan\theta} + \dfrac{\cos\theta}{\cot\theta} = \cos\theta + \sin\theta.$

Solution: $\quad \dfrac{\sin\theta}{\tan\theta} + \dfrac{\cos\theta}{\cot\theta} = \dfrac{\sin\theta}{\dfrac{\sin\theta}{\cos\theta}} + \dfrac{\cos\theta}{\dfrac{\cos\theta}{\sin\theta}}$

$$= \sin\theta\,\frac{\cos\theta}{\sin\theta} + \cos\theta\,\frac{\sin\theta}{\cos\theta}$$

$$= \cos\theta + \sin\theta$$

EXAMPLE 46 Prove $\dfrac{\cos A \csc A}{\cot^2 A} = \tan A.$

Solution: $\quad \dfrac{\cos A \csc A}{\cot^2 A} = \dfrac{\cos A \csc A}{\dfrac{1}{\tan^2 A}}$

$$= \cos A \csc A \tan^2 A$$

$$= \tan A(\cos A \csc A \tan A)$$

$$= \tan A\,\frac{\cos A \csc A \sin A}{\cos A}$$

$$= \tan A(\csc A \sin A)$$

$$= \tan A(1)$$

$$= \tan A$$

EXAMPLE 47 Prove $\sec B - \cos B = \dfrac{\tan B}{\csc B}.$

Solution: $\quad \sec B - \cos B = \dfrac{1}{\cos B} - \cos B$

$$= \frac{1 - \cos^2 B}{\cos B}$$

$$= \frac{\sin^2 B}{\cos B}$$

$$= \sin B\,\frac{\sin B}{\cos B}$$

$$= \sin B \tan B$$

$$= \frac{1}{\csc B}\tan B$$

$$= \frac{\tan B}{\csc B}$$

PROBLEM SET 5-20

Using the trigonometric identities (1) through (8) find the value of the sin, cos, tan, and cot, unless given. The quadrant in which θ lies is given.

1. $\cos \theta = \frac{5}{13}, 0° < \theta < 90°$

2. $\tan \theta = \sqrt{15}, 180° < \theta < 270°$

3. $\sin \theta = \frac{\sqrt{2}}{2}, 90° < \theta < 180°$

4. $\cot \theta = \frac{1}{-2}, 270° < \theta < 360°$

5. $\cos \theta = \frac{\sqrt{3}}{2}, 0° < \theta < 90°$

6. $\sin \theta = \frac{1}{2}, 90° < \theta < 180°$

7. $\cot \theta = \sqrt{3}, 270° < \theta < 360°$

8. $\tan \theta = \frac{-\sqrt{3}}{3}, 90° < \theta < 180°$

9. $\sin \theta = \frac{\sqrt{2}}{2}, 180° < \theta < 270°$

10. $\cos \theta = \frac{\sqrt{8}}{2}, 0° < \theta < 90°$

PROBLEM SET 5-21

Prove the following identities.

1. $\dfrac{\sec \theta - \cos \theta}{\sin^2 \theta} = \sec \theta$

2. $\dfrac{\cos \theta \tan \theta - \sin^2 \theta}{1 - \sin \theta} = \sin \theta$

3. $(\sec \theta - 1)(\sin \theta + \sin \theta \cos \theta) = \sin^2 \theta \tan \theta$

4. $(\sin \theta - \cos \theta)(\cot \theta + 1) - \sin \theta = -\cot \theta \cos \theta$

5-22
Oblique Triangles

Many times in mathematical and technical fields problems arise in which *oblique triangles* (a triangle which does not contain a right angle) occur. Therefore it is necessary that we have other means for solving triangles other than the six trigonometric functions for right triangles.

We will now consider two commonly used methods for solving oblique triangles. However, the right triangle and the functions involving right triangles will be used in these solutions.

5-23
Law of Sines

The Law of Sines states that in any triangle ABC,

$$\frac{a}{\sin A} = \frac{b}{\sin B} = \frac{c}{\sin C}$$

To prove this let us consider the triangle ABC, and the altitude CD. In $\triangle ACD$

$$\sin A = \frac{h}{b}$$

or

$$b \sin A = h$$

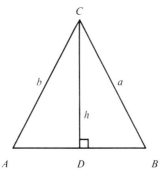

Figure 5-52

In $\triangle BCD$,

$$\sin B = \frac{h}{a}$$

or

$$a \sin B = h$$

Since h is the same in both cases $a \sin B = b \sin A$. Dividing both sides by $(\sin A)(\sin B)$

$$\frac{a \sin B}{\sin A \sin B} = \frac{b \sin A}{\sin A \sin B}$$

$$\frac{a}{\sin A} = \frac{b}{\sin B}$$

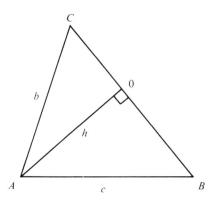

Figure 5-53

Suppose you draw the altitude AO in triangle AOC

$$\sin C = \frac{h'}{b}$$

or

$$b \sin C = h'$$

In triangle AOB

$$\sin B = \frac{h'}{c}$$

or

$$c \sin B = h'$$

Since h' is the same in both cases

$$b \sin C = c \sin B$$

Dividing both sides by $(\sin B)(\sin C)$

$$\frac{b \sin C}{\sin B \sin C} = \frac{c \sin B}{\sin B \sin C}$$

$$\frac{b}{\sin B} = \frac{c}{\sin C}$$

If

$$\frac{a}{\sin A} = \frac{b}{\sin B}$$

and

$$\frac{b}{\sin B} = \frac{c}{\sin C}$$

then

$$\frac{a}{\sin A} = \frac{b}{\sin B} = \frac{c}{\sin C}$$

This result is known as the *Law of Sines*, and can be used to solve triangles provided enough information is given to solve either of the following proportions:

$$(1) \quad \frac{a}{\sin A} = \frac{b}{\sin B}$$

$$(2) \quad \frac{a}{\sin A} = \frac{c}{\sin C}$$

$$(3) \quad \frac{b}{\sin B} = \frac{c}{\sin C}$$

Remember to solve a proportion like these you must know three of the four factors.

EXAMPLE 48   If $a = 10$, $\angle A = 45°$, and $\angle B = 50°$, find c.

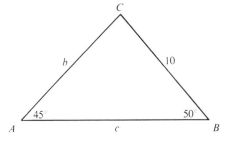

Figure 5-54

Solution:
$$\angle C = 180° - (\angle B + \angle A)$$
$$= 180° - (50° + 45°)$$
$$= 180° - 95°$$
$$= 85°$$

$$\frac{a}{\sin A} = \frac{c}{\sin C}$$

$$\frac{10}{\sin 45°} = \frac{c}{\sin 85°}$$

$$\frac{10 \sin 85°}{\sin 45°} = c$$

$$\frac{10(.9962)}{.7071} = c$$

$$\frac{9.962}{.7071} = c$$

$$14 = c$$

Sometimes it is possible to have two or no solutions with the information that is given. It is best to draw a triangle to scale before starting your work to determine the number of solutions.

5-24

Law of Cosines

The Law of Cosines states that in any triangle the square of one side is equal to the sum of the squares of the other two sides diminished by twice the product of the other two sides and the cosine of their included angle.

Stated algebraically for any triangle ABC with sides a, b, and c:

(1) $a^2 = b^2 + c^2 - 2bc \cos A$

(2) $b^2 = a^2 + c^2 - 2ac \cos B$

(3) $c^2 = a^2 + b^2 - 2ab \cos C$

The proof of the Law of Cosines is beyond the scope of the student at this time and will be given in appendix C.

PROBLEM SET 5-22

Determine the unknown sides and angles for the following triangles.

	side a	side b	side c	angle A	angle B	angle C
1.	36			28°	65°	
2.	60		156	20°		
3.		56.2	63.9	71.5°		
4.		40	40		80°	
5.		537	428	32.6°		
6.	28	21			36.9°	
7.	50		53.7		40°	
8.	40	3				101°
9.	12	8	14			
10.	.624	.663			47°	

5-25

Graphs of the Trigonometric Functions

One way of comparing trigonometric functions is by graphs. Graphs show pictorially how the functions differ and hence are valuable in many technical areas.

The graphs will be constructed on a rectangular coordinate system, and the angles will be expressed in radians.

5-26

$$y = a \sin x \text{ and } y = a \cos x$$

We will begin by constructing $y = \sin x$.

x	0	$\dfrac{\pi}{6}$	$\dfrac{\pi}{3}$	$\dfrac{\pi}{2}$	$\dfrac{2\pi}{3}$	$\dfrac{5\pi}{6}$	π	$\dfrac{7\pi}{6}$	$\dfrac{4\pi}{3}$	$\dfrac{3\pi}{2}$	$\dfrac{5\pi}{3}$	$\dfrac{11\pi}{6}$	2π
y	0	.5	.87	1	.87	.5	0	$-.5$	$-.87$	-1	$-.87$	$-.5$	0

Plotting these points we obtain the graph depicted by Figure 5-55.

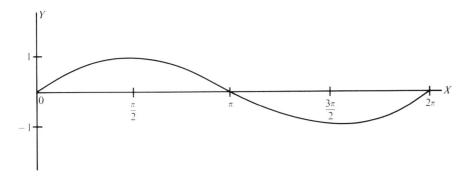

Figure 5-55

This table shows the values for $y = \cos x$

x	0	$\dfrac{\pi}{6}$	$\dfrac{\pi}{3}$	$\dfrac{\pi}{2}$	$\dfrac{2\pi}{3}$	$\dfrac{5\pi}{6}$	π	$\dfrac{7\pi}{6}$	$\dfrac{4\pi}{3}$	$\dfrac{3\pi}{2}$	$\dfrac{5\pi}{3}$	$\dfrac{11\pi}{6}$	2π
y	1	.87	.5	0	$-.5$	$-.87$	-1	$-.87$	$-.5$	0	.5	.87	1

Figure 5-56 shows the graph of these points.

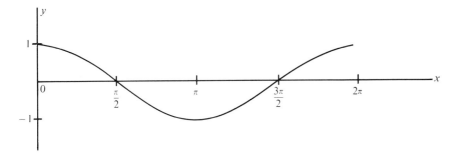

Figure 5-56

What about the graph of $y = a \sin x$? or $y = a \cos x$? You will see from the following examples that the value of y in $y = a \sin x$ is a times the value of y in $y = \sin x$. The same will be true of the values of y in $y = a \cos x$ and $y = \cos x$.

EXAMPLE 49   Graph $y = 2 \sin x$.

Solution: Plot the following points:

x	0	$\frac{\pi}{6}$	$\frac{\pi}{3}$	$\frac{\pi}{2}$	$\frac{2\pi}{3}$	$\frac{5\pi}{6}$	π	$\frac{7\pi}{6}$	$\frac{4\pi}{3}$	$\frac{3\pi}{2}$	$\frac{5\pi}{3}$	$\frac{11\pi}{6}$	2π
y	0	1	1.73	2	1.73	1	0	-1	-1.73	-2	-1.73	-1	0

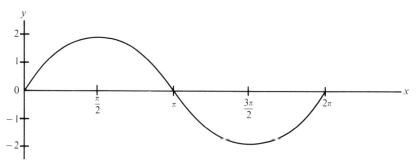

Figure 5-57

EXAMPLE 50   Graph $y = 3 \cos x$.

Solution:

x	0	$\frac{\pi}{6}$	$\frac{\pi}{3}$	$\frac{\pi}{2}$	$\frac{2\pi}{3}$	$\frac{5\pi}{6}$	π	$\frac{7\pi}{6}$	$\frac{4\pi}{3}$	$\frac{3\pi}{2}$	$\frac{5\pi}{3}$	$\frac{11\pi}{6}$	2π
y	3	2.6	1.5	0	-1.5	-2.6	-3	-2.6	-1.5	0	1.5	2.6	3

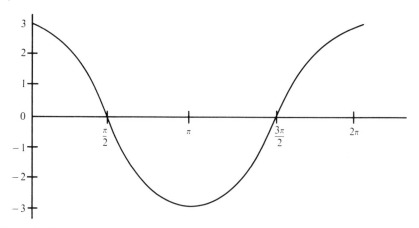

Figure 5-58

EXAMPLE 51 Sketch $y = -2 \sin x$.

Solution: Find the high and low points and where it crosses the x-axis.

x	0	$\dfrac{\pi}{2}$	π	$\dfrac{3\pi}{2}$	2π
y	0	-2	0	2	0

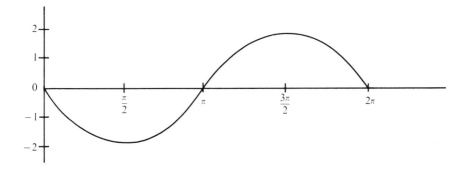

Figure 5-59

    It should be noted that each function was in the form of $y = a \sin x$ and reached a *height* of a. Also, notice that had each graph been extended in either direction, it would have repeated the same y values each 2π units along the x-axis because $\sin x = \sin (x + 2\pi) = \sin (x + n \cdot 2\pi)$, where $n = 0,1,2,3$ and so forth. Therefore we say that a function like $y = a \sin x$ has a *height* or *amplitude* of a units and its *period* or *wave-length* (the distance between the points along the x-axis in which the values of y repeat) is 2π units.

    We can generalize the same information about the *amplitude* and *period* of $y = a \cos x$.

EXAMPLE 52 What is the amplitude and period of the following?

$$\text{(a) } y = -2 \sin x$$

$$\text{(b) } y = \tfrac{3}{2} \cos x$$

$$\text{(c) } y = a \sin x$$

$$\text{(d) } y = b \cos x$$

Solution: (a) -2; 2π (b) $\tfrac{3}{2}$; 2π

                    (c) a; 2π (d) b; 2π

One can see from Figures 5-55, 5-56, 5-57, and 5-58, that it is necessary to have only the high and low points and the point where the equation crosses the x-axis in order to draw a rough sketch of the graph.

The graph of $y = \tan \theta$, $y = \cot \theta$, $y = \sec \theta$, and $y = \csc \theta$ which are drawn in the examples below, are not used as much as $y = \sin \theta$ and $y = \cos \theta$.

EXAMPLE 53   Graph $y = \tan \theta$.

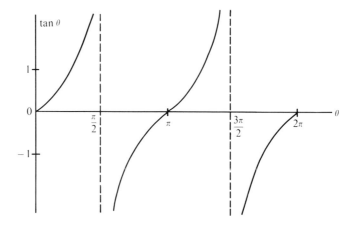

Figure 5-60

EXAMPLE 54   Graph $y = \cot \theta$.

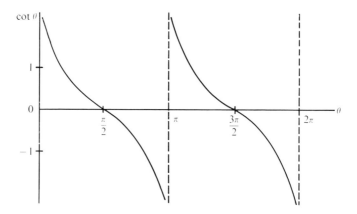

Figure 5-61

EXAMPLE 55   Graph $y = \sec \theta$.

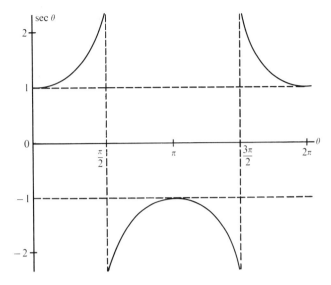

Figure 5-62

EXAMPLE 56   Graph $y = \csc \theta$.

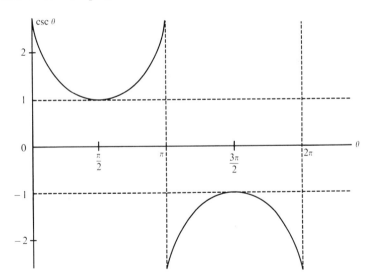

Figure 5-63

It can be seen from the previous examples that the tangent and cotangent curves have a period of π units, but the secant and cosecant curves have a period of 2π units.

EXAMPLE 57   Sketch $y = 2 \cos \theta$.

Another way to sketch a function like this other than finding the high and low points and the points of intersection on the x-axis is as follows. First sketch $y = \cos \theta$. Since $y = \cos \theta$ crosses the x-axis at $\dfrac{\pi}{2}$ and $\dfrac{3\pi}{2}$, $y = 2 \cos \theta$ will also cross here. Therefore, we need only to begin at 2, go through $\dfrac{\pi}{2}$ and continue in the same general direction to -2. Then go back through $\dfrac{3\pi}{2}$ and up to 2.

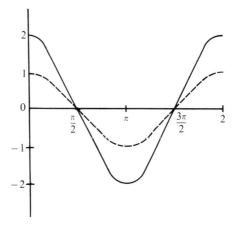

Figure 5-64

EXAMPLE 58   Sketch $y = -2 \cos \theta$.

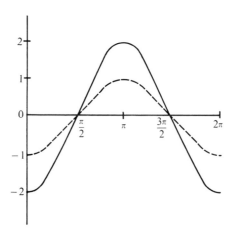

Figure 5-65

First sketch $y = -\cos \theta$, then sketch $y = -2 \cos \theta$.

5-28
$$y = a \sin bx \text{ and } y = a \cos bx$$

When we see a function like $y = \sin 2x$ or $y = 2 \cos 3x$, can we tell anything about the amplitude or period? Is there some way that we can pre-determine these values and sketch their graph?

Let us now graph $y = \sin 2x$. To find values for y we first assign values to x, next we double that value of x, then find the sine of the result we get from the doubling of x. For example, if $x = \frac{\pi}{3}$, then $2x = \frac{2\pi}{3}$ and $\sin \frac{2\pi}{3} = .87$. Therefore, if $y = \sin 2x$ and $x = \frac{\pi}{3}$ then $y = .87$.

We will now make a table for $y = \sin 2x$.

x	0	$\frac{\pi}{12}$	$\frac{\pi}{6}$	$\frac{\pi}{4}$	$\frac{\pi}{3}$	$\frac{5\pi}{12}$	$\frac{\pi}{2}$	$\frac{7\pi}{12}$	$\frac{2\pi}{3}$	$\frac{3\pi}{4}$	$\frac{5\pi}{6}$	$\frac{11\pi}{12}$	π
y	0	.5	.87	1	.87	.5	0	$-.5$	$-.87$	-1	$-.87$	$-.5$	0

Graphing these points we get the following.

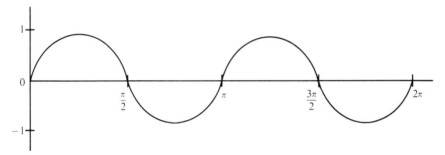

Figure 5-66

In comparing the graph of $y = \sin 2x$ in Figure 5-66 with the graph of $y = \sin x$ in Figure 5-67 we see that the amplitudes are the same but the period of $y = \sin 2x$ equals π, which is one-half the period of $y = \sin x$. Another way of comparing the periods is to say that everything in $y = \sin 2x$ is happening twice as fast as in $y = \sin x$.

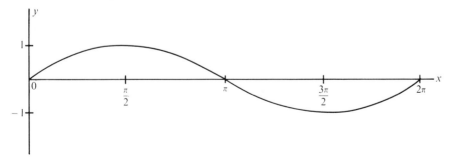

Figure 5-67

Let us now graph $y = 2 \sin 2x$ and compare it with $y = 2 \sin x$. First we make a table for $y = 2 \sin 2x$ and graph the points.

x	0	$\dfrac{\pi}{12}$	$\dfrac{\pi}{6}$	$\dfrac{\pi}{4}$	$\dfrac{\pi}{3}$	$\dfrac{5\pi}{12}$	$\dfrac{\pi}{2}$	$\dfrac{7\pi}{12}$	$\dfrac{2\pi}{3}$	$\dfrac{3\pi}{4}$	$\dfrac{5\pi}{6}$	$\dfrac{11\pi}{12}$	π
y	0	1	1.7	2	1.7	1	0	-1	-1.7	-2	-1.7	-1	0

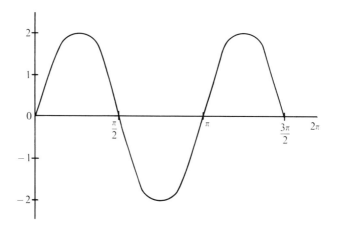

Figure 5-68

If we compare $y = 2 \sin 2x$ in Figure 5-68 with $y = 2 \sin x$ in Figure 5-69, we see that the amplitudes are equal, but the period of $y = 2 \sin 2x$ is π, and the period of $y = 2 \sin x$ is 2π.

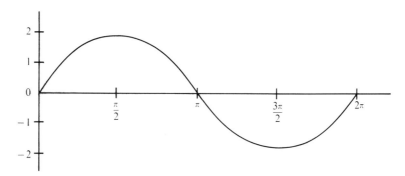

Figure 5-69

We can generalize our observations of the previous examples.

The graph of the sine wave $y = a \sin bx$ has amplitude a with period $\dfrac{2\pi}{b}$.

Had we drawn similar graphs of the cosine function we could have made the following generalization about $y = a \cos bx$.

The graph of the cosine wave $y = a \cos bx$ has amplitude a with period $\dfrac{2\pi}{b}$.

5-29
$y = a \sin (bx + c)$ and $y = a \cos (bx + c)$

Consider the function $y = \sin \left(x + \dfrac{\pi}{6} \right)$. When $x = 0$, $y = \sin \left(0 + \dfrac{\pi}{6} \right) = \sin \dfrac{\pi}{6} = .5$. When $x = \dfrac{\pi}{6}$, $y = \sin \left(\dfrac{\pi}{6} + \dfrac{\pi}{6} \right) = \sin \dfrac{\pi}{3} = .87$. If we let x vary from 0 to 2π we get the following values for y.

x	0	$\dfrac{\pi}{6}$	$\dfrac{\pi}{3}$	$\dfrac{\pi}{2}$	$\dfrac{2\pi}{3}$	$\dfrac{5\pi}{6}$	π	$\dfrac{7\pi}{6}$	$\dfrac{4\pi}{3}$	$\dfrac{3\pi}{2}$	$\dfrac{5\pi}{3}$	$\dfrac{11\pi}{6}$	2π
y	.5	.87	1	.87	.5	0	-5	$-.87$	-1	$-.87$	$-.5$	0	.5

If we graph these points, as in Figure 5-70, and also draw the graph of $y = \sin x$ on the same axis we can see how the two functions compare.

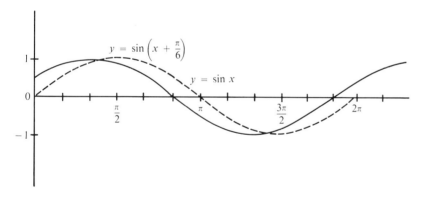

Figure 5-70

In Figure 5-70 we see that the graph of $y = \sin \left(x + \dfrac{\pi}{6} \right)$ and $y = \sin x$ are identical except $y = \sin \left(x + \dfrac{\pi}{6} \right)$ has been shifted $\dfrac{\pi}{6}$ units to the left of $y = \sin x$.

From Figure 5-70, we can say that c, in $y = a \sin (bx + c)$, will shift the curve of $y = a \sin bx$ either c units to the left if $c > 0$ or c units to the right if $c < 0$. The amount of shift is called *phase displacement* and is found by $\dfrac{c}{b}$.

Therefore, in generalizing the above we say that to sketch the curves of the functions $y = a \sin(bx + c)$ or $y = a \cos(bx + c)$:

(1) a = amplitude

(2) $\dfrac{2\pi}{b}$ = period

(3) $\dfrac{c}{b}$ = phase displacement, and will be $\begin{cases} \text{left, if } \dfrac{c}{b} > 0 \\[2mm] \text{right, if } \dfrac{c}{b} < 0 \end{cases}$

EXAMPLE 59 Find the amplitude, period, phase displacement, and sketch the graph of $y = \sin\left(x + \dfrac{\pi}{2}\right)$.

Solution:

(1) amplitude = a = 1

(2) period = $\dfrac{2\pi}{b} = \dfrac{2\pi}{1} = 2\pi$

(3) phase displacement = $\dfrac{c}{b} = \dfrac{\dfrac{\pi}{2}}{1} = \dfrac{\pi}{2}$

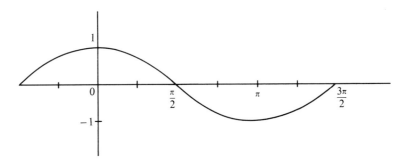

Figure 5-71

EXAMPLE 60 Sketch the graph of $y = 2\sin\left(x - \dfrac{\pi}{3}\right)$.

Solution:

(1) a = 2

(2) $\dfrac{2\pi}{b} = \dfrac{2\pi}{1} = 2\pi$

(3) $\dfrac{c}{b} = \dfrac{-\dfrac{\pi}{3}}{1} = -\dfrac{\pi}{3}$

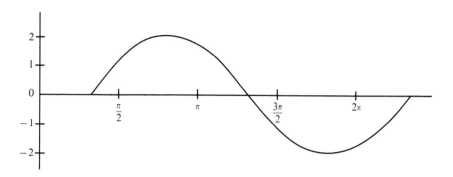

Figure 5-72

EXAMPLE 61   Sketch $y = \sin\left(\dfrac{x}{2} - \dfrac{\pi}{6}\right)$.

Solution:

(1) $a = 1$

(2) $\dfrac{2\pi}{b} = \dfrac{2\pi}{1/2} = 4\pi$

(3) $\dfrac{c}{b} = \dfrac{-\dfrac{\pi}{6}}{1/2} = -\dfrac{\pi}{12}$

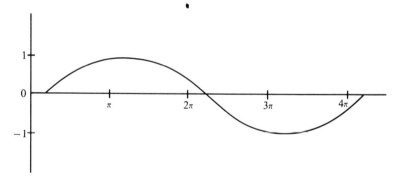

Figure 5-73

5-30

$$y = a \sin x + b \cos x$$

Many trigonometric functions involve a combination of two or more functions.

To find the graph of a function which is the sum of two other functions, first graph each function separately and then add the y-coordinates graphically.

EXAMPLE 62   Sketch $y = x + \cos x$.

Solution: First draw the graph of $y = x$. Next draw the graph of $y = \cos x$, then add the y-coordinates together.

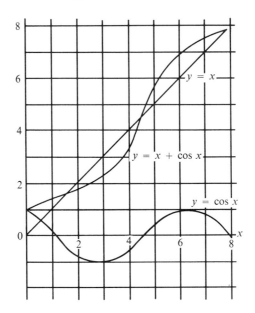

Figure 5-74

EXAMPLE 63   Sketch $y = \sin x + \cos x$.

Solution: First draw $y = \sin x$. Next draw $y = \cos x$. Add the y-coordinates together making the graph of $y = \sin x + \cos x$.

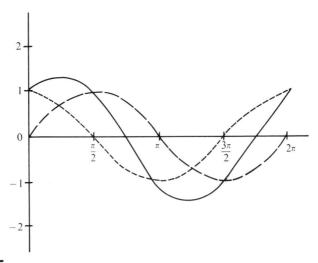

Figure 5-75

Graph the following equations.

1. $y = 2 \sin 2x$

2. $y = \sin 4x$

3. $y = 3 \sin 2x$

4. $y = \sin \frac{1}{2}x$

5. $y = \cos \frac{1}{2}x$

6. $y = 2 \sin 3x$

7. $y = 4 \sin 2x$

8. $y = 3 \sin 4x$

9. $y = \frac{1}{2} \cos x$

10. $y = \sin\left(x + \frac{\pi}{6}\right)$

11. $y = 2 \sin x + \cos x$

12. $y = \sin\left(2x + \frac{\pi}{3}\right)$

13. $y = \sin 2\left(x + \frac{\pi}{6}\right)$

14. $y = 2 \sin\left(\frac{1}{2}x + \frac{\pi}{2}\right)$

15. $y = 3 \cos\left(2x + \frac{\pi}{3}\right)$

6

Complex
Numbers

Thus far the only kind of numbers we have used have been real numbers. In this chapter, we shall be using imaginary numbers. *Imaginary numbers* is the term applied to square roots of negative numbers such as $\sqrt{-1}$, $\sqrt{-4}$, or $\sqrt{-7}$. In our work we shall let the symbol j equal $\sqrt{-1}$ which allows us to represent imaginary numbers without the minus sign.

EXAMPLE 1

$$\sqrt{-1} = j$$
$$\sqrt{-4} = \sqrt{4} \cdot \sqrt{-1} = 2\sqrt{-1} = 2j$$
$$\sqrt{-7} = \sqrt{7} \cdot \sqrt{-1} = \sqrt{7}\,j$$
$$\sqrt{-12} = \sqrt{12} \cdot \sqrt{-1} = 2\sqrt{3} \cdot \sqrt{-1} = 2\sqrt{3}\,j$$

Sometimes it is necessary to raise imaginary numbers to some power. Using the definitions that we know and the definition of j we can get the following.

$$j = \sqrt{-1} \qquad\qquad j^5 = j^4 \cdot j^1 = \sqrt{-1}$$
$$j^2 = -1 \qquad\qquad j^6 = j^4 \cdot j^2 = -1$$
$$j^3 = j^2 \cdot j^1 = -j^1 \qquad j^7 = j^4 \cdot j^3 = -j$$
$$j^4 = j^2 \cdot j^2 = 1 \qquad j^8 = j^4 \cdot j^4 = 1$$

Continuing with this it can be seen that the powers of j will repeat the same cycle of j, -1, $-j$, 1.

If we combine a real number and an imaginary number to form a number like $2 + 3j$, we call this a *complex number*. A *complex number* is a number in the form of $a + bj$ where a and b are real numbers and $j = \sqrt{-1}$. The complex number $a + bj$ has two parts: a, which is the real part and bj, which is the imaginary part. If $a = 0$ and $b \neq 0$, we have the *pure imaginary number*, bj. If $a \neq 0$ and $b = 0$, we have the real number, a. Thus, we see that the complex numbers include the real numbers. Complex numbers are neither positive nor negative but the real and imaginary parts can be positive or negative.

PROBLEM SET 6-1

Express the following numbers using the j operator.

1. $\sqrt{-16}$ 2. $\sqrt{-81}$

3. $\sqrt{-8}$ 4. $-\sqrt{-49}$

5. $\sqrt{-\frac{9}{5}}$ 6. $\sqrt{-\frac{25}{32}}$

7. $\sqrt{-\dfrac{E^2}{R}}$ 8. $\sqrt{-\dfrac{1}{w^2}}$

9. $-\sqrt{-\dfrac{4B}{2}}$ 10. $\sqrt{-\dfrac{12x^2}{45y}}$

6-2
Properties of Complex Numbers

Complex numbers follow the operations of algebra which were defined in Chapter 2. Most of the properties of complex numbers are stated below. However only addition, multiplication, and division of complex numbers will be proven.

(1) Equality: $a + bj = c + dj$, if and only if $a = c$ and $b = d$.

(2) Addition: $(a + bj) + (c + dj) = (a + c) + (b + d)j$.

Proof: $(a + bj) + (c + dj) = a + bj + c + dj$
$$= a + c + bj + dj$$
$$= (a + c) + (b + d)j$$

(3) Multiplication: $(a + bj)(c + dj) = (ac - bd) + (ad + bc)j$.

Proof: $(a + bj)(c + dj) = ac + adj + bcj + bdj^2$

$$= ac + bdj^2 + adj + bcj$$

$$= (ac - bd) + (ad + bc)j$$

The following examples show how the above properties are used.

EXAMPLE 2   Find x and y if $4x + 3yj = 24 - 15j$.

Solution: $4x + 3yj = 24 - 15j$

$$4x = 24 \qquad 3y = -15$$

$$x = 6 \qquad y = -5$$

EXAMPLE 3   Find the sum of $(2 - 6j)$ and $(1 + 3j)$.

Solution: $(2 - 6j) + (1 + 3j) = (2 + 1) + (-6 + 3)j$

$$= 3 - 3j$$

EXAMPLE 4   Find the product of $(1 + 4j)$ and $(3 - 2j)$.

Solution: $(1 + 4j)(3 - 2j) = (1 \cdot 3 + 4 \cdot 2) + (-2 + 12)j$

$$= (3 + 8) + 10j$$

$$= 11 + 10j$$

(4) Conjugate: The complex number $a - bj$ is called the conjugate of $a + bj$.

EXAMPLE 5   Find the conjugates of the following complex numbers.

(a) $3 + 2j$ (b) $4 - j$

(c) 2 (d) $3j$

Solution: (a) $3 - 2j$ (b) $4 + j$

(c) 2 (d) $-3j$

(5) Division: $\dfrac{a + bj}{c + dj} = \dfrac{ac + bd}{c^2 + d^2} + \dfrac{bc - ad}{c^2 + d^2}j$.

Proof: The quotient of two complex numbers such as $\dfrac{a + bj}{c + dj}$ can be found by multiplying both the numerator and denominator by the conjugate of the denominator.

$$\frac{a + bj}{c + dj} = \frac{a + bj}{c + dj} \cdot \frac{c - dj}{c - dj}$$

$$= \frac{ac + adj + bcj - bdj^2}{c^2 - cdj + cdj - d^2j^2}$$

$$= \frac{ac + bd + bcj - adj}{c^2 + d^2}$$

$$= \frac{(ac + bd) + (bc - ad)j}{c^2 + d^2}$$

$$= \frac{ac + bd}{c^2 + d^2} + \frac{bc - ad}{c^2 + d^2}j$$

EXAMPLE 6 Find $\dfrac{6 + 2j}{3 - j}$.

Solution:

$$\frac{6 + 2j}{3 - j} = \frac{(6 + 2j)(3 + j)}{(3 - j)(3 + j)}$$

$$= \frac{18 + 6j + 6j + 2j^2}{9 + 3j - 3j - j^2}$$

$$= \frac{18 + 12j - 2}{9 + 1}$$

$$= \frac{16 + 12j}{10}$$

$$= \frac{8}{5} + \frac{6}{5}j$$

You will notice in the above example that the answer was in the form of $a + bj$. All answers should be written in this form, and will be given in this form throughout the text.

PROBLEM SET 6-2

Express the following numbers in the form, $a + bj$. Also write the conjugate of each.

1. $5 + \sqrt{-9}$

2. $-3 - \sqrt{-12}$

3. $j^2 + \sqrt{-8}$

4. $\sqrt{36} + \sqrt{-36}$

5. $j^3 - \sqrt{4}$

6. $-\sqrt{49} - \sqrt{-54}$

7. $\sqrt{-27} + \sqrt{25}$

8. $\sqrt{(-2)^2} + \sqrt{-2}$

9. $(\sqrt{-4})^2 - \sqrt{-8}$

10. $-(\sqrt{-6}) + (2\sqrt{2})^2$

Determine the quotient for the following:

1. $\dfrac{12}{1 + 2j}$

2. $\dfrac{5 - 6j}{3 + 4j}$

3. $\dfrac{3}{1 + 3j}$

4. $\dfrac{3 + 8j}{2 + 4j}$

5. $\dfrac{1 + j}{1 - j}$

6. $\dfrac{7 + 2j}{3 - 4j}$

7. $\dfrac{8 + 2j}{j}$

8. $\dfrac{3j}{4j - 5}$

9. $\dfrac{j^3 + j}{1 - j^3}$

10. $\dfrac{6 + \sqrt{-4}}{\sqrt{9} - \sqrt{-9}}$

6-3
Graphical Representation of Complex Numbers

Since complex numbers are made up of two parts it is necessary to represent them graphically in a different way. For this we shall use a set of axes similar to the rectangular coordinate system except we shall call the x-axis the axis of reals and the y-axis the axis of imaginaries, and shall label these as r and j, respectively, as shown in Figure 6.1.

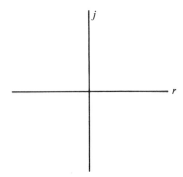

Figure 6-1

We shall call our representation the *complex plane* with the unit of measure on the r-axis being 1 and the unit of measure on the j-axis being j.

For example, in graphing the complex number $5 + 3j$ we shall go 5 units to the right of the axis of reals and 3 units up from the axis of imaginaries. The point A in Figure 6.2 represents the complex number $5 + 3j$.

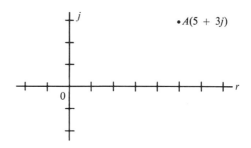

Figure 6-2

It is common to let a complex number represent a vector, thus the complex number $(3 - 2j)$ represents the vector OP in Figure 6-3.

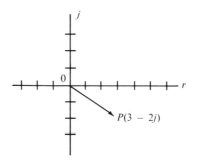

Figure 6-3

6-4
Graphical Addition and Subtraction of Complex Numbers

Two complex numbers, $a + bj$ and $c + dj$, can be added graphically as shown in Figure 6-4.

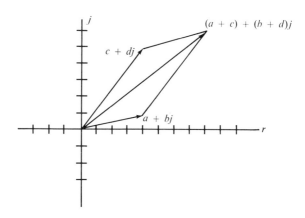

Figure 6-4

The sum is the vector represented as the diagonal of the parallelogram drawn by using the vectors made by the two complex numbers as two sides of the parallelogram.

EXAMPLE 7   Add $(1 + 2j)$ and $(3 - 3j)$ graphically.

Solution: Draw the vectors representing $(1 + 2j)$ and $(3 - 3j)$, and complete the parallelogram. The complex number represented by $(4 - j)$ will be the sum.

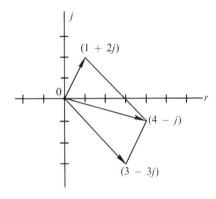

Figure 6-5

To subtract $(c + dj)$ from $(a + bj)$ graphically, add $(-c - dj)$ to $(a + bj)$.

EXAMPLE 8   Subtract $(3 - 2j)$ from $(1 - 3j)$.

Solution: Change $(3 - 2j)$ to $(-3 + 2j)$ and add this to $(1 - 3j)$. Therefore, the sum is $(-2 - j)$.

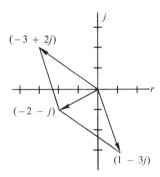

Figure 6-6

Perform the indicated operations graphically.

1. $(3 - 7j) + (4 + 5j)$ 2. $(-2 + j) + (6 + 3j)$

3. $(7 - 8j) - (2 + 3j)$ 4. $(\sqrt{9} + \sqrt{-9}) + (j^2 - \sqrt{-4})$

5. $(2\sqrt{4} + 3j^3) - (\sqrt{16} + 2j)$ 6. $(\sqrt{25} + j^5) + (j^3)$

7. $(\sqrt{-1} \cdot j + j^3) + (j^4 + \sqrt{-1})$ 8. $(\sqrt{-12} + \sqrt{12}) - (-\sqrt{-7} + \sqrt{4})$

6-5
Polar Representation of Complex Numbers

By plotting the complex number $a + bj$ on a complex plane it is possible to write the complex number in a different form. Let r equal the magnitude of the vector and θ equal the angle formed by the vector and the r-axis and use the following relations.

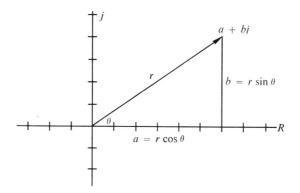

Figure 6-7

$$a = r \cos \theta \qquad r^2 = a^2 + b^2$$

$$b = r \sin \theta \qquad \tan \theta = \frac{b}{a}$$

$$a + bj = r \cos \theta + rj \sin \theta$$
$$= r(\cos \theta + j \sin \theta)$$

The right side, $r(\cos \theta + j \sin \theta)$, is called the *polar form* of the complex number. Sometimes it is called the *trigonometric form* and may be abbreviated as $(r \text{ cis } \theta)$ or $r\angle\theta$. We will use the abbreviation $r\angle\theta$ to mean $r(\cos \theta + j \sin \theta)$. The *magnitude* (sometimes called *modulus* or *absolute value*) of a complex number is the distance r. The angle θ is called the *amplitude* or *argument* of the complex number.

The expression $a + bj$ is called the *algebraic form* or *rectangular form* of a complex number.

In changing from algebraic form to polar form it is best to plot the complex number on a complex plane to determine the angle θ.

EXAMPLE 9   Express each of the following in polar form.

$$\text{(a) } 3 + 3j \qquad \text{(b) } 1 - j$$
$$\text{(c) } -4j \qquad \text{(d) } 5$$

Solution: (a) Plot the number $(3 + 3j)$ on a complex plane. If $3 + 3j = a + bj$ then $a = 3, b = 3$.

$$r^2 = a^2 + b^2 \qquad \tan \theta = \frac{b}{a}$$
$$= 3^2 + 3^2 \qquad \tan \theta = \frac{3}{3}$$
$$= 9 + 9 \qquad \tan \theta = 1$$
$$= 18 \qquad \theta = 45°, 225°$$
$$r = 3\sqrt{2}$$

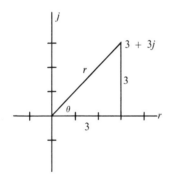

Figure 6-8

Since $3 + 3j$ is in the first quadrant $\theta = 45°$. Therefore,

$$3 + 3j = 3\sqrt{2} \, (\cos 45° + j \sin 45°)$$
$$= 3\sqrt{2} \, \angle 45°$$

(b) Plot the number $(1 - j)$ on a complex plane. If $1 - j = a + bj$ then $a = 1, b = -1$.

$$r^2 = a^2 + b^2 \qquad \tan \theta = \frac{-1}{1}$$
$$= 1^2 + 1^2 \qquad = -1$$
$$= 1 + 1 \qquad \theta = 135° \text{ or } 315°$$
$$= 2$$
$$r = \sqrt{2}$$

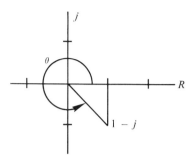

Figure 6-9

Since $1 - j$ is in the fourth quadrant, $\theta = 315°$. Therefore,

$$1 - j = \sqrt{2}\,(\cos 315° + j \sin 315°)$$

$$= \sqrt{2}\,\angle\,315°$$

(c) Plot $-4j$ on a complex plane. If $-4j = a + bj$ then $a = 0$, $b = -4$.

$$r^2 = a^2 = b^2 \qquad \tan \theta = \frac{-4}{0}$$

$$= 0 + (-4)^2 \qquad \tan \theta = ——$$

$$= 0 + 16 \qquad\quad \theta = 90°,\ 270°$$

$$r = 4$$

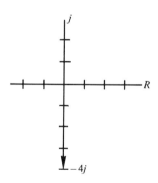

Figure 6-10

We can see by the drawing that $\theta = 270°$. Therefore,

$$-4j = 4(\cos 270° + j \sin 270°)$$

$$= 4\,\angle\,270°$$

(d) Plot 5 on a complex plane. If $5 = a + bj$ then $a = 5$, $b = 0$.

$$r^2 = a^2 + b^2 \qquad \tan \theta = \frac{0}{5}$$

$$= 5^2 + 0^2 \qquad\qquad = 0$$

$$= 25 \qquad\qquad \theta = 0°, 360°$$

$$r = 5$$

Therefore, $5 = 5(\cos 0° + j \sin 0°)$

$$= 5\angle 0°$$

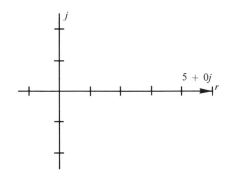

Figure 6-11

PROBLEM SET 6-5

Find the sum of the following complex numbers. Express the results in polar form, $r\angle\theta$.

1. $2 + 2j$ 2. $3 + 5j$
    $3 + j$ $1 + 4j$

3. $0 + 6j$ 4. $25 + 19j$
    $5 + 2j$ $11 - 4j$

5. $-40 - 27j$ 6. $123 + 92j$
    $27 + 53j$ $-50 + 14j$

7. $43 + 136j$ 8. $589 + 162j$
    $-273 + 320j$ $345 + 667j$

9. $-424 - 376j$ 10. $-949 + 768j$
    $-291 - 667j$ $833 - 594j$

Find the product of the following complex numbers. Express the results in polar form, $r\angle\theta$.

1. $7 + 9j$
    $4 - 2j$

2. $16 - 11j$
    $9 + 18j$

3. $-20 + 20j$
    $-32 - 6j$

4. $77 - 6j$
    $21 - 44j$

5. $30 + 48j$
    $-12 + 17j$

6. $-56 - 38j$
    $26 + 45j$

7. $.25 + .60j$
    $.47 - .18j$

8. $-.019 - .278j$
    $.345 + .116j$

6-6
Multiplication and Division of Complex Numbers In Polar Form

To find the product of two complex numbers in polar form, multiply the two magnitudes and add the two angles.

$$\{r_1(\cos\theta_1 + j\sin\theta_1)\}\{r_2(\cos\theta_2 + j\sin\theta_2)\}$$

$$= r_1r_2\{\cos(\theta_1 + \theta_2) + j\sin(\theta_1 + \theta_2)\}$$

The proof of this is as follows:

$$[r_1(\cos\theta_1 + j\sin\theta_1)][r_2(\cos\theta_2 + j\sin\theta_2)]$$

$$= r_1r_2(\cos\theta_1 + j\sin\theta_1)(\cos\theta_2 + j\sin\theta_2)$$

$$= r_1r_2(\cos\theta_1\cos\theta_2 + j\sin\theta_2\cos\theta_1 + j\sin\theta_1\cos\theta_2 + j^2\sin\theta_1\sin\theta_2)$$

$$= r_1r_2(\cos\theta_1\cos\theta_2 - \sin\theta_1\sin\theta_2 + j\sin\theta_2\cos\theta_1 + j\sin\theta_1\cos\theta_2)$$

$$= r_1r_2[(\cos\theta_1\cos\theta_2 - \sin\theta_1\sin\theta_2) + j(\sin\theta_2\cos\theta_1 + \sin\theta_1\cos\theta_2)]$$

$$= r_1r_2[\cos(\theta_1 + \theta_2) + j\sin(\theta_1 + \theta_2)]$$

To find the quotient of two complex numbers in polar form, divide the two magnitudes and subtract the two angles.

$$\frac{r_1(\cos\theta_1 + j\sin\theta_1)}{r_2(\cos\theta_2 + j\sin\theta_2)} = \frac{r_1}{r_2}\{\cos(\theta_1 - \theta_2) + j\sin(\theta_1 - \theta_2)\}$$

The proof of this is as follows:

$$\frac{r_1(\cos \theta_1 + j \sin \theta_1)}{r_2(\cos \theta_2 + j \sin \theta_2)}$$

$$= \frac{r_1}{r_2} \left[\frac{\cos \theta_1 + j \sin \theta_1}{\cos \theta_2 + j \sin \theta_2} \cdot \frac{\cos \theta_2 - j \sin \theta_2}{\cos \theta_2 - j \sin \theta_2} \right]$$

$$= \frac{r_1}{r_2} \left[\frac{\cos \theta_1 \cos \theta_2 - j \sin \theta_2 \cos \theta_1 + j \sin \theta_1 \cos \theta_2 - j^2 \sin \theta_1 \sin \theta_2}{\cos^2 \theta_2 - j \sin \theta_2 \cos \theta_2 + j \sin \theta_2 \cos \theta_2 - j^2 \sin^2 \theta_2} \right]$$

$$= \frac{r_1}{r_2} \left[\frac{(\cos \theta_1 \cos \theta_2 + \sin \theta_1 \sin \theta_2) + j(\sin \theta_1 \cos \theta_2 - \cos \theta_1 \sin \theta_2)}{\cos^2 \theta_2 + \sin^2 \theta_2} \right]$$

$$= \frac{r_1}{r_2} \left[\frac{\cos (\theta_1 - \theta_2) + j \sin (\theta_1 - \theta_2)}{1} \right]$$

$$= \frac{r_1}{r_2} [\cos (\theta_1 - \theta_2) + j \sin (\theta_1 - \theta_2)]$$

EXAMPLE 10   Find the product of

$$3(\cos 75° + j \sin 75°) \quad \text{and} \quad 4(\cos 50° + j \sin 50°)$$

Solution:

$$3(\cos 75° + j \sin 75°) \cdot 4(\cos 50° + j \sin 50°)$$

$$= 3 \cdot 4\{\cos (75° + 50°) + j \sin (75° + 50°)\}$$

$$= 12(\cos 125° + j \sin 125°)$$

EXAMPLE 11   Find the quotient of

$$6(\cos 100° + j \sin 100°) \quad \text{and} \quad 2(\cos 80° + j \sin 80°)$$

Solution:

$$\frac{6(\cos 100° + j \sin 100°)}{2(\cos 80° + j \sin 80°)} = \frac{6}{2} \{\cos (100° - 80°) + j \sin (100° - 80°)\}$$

$$= 3(\cos 20° + j \sin 20°)$$

EXAMPLE 12   Find the product of

$$5(\cos 300° + j \sin 300°) \quad \text{and} \quad 2(\cos 180° + j \sin 180°)$$

Solution:

$$5(\cos 300° + j \sin 300°) \cdot 2(\cos 180° + j \sin 180°)$$

$$= 5 \cdot 2\{\cos (300° + 180°) + j \sin (300° + 180°)\}$$

$$= 10(\cos 480° + j \sin 480°)$$

$$= 10(\cos 120° + j \sin 120°)$$

Perform the indicated operations on the following complex numbers.

1. $2(\cos 30° + j \sin 30°) \cdot 3(\cos 45° + j \sin 45°)$

2. $6(\cos 55° + j \sin 55°) \cdot 4(\cos 10° + j \sin 10°)$

3. $(\cos 20° + j \sin 20°) \cdot 7(\cos 120° + j \sin 120°)$

4. $5(\cos 80° + j \sin 80°) \cdot 8(\cos 270° + j \sin 270°)$

5. $\sqrt{9}\,(\cos 27° + j \sin 27°) \cdot \sqrt{16}\,(\cos 340° + j \sin 340°)$

6. $\dfrac{6(\cos 40° + j \sin 40°)}{2(\cos 60° + j \sin 60°)}$

7. $\dfrac{5(\cos 70° + j \sin 70°)}{(\cos 8° + j \sin 8°)}$

8. $\dfrac{\sqrt{18}\,(\cos 66° + j \sin 66°)}{\sqrt{8}\,(\cos 150° + j \sin 150°)}$

9. $\dfrac{(\cos 37° + j \sin 37°)}{\sqrt{49}\,(\cos 180° + j \sin 180°)}$

10. $\dfrac{\cos 90° + j \sin 90°}{\cos 450° + j \sin 450°}$

6-7
Powers and Roots of Complex Numbers

De Moivre's theorem says that

$$[r(\cos \theta + j \sin \theta)]^n = r^n(\cos n\theta + j \sin n\theta)$$

This theorem is true for all real numbers n and can be used to find roots of complex numbers. Thus,

$$\sqrt[n]{r(\cos \theta + j \sin \theta)} = \{r(\cos \theta + j \sin \theta)\}^{1/n}$$

$$= \sqrt[n]{r}\left(\cos \frac{\theta + K \cdot 360°}{n} + j \sin \frac{\theta + K \cdot 360°}{n}\right)$$

$$\text{where } K = 0, 1, 2, \ldots, n - 1$$

The reason we say $\cos \dfrac{\theta + K \cdot 360°}{n}$ and $\sin \dfrac{\theta + K \cdot 360°}{n}$ is because $\cos \theta$ and $\sin \theta$ are periodic functions making $\cos \theta = \cos(\theta + K \cdot 360°)$ and $\sin \theta = \sin(\theta + K \cdot 360°)$. Therefore, $\dfrac{\theta + K \cdot 360°}{n}$ takes on distinct values as K takes on the values $0, 1, 2, \ldots, n - 1$.

EXAMPLE 13 Find $(1 - j)^{10}$.

Solution: Change $1 - j$ to $\sqrt{2} (\cos 315° + j \sin 315°)$.

$$\begin{aligned}
(1 - j)^{10} &= \{\sqrt{2} (\cos 315° + j \sin 315°)\}^{10} \\
&= (\sqrt{2})^{10} (\cos 10 \cdot 315° + j \sin 10 \cdot 315°) \\
&= 32(\cos 3150° + j \sin 3150°) \\
&= 32(\cos 270° + j \sin 270°) \\
&= 32\{0 + j(-1)\} \\
&= -32j
\end{aligned}$$

EXAMPLE 14 Find the three cube roots of $8j$.

Solution: $\sqrt[3]{8j} = \sqrt[3]{8(\cos 90° + j \sin 90°)}$

$$= \sqrt[3]{8} \left(\cos \frac{90° + K \cdot 360°}{3} + j \sin \frac{90° + K \cdot 360°}{3} \right)$$

Letting r_1, r_2, and r_3 be the three roots when $K = 0$, $K = 1$, $K = 2$ respectively, we have:

$$\begin{aligned}
r_1 &= \sqrt[3]{8} \left(\cos \frac{90° + 0 \cdot 360°}{3} + j \sin \frac{90° + 0 \cdot 360°}{3} \right) \\
&= 2(\cos 30° + j \sin 30°) \\
&= 2 \left(\frac{\sqrt{3}}{2} + j \left(\frac{1}{2} \right) \right) \\
&= \sqrt{3} + j
\end{aligned}$$

$$\begin{aligned}
r_2 &= \sqrt[3]{8} \left(\cos \frac{90° + 1 \cdot 360°}{3} + j \sin \frac{90° + 1 \cdot 360°}{3} \right) \\
&= 2(\cos 150° + j \sin 150°) \\
&= 2 \left\{ -\frac{\sqrt{3}}{2} + j \left(\frac{1}{2} \right) \right\} \\
&= -\sqrt{3} + j
\end{aligned}$$

$$\begin{aligned}
r_3 &= \sqrt[3]{8} \left(\cos \frac{90° + 2 \cdot 360°}{3} + j \sin \frac{90° + 2 \cdot 360°}{3} \right) \\
&= 2(\cos 270° + j \sin 270°) \\
&= 2\{0 + j(-1)\} \\
&= -2j
\end{aligned}$$

Evaluate the following.

1. $(1 + j)^7$ 2. $(3 - 3j)^5$

3. $(2 + 2j)^4$ 4. $(-27 + j)^{1/3}$

5. $(-2 + j)^{10}$ 6. $(-2 - 5j)^{15}$

7. $(\frac{1}{2} + 3j)^{1/2}$ 8. $\{2(\cos 30 + j \sin 30)\}^8$

9. $(1 - j)^3$ 10. $\left(\frac{2}{5} - \frac{j}{2}\right)^{1/4}$

6-8
Application of Complex Numbers

Complex numbers find application in many scientific and technical problems. In physics or applied mechanics, vector quantities such as displacement, velocity, acceleration, or force may be described using complex numbers. The analysis of alternating-current electrical problems makes extensive use of complex numbers, or the *j operator*.

A positive vector, z, may be viewed as being made up of a horizontal and vertical component. If we define the horizontal component as lying along the x-axis or *real* axis and the vertical component as lying along the y-axis or *imaginary* axis, and use the *j operator* to identify the vertical component, then we can express the vector z as a complex number.

$$z = x + jy$$

The *j operator* indicates that the y component lies along the vertical or imaginary axis. A positive j means the vertical component is projected upward, while a negative sign before the j means the vertical component is projected downward.

In electrical problems different symbols are generally used such that the equation looks like:

$$z = r + jx$$

where

z = total impedance

r = pure resistance

x = reactance

6-9
Technical Example Using Complex Numbers (Electrical)

Figure 6-12 represents a differentiating circuit. Suppose we wish to determine the total impedance of the circuit, the voltage across the resistor and the capacitor, and the phase angle between the voltage and current in the circuit.

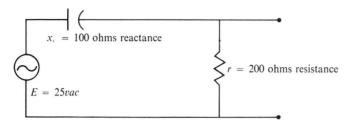

Figure 6-12

The total impedance of the circuit is found by

$$z = r - jx_c$$
$$z = 200 - j100$$
$$|z| = \sqrt{(200)^2 + (-100)^2}$$
$$= \sqrt{40,000 + 10,000}$$
$$= \sqrt{50,000}$$
$$|z| = 223.6 \text{ ohms impedance}$$

The negative sign associated with the j operator implies that the imaginary component is directed downward. In an electrical circuit this means that the voltage lags the current by 90° which is indeed the case in a capacitor.

The phase angle of the circuit is found by

$$\tan \theta = \frac{-jx_c}{r}$$
$$= \frac{-100}{200}$$
$$\tan \theta = -.5$$
$$\theta = \tan^{-1}(-.5)$$
$$\theta = -26.5°$$

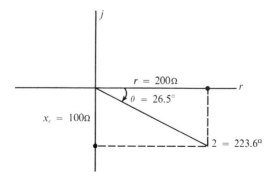

Figure 6-13

The circuit impedance, $z = r - jx_c$, may be written as

$$z = 223.6\angle -26.5°$$

which is the polar form for representing a complex number.

The total current in the circuit is found by dividing the voltage by the impedance.

$$I = \frac{E}{Z}$$

$$= \frac{25 \text{ volts}}{223.6\angle -26.5° \text{ ohms}}$$

$$I = .112 \text{ amps } \angle +26.5°$$

The positive sign associated with the phase angle indicates that the current leads the voltage by 90°.

The voltage across the resistor is found by multiplying the current times the resistance.

$$E_r = Ir = (.112 \text{ amps})(200 \text{ ohms}) = 22.4 \text{ volts}$$

The voltage across the capacitor is found by multiplying the current times the reactance.

$$E_{x_c} = IX_c = (.112 \text{ amps})(100 \text{ ohms}) = 11.2 \text{ volts}$$

Notice that the sum of the voltage across the resistor and capacitor is *not* equal to the applied voltage. This is due to the phase difference.

6-10
Technical Example Using Complex Numbers (Mechanical)

The following forces act on a point.

(a) 50 lbs at 53°
(b) 70 lbs at 45°
(c) 100 lbs at 180°
(d) 30 lbs at 270°

We can determine the magnitude and direction of the resultant force by the use of complex numbers. Our first step is to express each force as a complex number as follows.

(a) $30 + 40j$
(b) $50 + 50j$
(c) $-100 + 0j$
(d) $0 - 30j$

Adding these four complex numbers gives a result of $-20 + 60j$. Converting this to polar form we get:

$$R = -20 + 60j = 63\angle 108.5°$$

This tells us that the four original forces can be replaced by this single resultant force and produce the same net force on point A in Figure 6.14.

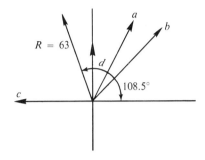

Figure 6-14

7

Logarithms

The *logarithm* of a number is the exponent, which when placed on a given base produces the number. For example, the logarithm of 25 to the base 5 is 2 because $5^2 = 25$. Another way of expressing this is $\log_5 25 = 2$.

EXAMPLE 1   Express the following in words:

$$(1)\ \log_3 9 = 2$$
$$(2)\ \log_2 16 = 4$$
$$(3)\ \log_7 49 = 2$$

Solution: (1) the logarithm of 9 to the base 3 equals 2.

(2) the logarithm of 16 to the base 2 equals 4.

(3) the logarithm of 49 to the base 7 equals 2.

An expression like $\log_3 9 = 2$ is called a *logarithmic equation*. An expression like $3^2 = 9$ is called an *exponential equation*. A logarithmic equation and an exponential equation say the same thing but in two different ways. In general:

$$\text{if}\quad x^n = y \qquad \text{then}\quad \log_x y = n$$

or

$$\text{if}\quad \log_x y = n \qquad \text{then}\quad x^n = y$$

239

EXAMPLE 2   Express the following in exponential form.

$$(1) \log_4 64 = 3$$
$$(2) \log_4 16 = 2$$
$$(3) \log_5 625 = 4$$

Solution:

$$(1) \ 4^3 = 64$$
$$(2) \ 4^2 = 16$$
$$(3) \ 5^4 = 625$$

EXAMPLE 3   Express the following in logarithmic form.

$$(1) \ 5^3 = 125$$
$$(2) \ 3^4 = 81$$
$$(3) \ (\tfrac{1}{2})^2 = \tfrac{1}{4}$$

Solution:

$$(1) \log_5 125 = 3$$
$$(2) \log_3 81 = 4$$
$$(3) \log_{1/2} \tfrac{1}{4} = 2$$

EXAMPLE 4   Find x if $\log_x 121 = 2$.

Solution: Change $\log_x 121 = 2$ to exponential form and solve for x.

$$x^2 = 121$$
$$x = \sqrt{121}$$
$$x = 11$$

EXAMPLE 5   Find y, if $\log_{49} y = \tfrac{1}{2}$.

Solution: Change $\log_{49} y = \tfrac{1}{2}$ to exponential form and solve for y.

$$49^{1/2} = y$$
$$\sqrt{49} = y$$
$$7 = y$$

PROBLEM SET 7-1

Express the following in exponential form.

1. $\log_{10} 1,000 = 3$
2. $\log_7 49 = 2$
3. $\log_4 64 = 3$
4. $\log_{10} 1 = 0$
5. $\log_{25} 5 = .5$
6. $\log_{10} 10,000 = 4$
7. $\log_8 512 = 3$
8. $\log_3 243 = 5$
9. $\log_e y = x$
10. $\log_x X = 1$

Express the following equations in logarithmic form.

1. $10^2 = 100$
2. $5^3 = 125$
3. $2^5 = 32$
4. $3^4 = 81$
5. $4^0 = 1$
6. $10^6 = 1,000,000$
7. $16^{.5} = 4$
8. $27^{1/3} = 3$
9. $2^8 = 256$
10. $U^v = t$

Determine the value of X.

1. $2^x = 8$
2. $10^x = 1$
3. $3^x = 81$
4. $25^x = 5$
5. $4^x = 256$
6. $X = \log_3 27$
7. $\log_4 X = 3$
8. $4^x = \sqrt{4}$
9. $X^4 = 10,000$
10. $X^6 = 64$

7-2
Properties of Logarithms

Since $\log_x y = n$ and $x^n = y$ have the same meaning, let us examine $x^n = y$ a little further.

Suppose x is negative and n is fractional, what type of number is y? For example, let $x = -4$, and $n = \frac{1}{2}$, then

$$-4^{1/2} = y$$

If we express this with radicals we have

$$\sqrt{-4} = y$$

A number like $\sqrt{-4}$ is called an *imaginary number*.

We can see from this example that if x is negative and n is fractional the result may be an imaginary number.

If we let x equal 1 and n equal any number the result will equal 1, since 1 to any power is 1, for example,

$$1^2 = 1 \qquad 1^{1/2} = 1$$

$$1^3 = 1 \qquad 1^5 = 1$$

Therefore, to avoid y becoming imaginary or always equal to 1, we require the base x to be positive and different from 1.

Since working with logarithms requires working with exponents, before we study the properties of logarithms we shall review the following laws of exponents.

(1) $a^m \cdot a^n = a^{m+n}$

(2) $\dfrac{a^m}{a^n} = a^{m-n}$ if $m > n$ and $a \neq 0$

$\dfrac{a^m}{a^n} = \dfrac{1}{a^{n-m}} = a^{m-n}$ if $m < n$ and $a \neq 0$

(3) $(a^m)^n = a^{mn}$

(4) $(ab)^m = a^m b^m$

(5) $(a)^0 = 1$

We shall make use of the laws of exponents to prove the properties of logarithms.

(1) Product Property   The logarithm of the product of two or more numbers equals the sum of the logarithms of the numbers.

$$\log_x (ab) = \log_x a + \log_x b$$

Proof: Let $\log_x a = m$ and $\log_x b = n$. Changing each to exponential form produces

$$x^m = a \quad \text{and} \quad x^n = b$$
$$ab = x^m \cdot x^n$$
$$= x^{m+n}$$

Changing $ab = x^{m+n}$ to logarithmic form we get:

$$\log_x (ab) = m + n$$
$$= \log_x a + \log_x b$$

(2) Quotient Property   The logarithm of the quotient of two numbers equals the logarithm of the numerator minus the logarithm of the denominator.

$$\log_x \left(\frac{a}{b}\right) = \log_x a - \log_x b$$

Proof: Let $\log_x a = m$ and $\log_x b = n$. Changing each to exponential form we get $x^m = a$ and $x^n = b$.

$$\frac{a}{b} = \frac{x^m}{x^n}$$
$$= x^{m-n}$$

Changing $\dfrac{a}{b} = x^{m-n}$ to logarithmic form we get:

$$\log_x \frac{a}{b} = m - n$$
$$= \log_x a - \log_x b$$

(3) Power Property   The logarithm of the rth power of a number equals r times the logarithm of the number.

$$\log_x (a^r) = r \cdot \log_x a$$

Proof: Let $\log_x a = m$. Changing $\log_x a = m$ to exponential form we get $x^m = a$.

$$a^r = (x^m)^r$$
$$= x^{rm}$$

Changing $a^r = x^{r \cdot m}$ to logarithmic form we get:

$$\log_x (a^r) = r \cdot m$$
$$= r \log_x a$$

The following examples show how the properties of logarithms are used.

EXAMPLE 6 $\quad \log_5 (4 \cdot 3) = \log_5 4 + \log_5 3$

EXAMPLE 7 $\quad \log_2 (7 \cdot 8) = \log_2 7 + \log_2 8$

EXAMPLE 8 $\quad \log_4 \frac{10}{3} = \log_4 10 - \log_4 3$

EXAMPLE 9 $\quad \log_8 \frac{6}{11} = \log_8 6 - \log_8 11$

EXAMPLE 10 $\quad \log_6 (10^3) = 3 \log_6 10$

EXAMPLE 11 $\quad \log_4 \sqrt{5} = \log_4 (5)^{1/2}$
$$= \tfrac{1}{2} \log_4 5$$

EXAMPLE 12 $\quad \log_4 \sqrt[3]{25} = \log_4 (25^{1/3})$
$$= \tfrac{1}{3} \log_4 25$$

7-3
Types of Logarithms

The two most common types of logarithms are the *common* and the *natural*. Common logarithms are logarithms to the base 10. Natural logarithms are to the base e which is approximately 2.71828.

As we continue to work with logarithms unless a particular base is specified, it will be understood that we are working with either base 10 or base e. To distinguish between these two we shall use *log* to mean common logarithms and *ln* to mean natural logarithms. For example,

$$\log 5 \text{ means } \log_{10} 5$$
$$\ln 8 \text{ means } \ln_e 8$$

We know that the logarithm of a number is an exponent of a particular base which will give the number. For example,

$$\log 1 = 0 \quad \text{because} \quad 10^0 = 1$$
$$\log 10 = 1 \quad \text{because} \quad 10^1 = 10$$

EXAMPLE 13   What is log 6?

Solution: If the $\log 1 = 0$ and $\log 10 = 1$, then $\log 6$ would be between 0 and 1.

$$10^0 = 1$$
$$10^? = 6$$
$$10^1 = 10$$

It will be shown later from Table 4, that $\log 6 = .77815$.

EXAMPLE 14   Find log 37.

Solution: Our problem is to find which exponent of 10 will equal 37. In other words

$$10^? = 37$$

Since

$$10^1 = 10$$

and

$$10^2 = 100$$

then the exponent which we are looking for must be somewhere between 1 and 2.

$$10^{1.56820} = 37$$
$$\log 37 = 1.56820$$

A logarithm has two parts. The whole number is called the characteristic, and the decimal number is called the mantissa. In Example 13 where $\log 6 = .77815$,

0 is the characteristic and .77815 is the mantissa. In Example 14 where log 37 = 1.5682, 1 is the characteristic and .56820 is the mantissa.

The characteristic and mantissa of the logarithm are completely different and are found in different ways. To find the characteristic and mantissa of the logarithm of a number consider the following example.

EXAMPLE 15   Find log 154.

Solution: First write the number in scientific notation.

$$154 = 1.54 \times 10^2$$

The characteristic is the exponent of the 10, which in this case is 2.

To find the mantissa, look in Table 4 in the appendices and find the number 154 in the left-hand column N.

To the right of 154 you will find the number 752. This number is preceded by the number 18, making the complete mantissa .18752. Each of the numbers is understood to have a decimal point in front, therefore the complete mantissa for 154 should be .18752. Therefore,

$$\log 154 = 2.18752$$

You will notice that most of the numbers in Table 4 are only three place numbers and should be five place numbers. They should be preceded by the two numbers which can be found to the left of the number under the 0 column. If an asterisk appears on the number in question, this means to prefix the two numbers on the following line, under the 0 column, as the first two numbers of the mantissa. The following example shows how the asterisks are used.

EXAMPLE 16   Find log 4074.

Solution: First write 4074 in scientific notation.

$$4074 = 4.074 \times 10^3$$

The exponent, 3, of 10 is the characteristic. In Table 4 find the number 407 under the column N, and go to the column headed by 4. In this column you find the number 002. Since this number has an asterisk you should prefix the two-place number 61 which is found in the next line under the 0 column, making the complete mantissa .61002. Therefore,

$$\log 4074 = 3.61002$$

EXAMPLE 17   Find log .014.

Solution: Write .014 in scientific notation.

$$.014 = 1.4 \times 10^{-2}$$

The characteristic is -2, and the mantissa is .14613. Therefore,

$$\log .014 = -2 + .14613$$
$$= -1.85387$$

In our previous discussions we stated that we wanted the logarithms to be positive and in this example we find the log .014 to be negative. To overcome this we write the characteristic (-2) as $(8 - 10)$, placing the 8 in front of .14613 and the -10 at the end. Therefore,

$$\log .014 = 8.14613 - 10$$

Since Table 4, contains logarithms of four place numbers only, you may wonder what to do with a number which has more than four places such as 168.149. We can use Table 4, to find logarithms of five-place numbers by interpolation. If the number has more than five places it is necessary that it be rounded to five places.

EXAMPLE 18   Find log 168.149.

Solution: Rounding 168.149 to five places we get 168.15. Writing 168.15 in scientific notation we get 1.6815×10^2. The characteristic is 2, but the mantissa can not be found directly from Table 4. However it is somewhere between 168.10 and 168.20.

$$10\left[\begin{array}{c} 5\left[\begin{array}{c}\log 168.10 = 2.22557 \\ \log 168.15 = ? \end{array}\right. \\ \log 168.20 = 2.22583 \end{array}\right] 26$$

The diagrams shows that 168.15 is $\frac{5}{10}$ between 168.10 and 168.20, therefore the log 168.15 is $\frac{5}{10}$ between the log 168.10 and log 168.20.

Subtracting 2.22557 from 2.22583 we get 26. Now find $\frac{5}{10}$ of 26.

$$\frac{5}{10} \times 26 = 13.$$

Add 13 to the last digit of 2.22557

$$\begin{array}{r} 2.22557 \\ 13 \\ \hline 2.22570 \end{array}$$

Therefore,

$$\log 168.15 = 2.22570.$$

EXAMPLE 19   Find log .12693.

Solution:

$$10\left[\begin{array}{c} \log .12700 = 9.10380 - 10 \\ 3\left[\begin{array}{c}\log .12693 = ? \\ \log .12690 = 9.10346 - 10\end{array}\right. \end{array}\right] 34$$

$$\frac{3}{10} \times 34 = \frac{102}{10} = 10.2$$

Add 10.2 and 9.10346 − 10.

$$9.10346 - 10$$
$$\underline{10}$$
$$9.10356 - 10$$

Therefore,

$$\log .12693 = 9.10356 - 10$$

Instead of using interpolation to find this answer, we can use the proportional parts table to the right of each page. It is necessary that you determine mentally the position ($\frac{3}{10}$) of .12693 between .12690 and .12700, and that you determine the difference (34) of .10346 and .10380. Find the proportional parts table headed with 34, locate 3 in the left hand column. The number corresponding to 3 under 34 is the number 10.2. Round 10.2 to 10 and add to 9.10346 − 10.

EXAMPLE 20   Find log 9463.1.

Solution: The characteristic is 3. The mantissa of log 9463.0 is .97603 and the mantissa of log 9464.0 is .97607. This difference is 4 and since 9463.1 is $\frac{1}{10}$ of the way between 9463.0 and 9464.0 we find the number .4 in the proportional parts table adjacent to 1. Round .1 to 0 and add to .97603. This will be the mantissa of log 9463.1. Therefore,

$$\log 9463.1 = 3.97603$$

PROBLEM SET 7-4

Find the logarithm of the following numbers.

1. 2	2. 100	3. 270	4. 3.14
5. 536	6. .005	7. 5,976	8. 1.247
9. 24,725	10. .00028	11. 4×10^{-8}	12. 6.74×10^{10}
13. 7,643,000	14. 1,000,000	15. 27.6×10^{-5}	16. 6.28
17. 2.736×10^{-4}	18. 56,238	19. .00683	20. 482,600

7-6
Antilogarithms

If you are given the logarithm of a number and asked to find the number, this is the same as finding the antilogarithm of a number.

EXAMPLE 21 $\log N = 4.43807$, find N.

Solution: Look in the mantissas for .43807. This corresponds to the number 2742. However, this should be thought of as 2.742×10^4. If we move the decimal point 4 places to the right we find,

$$N = 27420.$$

EXAMPLE 22 $\log N = 1.76668$, find N.

Solution:

$$10 \begin{bmatrix} \text{—}\log 58.440 = 1.76671 \text{—} \\ \quad \log N = 1.76668 \text{—} \\ \text{—}\log 58.430 = 1.76664 \text{—} \end{bmatrix} \begin{matrix} \\ 4 \end{matrix} \quad 7$$

We see that the N is $\frac{4}{7}$ between 58.430 and 58.440. $\frac{4}{7} \times 10 = \frac{40}{7} = 5.9$. Therefore,

$$N = 58.436$$

The table of proportional parts can be used instead of interpolation. To do this find the difference (7) between 1.76664 and 1.76671. In the proportional parts table headed 7, look directly under the 7 until you find 4 (or the closest number to 4) which is the difference between 1.76664 and 1.76668. Since 4.2 is the closest number to 4, look to the left of 4.2 and you will see the number 6. This is the number to be added to 58.430.

If we agree that every number under column N has a decimal point after the first number, we can find logarithms without thinking of scientific notation. To find log 601, count the places the decimal point moves to change to 6.01. Since the decimal point moves 2 places, the number 2 becomes the characteristic. The mantissa for log 6.01 is .77887. Therefore,

$$\log 601 = 2.77886$$

This same reasoning can be applied in finding antilogarithms. For example, to find N if $\log N = 2.83104$, first find the number 6.777 which corresponds to the number .83104. Since the characteristic is 2, move the decimal point two places to the right. Therefore,

$$N = 677.7$$

PROBLEM SET 7–5

Determine the antilogarithm of the following logarithms.

1. .3010	2. 1.6990
3. 2.8451	4. .9294
5. 5.1761	6. 7.4771 − 10
7. 4.7160	8. 1.9777
9. 8.5315 − 10	10. 9.7993 − 10

To make use of the properties of logarithms in solving problems consider the following examples.

EXAMPLE 23   Find $(1.287)(3.42)$ by using logarithms.

Solution: $N = (1.287)(3.42)$

Taking log of both sides,

$$\log N = \log (1.287)(3.42)$$

Since $\log ab = \log a + \log b$,

$$\log N = \log 1.287 + \log 3.42$$
$$\log N = .10958 + .53403$$
$$\log N = .64361$$

Solving for N, we find,

$$N = 4.4016$$

EXAMPLE 24   Find N, if $N = \dfrac{.01036}{23.184}$.

Solution: $N = \dfrac{.01036}{23.184}$

Taking log of both sides,

$$\log N = \log \frac{.01036}{23.184}$$

Since $\log \dfrac{a}{b} = \log a - \log b$,

$$\log N = \log .01036 - \log 23.184$$
$$\log N = (8.01536 - 10) - 1.36519$$
$$\log N = 6.65017 - 10$$

Solving for N, we find,

$$N = .00044686$$

EXAMPLE 25   Find N, if $N = \sqrt[3]{147}$.

Solution: Change $\sqrt[3]{147}$ to $(147)^{1/3}$, and use the power property of logarithms.

$$N = (147)^{1/3}$$

Taking the log of both sides,

$$\log N = \log (147)^{1/3}$$

Since $\log a^n = n \log a$,

$$\log N = \tfrac{1}{3} \log 147$$
$$\log N = \tfrac{1}{3}(2.16732)$$
$$\log N = .72244$$
$$N = 5.2774$$

EXAMPLE 26   Find N, if $N = \sqrt[5]{.01027}$.

Solution: $N = \sqrt[5]{.01027}$

Changing $\sqrt[5]{.01027}$ to $(.01027)^{1/5}$,

$$N = (.01027)^{1/5}$$

Taking the log of both sides,

$$\log N = \log (.01027)^{1/5}$$
$$\log N = \tfrac{1}{5} \log (.01027)$$
$$\log N = \tfrac{1}{5}(8.01157 - 10)$$

Before we divide by 5, the -10 must be changed to a number such that when we divide the result will be -10, therefore we must change the -10 in the number $8.01157 - 10$ to -50. To do this we must add 40 to the 8 and -40 to the -10 thus $8.01157 - 10$ becomes $48.01157 - 50$, therefore

$$\log N = \tfrac{1}{5}(48.01157 - 50)$$
$$\log N = 9.60231 - 10$$
$$N = .40023$$

EXAMPLE 27   Find N, if $N = \dfrac{(341)(129)^2}{.102}$.

Solution: This problem involves all three properties and can be simplified as follows.

$$\log N = \frac{(341)(129)^2}{.102}$$
$$\log N = \log 341 + \log (129)^2 - \log .102$$
$$\log N = \log 341 + 2 \log 129 - \log .102$$
$$\log N = 2.53275 + 2(2.11059) - (9.00860 - 10)$$
$$\log N = 2.53275 + 4.22118 - (9.00860 - 10)$$
$$\log N = 6.75393 - (9.00860 - 10)$$
$$\log N = (16.75393 - 10) - (9.00860 - 10)$$
$$\log N = 7.74533$$
$$N = 55632000$$

   To use logarithms in working with signed numbers, first determine the sign of the answer before starting your work, then disregard all signs in the computation. The following example shows how this is done.

EXAMPLE 28   Find N, if $N = (-4.3764)^3$.

Solution: When we cube a negative number the result is negative also, therefore we will use logarithms to find $(4.3764)^3$ and prefix a minus sign on the answer.

$$\log N = \log (4.3764)^3$$

$$\log N = 3 \log (4.3764)$$

$$\log N = 3(.64112)$$

$$\log N = 1.92336$$

$$N = -83.822$$

PROBLEM SET 7-6

Complete the following using logarithms.

1. 60×19

2. 128×2.43

3. 3.14×6.28

4. $\dfrac{254}{37}$

5. $\dfrac{.025}{1,571}$

6. $(2 \times 10^{-3})(4 \times 10^4)$

7. $(3.6 \times 10^3)(8.12 \times 10^5)(2.93 \times 10^{-4})$

8. $\dfrac{(2.78)(153)}{1,695}$

9. $\dfrac{(4.37 \times 10^5)(6.87 \times 10^3)}{8.77 \times 10^6}$

10. $\dfrac{5 \times 10^{-6}}{(2.3 \times 10^4)(7.31 \times 10^5)}$

7-8
Application of Logarithms

Logarithms are widely used in technical problems having to do with sound level measurement. The Weber-Fechner law, an approximate law dealing with sound levels, states that a perceptible change in loudness is proportional to the logarithm of the ratio of the intensity of the existing sound to the intensity at the threshold of hearing. In equation form this is written as:

$$L = \log \frac{I}{I_t}$$

where:

L is the existing sensation of loudness

I is the intensity of existing sound

I_t is the intensity at the threshold of hearing (10^{-10} microwatts/cm^2)

The unit of loudness is the *bel* (in honor of Alexander Graham Bell). However, because the bel is too large for ordinary use in sound measurements the decibel is generally used. The decibel is one-tenth bel. The technical definition of the bel is quite involved. However, a satisfactory practical definition in terms of power is that increasing the power to the ear by ten causes a perceptible change in loudness of one bel or ten decibels (db).

$$L_{bels} = \log \frac{P_2}{P_1}$$

$$L_{db} = 10 \log \frac{P_2}{P_1}$$

In the above equations P_2 is the power which produces the intensity of the existing sound and P_1 is the power producing sound at the threshold of hearing. Since both have dimensions of watts, the ratio P_2 to P_1 is a dimensionless number (the watts cancel).

7-9

Technical Example Using Common Logarithms

One of the advantages of using logarithms is that the results are simply added. To illustrate this, suppose we wish to determine the power gain of the three-stage audio-amplifier shown in Figure 7-1. The total gain of the amplifier is the product

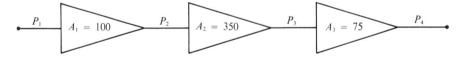

Figure 7-1

of the gain of each stage. The gain of the amplifier in Figure 7-1 is $A_1 \times A_2 \times A_3$ or $(100)(350)(75) = 2,625,000$ which means that P_4 is this many times greater than P_1. The total gain can also be expressed as the sum of the db gains for each stage as follows.

Stage 1:
$$db = 10 \log \frac{P_2}{P_1}$$
$$= 10 \log 100$$
$$= 10 \log 10^2$$
$$= 10(2 \log 10)$$
$$= 20 \log 10$$
$$= 20(1)$$
$$db = 20$$

Stage 2:
$$db = 10 \log \frac{P_3}{P_2}$$
$$= 10 \log 350$$
$$= 10 \log (10 \cdot 35)$$
$$= 10 \log 10 + \log 35$$
$$= 10(1 + 1.5441)$$
$$= 10(2.5441)$$
$$db = 25.441$$

Stage 3:
$$db = 10 \log \frac{P_4}{P_3}$$
$$= 10 \log 75$$
$$= 10 \log (10 \cdot 7.5)$$
$$= 10(\log 10 + \log 7.5)$$
$$= 10(1 + .8751)$$
$$= 10(1.8751)$$
$$db = 18.751$$

The total gain of the amplifier is the sum of the db gain of the individual stages.

Amplifier gain: $Ap = 20db + 25.441db + 18.751db$

$$Ap = 64.192db$$

The gain expressed in db would be particularly useful if one wished to plot the frequency response of the amplifier.

7-10

Graphs of $y = a^x$ and $y = \log ax$

An equation like $y = a^x$ is called an *exponential function*. If we interchange x and y we get the equation $x = a^y$ which is equivalent to $y = \log_a x$. If we let $a = 2$ in $y = a^x$ and $y = \log_a x$ and graph each we get the following.

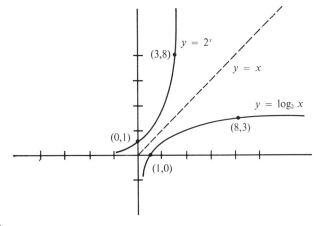

Figure 7-2

From Figure 7-2, it can be seen that $y = 2^x$ and $y = \log_2 x$ are symmetric to the line $y = x$. It can also be said that one is the image of the other with respect to $y = x$.

Functions like $y = 2^x$ and $y = \log_2 x$ are inverse functions.

It should be noted that if $x < 0$, $y = \log_2 x$ is not defined.

7-11

Logarithms to Base e and Other Bases

As previously mentioned another important number used as a base is e. Recall that e is approximately 2.71828, and that logarithms to base e are called natural logarithms (abbreviated ln).

There are times when it is useful to know how to find the logarithm of a number in a base other than 10. The most common base other than 10 is base e. The following shows how to change a logarithm in one base to a logarithm in some other base.

To prove this suppose:

(1) $\log_b n = y$

(2) $n = b^y$, changing to exponential form

(3) $\log_a n = \log_a b^y$, taking the logarithm of both sides to base a

(4) $\log_a n = y(\log_a b)$, by the power property

(5) $\log_a n = (\log_b n)(\log_a b)$, substituting $\log_b n$ for y

(6) $\dfrac{\log_a n}{\log_a b} = \log_b n$, dividing both sides by $\log_a b$

(7) or $\log_b n = \dfrac{\log_a n}{\log_a b}$, Symmetric Property

The following examples will show how this formula is used.

EXAMPLE 29 Find $\log_2 100$.

Solution:
$$\log_b n = \frac{\log_a n}{\log_a b}$$

$$\log_2 100 = \frac{\log_{10} 100}{\log_{10} 2}$$

$$\log_2 100 = \frac{2}{.30103}$$

Performing this division by logarithms, we have

$$\log 2 = 10.30103 - 10$$
$$-\log .30103 = \underline{9.47861 - 10}$$
$$.82242$$

The antilogarithm of .82242 is 6.6439, therefore

$$\log 100 = 6.6439$$

EXAMPLE 30   Find ln 16.

Solution: ln 16 = $\log_e$ 16, therefore:

$$\ln 16 = \log_e 16 = \frac{\log_{10} 16}{\log_{10} e}$$

or

$$\ln 16 = \frac{\log 16}{\log e}$$

Letting

$$e = 2.718$$

$$\ln 16 = \frac{\log 16}{\log 2.718}$$

$$\ln 16 = \frac{1.20412}{.43425}$$

$$\ln 16 = 2.7729$$

Logarithms to the base e can be found directly from Table 5. The following rules will be used for finding logarithms to base e.

To find the natural logarithm (ln) of a number which is $\frac{1}{10}$, $\frac{1}{100}$, $\frac{1}{1000}$, $\frac{1}{10000}$, and so forth of a number whose logarithm is given, subtract ln 10, 2 ln 10, 3 ln 10, 4 ln 10, and so forth, from the given logarithm.

To find the natural logarithm (ln) of a number which is 10, 100, 1000, 10000, and so forth, times a given number whose logarithm is given, add ln 10, ln 100, ln 1000, ln 10000 and so forth, to the given logarithm.

EXAMPLE 31   Find ln .0017.

Solution: ln .0017 = ln 1.7 − 3 ln 10

$$= .5306 - 6.9078$$

$$= -6.4772$$

EXAMPLE 32   Find ln 1500.

Solution: ln 1500 = ln 15 + 2 ln 10

$$= 2.7081 + 4.6052$$

In doing mathematical computation the same properties are true of natural logarithms as of common logarithms. The following example shows how these properties are applied.

EXAMPLE 33   Find N, if $N = \dfrac{14 \times \sqrt[3]{32}}{84}$.

Solution: $N = \dfrac{14 \times \sqrt[3]{32}}{84}$

$$\ln N = \ln \left(\frac{14 \times \sqrt[3]{32}}{84} \right)$$

$$= \ln 14 + \tfrac{1}{3} \ln 32 - \ln 84$$

$$= \ln 14 + \tfrac{1}{3}(\ln 3.2 + \ln 10) - (\ln 8.4 + \ln 10)$$

$$= 2.6391 + \tfrac{1}{3}(1.1632 + 2.3026) - (2.1282 + 2.3026)$$

$$= 2.6391 + \tfrac{1}{3}(3.4658) - (4.4308)$$

$$= 2.6391 + 1.1553 - 4.4308$$

$$\ln N = -.6364$$

$$N = .5292$$

7-11
Technical Example Using Natural Logarithms

Technical problems involving exponential functions can sometimes be simplified by the use of natural logarithms. Consider the following problem.

If, during a power failure, a home freezer warms up from $-10°F$ to $+2°F$ after standing for 3 hours in a room at $50°F$, how much longer will it take for the freezer to warm up to $32°F$?

The equation $T = T_0 e^{-kt}$ where

$$T_0 = \text{initial temperature difference}$$

$$T = \text{temperature difference at any time, } t$$

$$e = \text{exponential function}$$

$$t = \text{time in hours}$$

$$k = \text{constant}$$

will allow us to determine the solution to the problem.

Our first task is to determine the value of the constant, k. We can do this by substituting the known values into our equation. Doing this yields the following.

$$48 = 60e^{-k(3)}$$

$$\tfrac{48}{60} = e^{-k(3)}$$

$$e^{-3k} = 0.8$$

Now we need to take the natural logarithm of both sides of our equation.

$$\ln e^{-3k} = \ln .8$$
$$-3k(\ln e) = \ln .8$$
$$-3k(1) = \ln .8$$
$$-3k = -.223$$
$$k = \frac{-.223}{-3} = .0745$$

Going back to our original equation, we can substitute our set of known data, including k, and solve for time, t.

$$T_0 = T_0 e^{-kt}$$
$$18 = 60e^{-(.0745)t}$$
$$\tfrac{18}{60} = e^{-.0745t}$$
$$.3 = e^{-.0745t}$$

Taking the natural logarithm of both sides of our equation yields

$$\ln .3 = \ln e^{-.0745t}$$
$$\ln .3 = (-.0745t)(\ln e)$$
$$-1.204 = -.0745t$$
$$t = \frac{-1.204}{-.0745} = 16.2 \text{ hours total}$$

$$16.2 \text{ hrs} - 3 \text{ hrs} = 13.2 \text{ additional hours}$$

8

Algebra, Part 2

Factoring can be thought of as the reverse of multiplication. When two or more quantities are multiplied the result is called the product. Each of the quantities is called a factor. Thus in $3 \times 4 = 12$, 3 and 4 are called factors.

Sometimes the factors of a number can be given in several different ways. For example, if you are asked to find the factors of 24 you might say (3×8), (12×2), or $(3 \times 2 \times 4)$. Any of these would be correct.

As we continue our discussion of factoring we shall use the idea of multiplication and see its relation to factoring.

To multiply a polynomial by a monomial, multiply each term of the polynomial by the monomial.

EXAMPLE 1 Find the product of each of the following.

    (a) $2(3x - 4)$ (b) $2a(a + b)$

    (c) $4x(x^2 + 2x - 3)$ (d) $3x(x - y + 2)$

259

Solution: (a) $6x - 8$ (b) $2a^2 + 2ab$

(c) $4x^3 + 8x^2 - 12x$ (d) $3x^2 - 3xy + 6x$

PROBLEM SET 8-1

Find the product of each of the following.

1. $5a(3a + 4)$ 2. $e(4e + 7)$

3. $7r(2x + 3t + 5)$ 4. $xy(2x + 6y + 1)$

5. $8m^2(3n + 4m + 9mn)$ 6. $3ab(a + b + c)$

7. $\frac{1}{2}mv^2(2p + 6q + 8)$ 8. $9e(4^{1/2}e + 7)$

9. $7xy^2(2x^2y + xy + x)$ 10. $uv(4x + 9y + 3)$

11. $3^2e^2i^2(16e + 4^{1/2}i)$ 12. $8ab(5c + 6d + e)$

13. $5xyz(9x^2y + 4y^2)$

8-3
Factoring Out a Common Term

The easiest type of factoring involves eliminating a common term. To do this type of factoring, you must inspect the polynomial and decide what each term has in common. Divide each term of the polynomial by the common term and use the quotient for the other factor. Usually the common factor will be the largest quantity which is common to all terms. For example, to factor $(12x^2 + 6x)$ you could write $2(6x^2 + 3x)$ or $3x(4x + 2)$, but in neither case is the common term, 2 or $3x$, the largest factor which is common to both. We should factor $12x^2 + 6x$ as $6x(2x + 1)$. The largest common factor in a polynominal is usually called the *greatest common factor*.

EXAMPLE 2   Factor $24x^3 - 6x^2 + 18x$.

Solution: In looking at each term of $24x^3 - 6x^2 + 18x$, we see that $6x$ is the greatest common factor. Therefore if we divide $24x^3 - 6x^2 + 18x$ by $6x$, we get $4x^2 - x + 3$. Using this as the other factor we get:

$$24x^3 - 6x^2 + 18x = 6x(4x^2 - x + 3)$$

EXAMPLE 3   Factor $3a^2 - 9a + 6$.

Solution: The greatest common factor is 3. Therefore:

$$3a^2 - 9a + 6 = 3(a^2 - 3a + 2)$$

Factor each of the following.

1. $14x^2y + 21xy^2 + 35x^2y^2$
2. $17ei^2 + 34e^3i - 68e^2i$
3. $42a^2bc - 54a^2b^2c^2 - 63a^2bc^2$
4. $5u^3v^4w^7 + 16u^2v^3w^3 + 9u^4v^5w^4$
5. $39mn^3p + 52m^2np^2 - 61m^3n^3p^3$
6. $45rst - 180r^2st + 90r^2s^2t$
7. $33x^5y^2 + 55x^3y^4 - 77x^4y^6 + 121x^7y^3$
8. $70cd^5 - 28a^2b^3c^2d^2 + 56b^4c^3d^3e^2$
9. $72x^5y^4z^7 + 40w^3x^2 - 56x^3y^2 + 96x^4y^3z^5$
10. $21a^2b^5cd^3 + 56b^3c^3 + 42a^2d - 7ab^2c^4$

8-4
Multiplying the Sum and Difference of the Same Two Numbers

Suppose we wish to find the product of $(x + 1)(x - 1)$.

$$\begin{array}{r} x + 1 \\ \underline{x - 1} \\ -x - 1 \\ \underline{x^2 + x } \\ x^2 - 1 \end{array}$$

In this example each of the terms in the two binominals were the same. However, the signs of the second terms were different. This is what we mean by the sum and difference of the same two numbers.

EXAMPLE 4   Find the product of $(2x - 3)(2x + 3)$.

Solution:
$$\begin{array}{r} 2x - 3 \\ \underline{2x + 3} \\ 6x - 9 \\ \underline{4x^2 - 6x } \\ 4x^2 - 9 \end{array}$$

Again we have the product of the sum and difference of the same two numbers. Can you see any pattern to the result? If you examine the answer you will see that the first term is the square of the first term in either one of the factors and the second term of the answer is the square of the second term in either factor. Therefore we make the following statement or rule.

The product of the sum and difference of the same two numbers is the difference of the squares of the two terms of one of the factors.

It makes no difference as to which factor you wish to use, just square each term and place a minus sign between the two terms.

EXAMPLE 5   Find the product of each of the following.

$$\text{(a) } (a + b)(a - b) \qquad \text{(b) } (2x - y)(2x + y)$$

Solution: (a) $(a + b)(a - b) = a^2 - b^2$

(b) $(2x - y)(2x + y) = 4x^2 - y^2$

PROBLEM SET 8-3

Find the product of each of the following.

1. $(3x - y)(3x + y)$

2. $(5a + 3b)(5a - 3b)$

3. $(9e + 5i)(9e - 5i)$

4. $(2c - 6d)(2c + 6d)$

5. $(15m - 11n)(15m + 11n)$

6. $(13r + 9s)(13r - 9s)$

7. $(9^{1/2}x^2 - 5y)(9^{1/2}x^2 + 5y)$

8. $(16a^2 + 7b)(16a^2 - 7b)$

9. $(25^{1/4}x^2y + 4z)(25^{1/4}x^2y - 4z)$

10. $(36a^{1/2}b + 4^{1/2}c)(36a^{1/2}b - 4^{1/2}c)$

8-5
Factoring the Difference of Two Squares

We have just seen that if we multiply the sum and difference of the same two numbers the result is the difference of their squares. Therefore we can factor the difference of two squares. For example, if we wish to factor

$$4x^2 - 9$$

we should first determine if both terms are squares. Since this is the case, we find the square root of each. The square root of $4x^2$ is $2x$ and the square root of 9 is 3. We should start by placing two sets of parenthesis as follows:

$$(\quad)(\quad)$$

In each set of parenthesis we will place $2x$ and 3.

$$(2x \quad 3)(2x \quad 3)$$

Next we will insert a plus or minus sign in each parenthesis. It makes no difference whether we use a plus or minus sign in the first set of parenthesis, but we must use the opposite sign for the second parenthesis. Thus either of the following would be correct.

$$(2x + 3)(2x - 3)$$
$$(2x - 3)(2x + 3)$$

Therefore:

$$4x^2 - 9 = (2x + 3)(2x - 3)$$

EXAMPLE 6 Factor each of the following.

(a) $x^2 - y^2$ (b) $9x^2 - y^2$

(c) $16a^2 - 25y^2$ (d) $x^4 - (4x + 1)^2$

Solution: (a) $x^2 - y^2 = (x - y)(x + y)$

(b) $9x^2 - y^2 = (3x + y)(3x - y)$

(c) $16a^2 - 25y^2 = (4a - 5y)(4a + 5y)$

(d) $x^4 - (4x + 1)^2 = (x^2 - 4x - 1)(x^2 + 4x + 1)$

PROBLEM SET 8-4

Factor each of the following.

1. $a^2 - b^2$

2. $9e^2 - 4i^2$

3. $16u^2 - 25v^2$

4. $36m^4 - 9n^2$

5. $81x^4y^2 - 49z^2$

6. $25b^2c^2 - 100d^2$

7. $64r^4s^6 - 16t^2$

8. $49x^4z^8 - 121w^6$

9. $225n^{10} - 400p^8$

10. $169a^6b^4c^2 - 289d^2e^8$

8-6
Squaring a Binomial

Squaring a binomial is the same as multiplying the binomial by itself. For example,

$$(2x + 1)^2$$

could be found as follows.

$$
\begin{array}{r}
2x + 1 \\
2x + 1 \\
\hline
2x + 1 \\
4x^2 + 2x \\
\hline
4x^2 + 4x + 1
\end{array}
$$

Therefore:

$$(2x + 1)^2 = 4x^2 + 4x + 1$$

EXAMPLE 7 Square $(2x - 3y)$.

Solution:

$$
\begin{array}{r}
2x \;-\; 3y \\
2x \;-\; 3y \\
\hline
-6xy \;+\; 9y^2 \\
4x^2 \;-\; 6xy \\
\hline
4x^2 \;-\; 12xy \;+\; 9y^2
\end{array}
$$

Therefore:

$$(2x - 3y)^2 = 4x^2 - 12xy + 9y^2$$

EXAMPLE 8 Square $(3a + 4b)$.

Solution:

$$
\begin{array}{r}
3a \;+\; 4b \\
3a \;+\; 4b \\
\hline
12ab \;+\; 16b^2 \\
9a^2 \;+\; 12ab \\
\hline
9a^2 \;+\; 24ab \;+\; 16b^2
\end{array}
$$

Therefore:

$$(3a + 4b)^2 = 9a^2 + 24ab + 16b^2$$

Let us examine each of the above examples carefully and see if there is some kind of pattern to the answer. Notice in each case the first and last terms in the answer are the squares of the first and last terms of the problem. Also, the middle term of the answer is twice the product of the two terms of the problem. We now state this rule in a more formal manner.

To square a binomial, square the first and last terms of the binomial (the squares will become the first and last terms in the answer respectively and will both be positive). The middle term of the answer will be twice the product of the two terms of the binomial and will have the sign of the product.

Using this rule let us now solve $(3a - 2b)^2$. We begin by writing the problem and placing a set of parenthesis for the answer to be placed in.

$$(3a - 2b)^2 = (\qquad\qquad)$$

The first term in the binomial is $(3a)$ and the second term is $(-2b)$. Squaring $(3a)$ we get $(9a^2)$. Squaring $(-2b)$ we get $4b^2$. Placing these in the proper place we get

$$(3a - 2b)^2 = (9a^2 \qquad + 4b^2)$$

If we multiply the two terms $(3a)$ and $(-2b)$ we get $(-6ab)$. Doubling $(-6ab)$ we get $(-12ab)$ which is the middle term in the answer, thus:

$$(3a - 2b)^2 = (9a^2 - 12ab + 4b^2)$$

EXAMPLE 9 Simplify $(x + 4y)^2$.

Solution: Squaring the first term (x) we get x^2, squaring the second term $(4y)$ we get $16y^2$. Doubling the product of x and $4y$ gives $8xy$.
   Therefore:

$$(x + 4y)^2 = (x^2 + 8xy + 16y^2)$$

PROBLEM SET 8-5

Solve the following.

1. $(2x + 5)^2$ 2. $(8a + 3)^2$ 3. $(5cd + 7)^2$ 4. $(4m + 3n)^2$

5. $(9u - 3v)^2$ 6. $(8e + 3i)^2$ 7. $(6z^2 - 5w)^2$ 8. $(9x + 4)^2$

9. $(16^{1/2}x^2 - 9y^2)^2$ 10. $(\sqrt{25x^2y} - 8z^2)^2$

8-7
Factoring a Trinomial Which is a Perfect Square

In the previous examples involving squaring a binomial, each answer was a trinomial and each was a perfect square. By this we mean that the factors were identical. Now let us look at factoring a trinomial which is a perfect square. Keep in mind that factoring is "undoing" multiplication. Therefore we will do the reverse process of squaring a binomial.
   Consider the following example. Factor $25x^2 + 20x + 4$. We will examine each of the three terms to see if we have a trinomial which is a perfect square. The square root of $25x^2$ is $5x$, and the square root of 4 is 2. Next we shall see if $20x$ is twice the product of $5x$ and 2. $(5x)(2)$ equals $10x$. If we double $10x$ we get $20x$ which is the middle term. Therefore $25x^2 + 20x + 4$ is a perfect square and can be factored as:

$$25x^2 + 20x + 4 = (5x + 2)(5x + 2)$$
$$= (5x + 2)^2$$

PROBLEM SET 8-6

Factor each of the following.

1. $9x^2 + 6x + 1$ 2. $16a^2 + 48a + 36$

3. $4e^2 - 12e + 9$ 4. $49s^2 + 28s + 4$

5. $25t^2 - 10st + s^2$ 6. $36w^2 + 48xw + 16x^2$

7. $81b^2 + 18bc^2 + c^4$ 8. $64u^2 - 80uv + 25v^2$

9. $x^4 + 8x^2y + 16y^2$ 10. $m^4n^2 - 14m^2n + 49$

Multiplying $(x + a)(x + b)$ and $(ax + b)(cx + d)$

In finding the product of $(x + 1)(x + 2)$ or $(2x - 1)(x + 3)$ we can place one under the other and multiply as follows:

$$
\begin{array}{cc}
x + 1 & 2x - 1 \\
\underline{x + 2} & \underline{x + 3} \\
2x + 2 & 6x - 3 \\
\underline{x^2 + x} & \underline{2x^2 - x} \\
x^2 + 3x + 2 & 2x^2 + 5x - 3
\end{array}
$$

Thus:
$$(x + 1)(x + 2) = x^2 + 3x + 2$$
and
$$(2x - 1)(x + 3) = 2x^2 + 5x - 3$$

Let us examine each of the above examples. In the example

$$(x + 1)(x + 2) = x^2 + 3x + 2$$

we see that the first term in the answer is the product of the first terms in each binomial. The last term in the

$$(x + 1)(x + 2) = x^2 + 3x + 2$$

answer is the product of the last terms of the two binomials. To find the middle term multiply the two outside terms and then multiply the two inside terms and add the two products.

$$(x + 1)(x + 2) = x^2 + 3x + 2$$

Follow these same steps to find the product of $(2x - 1)(x + 3)$. Multiply $(2x)$ by (x) and get the first term in the answer.

$$(2x - 1)(x + 3) = 2x^2 + 5x - 3$$

Next multiply (-1) and (3) to get the last term in the answer.

$$(2x - 1)(x + 3) = 2x^2 + 5x - 3$$

Now multiply the two outside terms and the two inside terms and add the two product for the middle term in the answer.

$$(2x - 1)(x + 3) = 2x^2 + 5x - 3$$

EXAMPLE 10 Find the product of $(x - 3)(x + 2)$.

Solution:

$$(x - 3)(x + 2) = x^2$$

$$(x - 3)(x + 2) = x^2 \qquad - 6$$

$$(x - 3)(x + 2) = x^2 - x - 6$$

$$\begin{array}{c} -3x \\ \underline{2x} \\ -x \end{array}$$

Therefore:

$$(x - 3)(x + 2) = x^2 - x - 6$$

EXAMPLE 11 Find the product of $(3x - 4)(2x - 7)$.

Solution:

$$(3x - 4)(2x - 7) = 6x^2$$

$$(3x - 4)(2x - 7) = 6x^2 \qquad + 28$$

$$(3x - 4)(2x - 7) = 6x^2 - 29x + 28$$

$$\begin{array}{c} -8x \\ \underline{-21x} \\ -29x \end{array}$$

Therefore:

$$(3x - 4)(2x - 7) = 6x^2 - 29x + 28$$

PROBLEM SET 8-7

Find the product of the following.

1. $(x + 3)(x + 5)$
2. $(a + 2)(a - 7)$
3. $(e + 1)(e + 4)$
4. $(t - 6)(t - 2)$
5. $(x + 9)(x + 3)$
6. $(8y + 2)(4y - 6)$
7. $(7n - 5)(n + 3)$
8. $(4p - 11)(9p + 2)$
9. $(15s - 12)(7s - 23)$
10. $(42b - 26)(33b + 55)$

8-9
Factoring $x^2 + bx + c$ and $ax^2 + bx + c$

To factor a trinomial like $x^2 + bx + c$ or $ax^2 + bx + c$ it is very important to remember the multiplication in the previous section. For example if we are going to factor

$$x^2 - 3x + 2$$

we will start as follows:

$$x^2 - 3x + 2 = (\qquad)(\qquad)$$

We know that the product of the first two terms of the binomials is supposed to give the first term of the trinomial x^2. Since x times x is x^2 we can use these two factors of x^2 and have

$$x^2 - 3x + 2 = (x \qquad)(x \qquad)$$

Now we need to factor 2. The factors of 2 are (1 and 2) and (-1 and -2). Using these for the second terms we have the following.

$$x^2 - 3x + 2 = (x + 1)(x + 2) \tag{1}$$
$$= (x + 2)(x + 1) \tag{2}$$
$$= (x - 1)(x - 2) \tag{3}$$
$$= (x - 2)(x - 1) \tag{4}$$

Which of the above are true? If we multiply to see if we get the middle term $-3x$ we see that (3) and (4) are correct and (1) and (2) are incorrect. Therefore:

$$(x^2 - 3x + 2) = (x - 1)(x - 2)$$

Let us see what happens when we try to factor $2x^2 + 5x - 12$. If we use all the combinations of ways in which we can factor $2x^2$ and -12 and place the factors in their proper places we get:

$$2x^2 + 5x - 12 = (2x - 6)(x + 2)$$
$$= (2x + 6)(x - 2)$$
$$= (2x - 3)(x + 4)$$
$$= (2x + 3)(x - 4)$$
$$= (2x - 12)(x + 1)$$
$$= (2x + 12)(x - 1)$$
$$= (2x - 2)(x + 6)$$
$$= (2x + 2)(x - 6)$$
$$= (2x - 4)(x + 3)$$
$$= (2x + 4)(x - 3)$$
$$= (2x - 1)(x + 12)$$
$$= (2x + 1)(x - 12)$$

As far as the product of the first two terms of the answer equaling the first term in the problem $3x^2$ any one of the above might be true. The same is true of the last two terms of the binomials and the last term in the problem. But only one of the above twelve answers will give the middle term $5x$. By following the rules for finding the middle term we see that $(2x - 3)(x + 4)$ are the correct factors. Therefore:

$$2x^2 + 5x - 12 = (2x - 3)(x + 4)$$

EXAMPLE 12 Factor $3x^2 - x - 10$.

Solution: The factors of $3x^2$ are $(3x)$ and x. The factors of -10 are $(2$ and $-5)$ and $(-2$ and $5)$, $(1$ and $-10)$, and $(-1$ and $10)$. We must place these in the parentheses in such a manner that the sum of the products of the two outside terms and two inside terms are $(-x)$, thus

$$3x^2 - x - 10 = (3x + 5)(x - 2)$$

PROBLEM SET 8-8

Factor each of the following.

1. $x^2 + 6x + 8$
2. $2x^2 - 3x + 1$
3. $e^2 - 6e + 5$
4. $6t^2 + 12t + 6$
5. $r^2 + 10r + 21$
6. $5a^2 - 2a - 7$
7. $u^2 + 11u - 60$
8. $s^2 + 22s + 21$
9. $2y^2 + y - 6$
10. $i^2 - 7i - 60$
11. $12u^2 + 11u + 2$
12. $t^2 + 5t - 66$
13. $10a^2 - 27ab - 16b^2$
14. $c^2 - 17c + 72$
15. $3d^3 - 9d^2 - 30d$
16. $2x^2 - x - 36$
17. $v^2 - 15v - 54$
18. $6a^2 - 13a - 5$
19. $p^2 + 4p - 96$
20. $y^2 + 3y - 18$
21. $2d^2 + 15d - 8$
22. $b^2 + 11b + 28$
23. $4n^2 - 12n - 8$
24. $z^2 - 7z - 30$
25. $12w^2 - w - 1$
26. $m^2 - 18m - 40$
27. $r^2 - 12r + 35$
28. $3c^2 + 4c - 4$
29. $48h^2 + 14h - 45$
30. $3g^2 - 2g - 16$
31. $2r^2 - 15r - 27$
32. $6a^2 - a - 12$

8-10
Factoring $a^3 + b^3$ and $a^3 - b^3$

To factor $a^3 + b^3$ or $a^3 - b^3$ is called factoring the sum of two cubes or factoring the difference of two cubes. The factors of each should be committed to memory and are as follows.

$$a^3 + b^3 = (a + b)(a^2 - ab + b^2)$$
$$a^3 - b^3 = (a - b)(a^2 + ab + b^2)$$

These can be easily remembered if you follow this rule.

To factor the sum or difference of two cubes, the first factor will be a binomial whose terms are the cube roots of the original problem, and the sign will be the same as the original problem.

The second term is a trinomial whose first and last terms are squares of the two terms of the first factor, and the middle term is the product of the two terms in the first factor. The sign of the middle term will be negative for the sum of two cubes and positive for the difference of two cubes.

EXAMPLE 13   Factor $x^3 + 8^3$.

Solution: The cube root of x^3 is x and the cube root of 8 is 2. Therefore, the first factor is $(x + 2)$. The first and last terms of the second factor will be x and 4 respectively. For the middle term we multiply x and 2 which is $2x$. Therefore the middle term is $-2x$ because we are factoring the sum of two cubes. Hence, the complete second factor is $(x^2 - 2x + 4)$. Therefore:

$$x^3 + 8^3 = (x + 2)(x^2 - 2x + 4)$$

EXAMPLE 14   Factor $a^3 - 125^3$.

Solution: The cube root of a^3 is a and the cube root of 125 is 5. The first factor will be $(a - 5)$. The second factor will be $(a^2 + 5a + 25)$. Notice that the first and last terms are squares of each term in the first factor, and the middle term is the product of the two terms with a positive sign.

$$a^3 - 125^3 = (a - 5)(a^2 + 5a + 25)$$

PROBLEM SET 8-9

Factor each of the following.

1. $x^3 + 64$ 2. $e^3 + 27$

3. $v^3 - 125$ 4. $i^3 + 343$

5. $a^3 - 216$ 6. $8z^3 + 1$

7. $s^3 - 729$ 8. $27t^3 + 27$

9. $64r^3 - 512$ 10. $216y^3 + 343$

8-11
Factoring by Grouping

Sometimes it is necessary to group two or more terms together and to consider the group as a single term in order to factor. The following examples will show how this is done.

EXAMPLE 15 Factor $x^2 + 6x + 9 - y^2$.

Solution: Group the first three terms as:

$$(x^2 + 6x + 9) - y^2$$

Factor the group inside the parentheses.

$$(x + 3)^2 - y^2$$

Do you recognize this as the difference of two squares? Factor.

$$\{(x + 3) - y\}\{(x + 3) + y\}$$

Remove the parentheses.

$$\{x + 3 - y\}\{x + 3 + y\}$$

Therefore:

$$x^2 + 6x + 9 - y^2 = (x + 3 - y)(x + 3 + y)$$

EXAMPLE 16 Factor $x^2 - 4x + ax - 4a$.

Solution: Group as follows:

$$(x^2 - 4x) + (ax - 4a)$$

Take out the common factor of each.

$$x(x - 4) + a(x - 4)$$

Both of these have a common factor of $(x - 4)$. Factoring again we get:

$$(x - 4)(x + a)$$

Therefore:

$$x^2 - 4x + ax - 4a = (x - 4)(x + a)$$

It should be remembered that each expression should be factored completely and the answer should always be in the form of a product. If this is not the case then either the problem is not factored correctly or the problem cannot be factored.

PROBLEM SET 8-10

Factor each of the following.

1. $x^2 - 4x + 6xy - 24y$

2. $6a^2 - 10ab + 35cb - 21ac$

3. $x^3 + 2x^2 - 5x - 10$

4. $x^2 + 2x + 1 - y^2$

5. $h^3 + 3h^2 + 6h + 18$ 6. $7s^2r - 42sr^2 + 3s^2 - 18sr$

7. $5e^2i^3 + 30i^5 - 4e^2 - 24i^2$ 8. $a^2 + 4a + 4 - d^2 + 8d - 16$

8-12

Division

Division of an algebraic expression is basically the same as regular division. To show this we will use the following examples.

EXAMPLE 17   Divide $(x^3 + 2x^2 - 3x)$ by x.

Solution: Write this as a normal division problem.

$$x \overline{\smash{\big)}\ x^3 + 2x^2 - 3x}$$

Divide x^3 by x and place the quotient over the x^3 then multiply and subtract.

$$
\begin{array}{r}
x^2 \\
x \overline{\smash{\big)}\ x^3 + 2x^2 - 3x} \\
\underline{x^3 } \\
+ 2x^2
\end{array}
$$

Bring down the next term ($2x^2$) and continue by dividing $2x^2$ by x.

$$
\begin{array}{r}
x^2 + 2x \\
x \overline{\smash{\big)}\ x^3 + 2x^2 - 3x} \\
\underline{x^3 } \\
2x^2 \\
\underline{2x^2 } \\
- 3x
\end{array}
$$

Bring down the next term, $-3x$, and divide by x.

$$
\begin{array}{r}
x^2 + 2x \ - 3 \\
x \overline{\smash{\big)}\ x^3 + 2x^2 - 3x} \\
\underline{x^3 } \\
2x^2 \\
\underline{2x^2 } \\
- 3x \\
\underline{- 3x}
\end{array}
$$

Continue the division until it ends evenly or the variable in the remainder has a smaller exponent than the variable in the divisor.

EXAMPLE 18   Divide $8x^3 + 6x^2 - 3x + 4$ by $x + 1$.

Solution:

$$
\begin{array}{r}
8x^2 - 2x - 1 + \frac{5}{x+1} \\
x + 1 \overline{\smash{\big)}\ 8x^3 + 6x^2 - 3x + 4} \\
\underline{8x^3 + 8x^2} \\
-2x^2 - 3x \\
\underline{-2x^2 - 2x} \\
-x + 4 \\
\underline{-x - 1} \\
+5
\end{array}
$$

Notice how the remainder was written in the answer.

EXAMPLE 19   Divide $x^3 - 27$ by $x - 3$.

Solution:

$$x - 3 \overline{\smash{\big)}\ x^3 - 27}$$

Notice how the terms were written in the dividend. This was done because there will be some terms appearing in the solution which are not in the dividend.

$$
\begin{array}{r}
x^2 + 3x + 9 \\
x - 3 \overline{\smash{\big)}\ x^3 \qquad\quad - 27} \\
\underline{x^3 - 3x^2} \\
+3x^2 \qquad - 27 \\
\underline{+3x^2 - 9x} \\
9x - 27 \\
\underline{9x - 27}
\end{array}
$$

EXAMPLE 20   Divide $(-5x^2 + 6x^4 + 7x - 7x^3 - 10)$ by $(2x - 3)$.

Solution: You will notice in the solution that the dividend is rearranged so that we have descending powers of x.

$$
\begin{array}{r}
3x^3 + x^2 - x + 2 + \frac{-4}{2x-3} \\
2x - 3 \overline{\smash{\big)}\ 6x^4 - 7x^3 - 5x^2 + 7x - 10} \\
\underline{6x^4 - 9x^3} \\
2x^3 - 5x^2 \\
\underline{2x^3 - 3x^2} \\
-2x^2 + 7x \\
\underline{-2x^2 + 3x} \\
4x - 10 \\
\underline{4x - 6} \\
-4
\end{array}
$$

Perform the indicated division.

1. $\dfrac{5y^2 + 3y}{y}$

2. $\dfrac{4x^5 - 8x^3 + 16x}{4x}$

3. $\dfrac{6x^5 + 3x^4 - 9x^3 + 12x^2}{3x^2}$

4. $\dfrac{48r^2 - 36r + 84}{12}$

5. $\dfrac{t^8 - 6s^4t^6 + 6s^6t^4 - s^8t^2}{t^2}$

6. $\dfrac{\theta^2 + 12}{\theta - 4}$

7. $\dfrac{250Z^3 - 16w^3}{10Z - 4w}$

8. $\dfrac{10k^2 + 11k + 3}{2k + 1}$

9. $\dfrac{8 - 5v^2 - 18v + 21v^3}{3v - 4 + 7v^2}$

10. $\dfrac{\phi^3 + 27w^3 - \theta^3 + 9\theta\phi w}{\phi + 3w - \theta}$

11. $\dfrac{2h^2 - 2h^3 + 8h^4 + 27h - 30}{4h^2 + 3h - 6}$

12. $\dfrac{27c^2x^4 - 15c^2x^2 + 45cx^2 - 12c^2 + 46c - 42}{4c - 6 + 9cx^2}$

8-13
Fractions

Previously, addition and subtraction of fractions as well as reduction of fractions were discussed. Now we will take a closer look at these topics as well as multiplication and division of fractions, complex fractions, and fractional equations.

8-14
Reducing Fractions to Lowest Terms

Reducing fractions to lowest terms is nothing more than dividing both the numerator and denominator by the same term. Sometimes we can find the common term and reduce the fraction without very much difficulty. However, there are times when we must factor both the numerator and denominator in order to find the common factor of each.

EXAMPLE 21　　　Reduce $\dfrac{5x}{6x^2}$ to lowest terms.

Solution: We can divide both by x such as:

$$\frac{5x}{\underset{x}{\cancel{6x^2}}} = \frac{5}{6x}$$

EXAMPLE 22   Reduce $\dfrac{10a^2b^3}{15ab}$ to lowest terms.

Solution: First of all in this problem we will divide both the (10) and (15) by (5).

$$\frac{\overset{2}{\cancel{10}}\, a^2 b^3}{\underset{3}{\cancel{15}}\, a\, b}$$

Next we divide both (a^2) and (a) by (a).

$$\frac{\overset{2}{\cancel{10}}\ \overset{a}{\cancel{a^2}} b^3}{\underset{3\cdot 1}{\cancel{15}\ \cancel{a}\ b}}$$

Finally, divide both (b^3) and (b) by (b).

$$\frac{\overset{2\ \ a\ \ b^2}{\cancel{10}\ \cancel{a^2}\ \cancel{b^3}}}{\underset{3\cdot 1\cdot 1}{\cancel{15}\ \cancel{a}\ \cancel{b}}}$$

Multiply each part in the numerator and each part in the denominator.

$$\frac{\overset{2\ \ a\ \ b^2}{\cancel{10}\ \cancel{a^2}\ \cancel{b^3}}}{\underset{3\cdot 1\cdot 1}{\cancel{15}\ \cancel{a}\ \cancel{b}}} = \frac{2ab^2}{3}$$

Therefore:

$$\frac{10a^2b^3}{15ab} = \frac{2ab^2}{3}$$

EXAMPLE 23   Reduce $\dfrac{x^2 - 5x + 6}{x^2 - 9}$ to lowest terms.

Solution: Factor both the numerator and denominator and then cancel the common factors.

$$\frac{x^2 - 5x + 6}{x^2 - 9} = \frac{\cancel{(x - 3)}(x - 2)}{\cancel{(x - 3)}(x + 3)}$$

$$= \frac{x - 2}{x + 3}$$

Caution: Very often students make the mistake of cancelling out terms instead of factors which renders the answer incorrect. The following shows how this might be done.

$$\frac{\cancel{x^2} - 5x + \overset{2}{\cancel{6}}}{\cancel{x^2} - \cancel{9}} = \frac{-5x + 2}{-3}$$

Remember *in reducing fractions only common factors can be cancelled, not common terms.*

PROBLEM SET 8-12

Reduce the following fractions to their lowest terms.

1. $\dfrac{28e^2r}{12er^2}$

2. $\dfrac{48r^2s^5 + 7}{84r^2s^3 + 6}$

3. $\dfrac{120\lambda^8\beta^9\phi}{\lambda^3\beta^4\phi^5}$

4. $\dfrac{x^4 + 5x^2 - 6}{x^4 + 3x^2 - 18}$

5. $\dfrac{15n^7m^9}{21n^5m^4 + 27n^3m}$

6. $\dfrac{I^3 - 2I^2 + 4I - 8}{16 - I^4}$

7. $\dfrac{36a^2 - 25b^2}{6a^2 + 5ab}$

8. $\dfrac{4 - a^4}{8 + a^6}$

9. $\dfrac{K^2 - h^2}{h^2 + hK - 2K^2}$

10. $\dfrac{z^2 - 4 + 2v - vz}{2vz + 4 - v^2 - z^2}$

8-15
Multiplication of Fractions

In multiplying fractions, multiply the numerators (this will be the numerator of the answer), then multiply the denominators (this will be the denominator in the answer). Finally, reduce the fraction to lowest terms.

EXAMPLE 24   Find the product of $\dfrac{2x}{3x - 1} \cdot \dfrac{9x^2 - 3x}{x^2}$.

Solution: $\dfrac{2x}{3x - 1} \cdot \dfrac{9x^2 - 3x}{x^2} = \dfrac{2\cancel{x}}{\cancel{3x - 1}} \cdot \dfrac{3x(\cancel{3x - 1})}{\cancel{x}}$

$= 6x$

EXAMPLE 25 Find the product of $\dfrac{y - z}{x^2} \cdot \dfrac{x^2 - x}{(y - z)^2}$.

Solution: $\dfrac{y - z}{x^2} \cdot \dfrac{x^2 - x}{(y - z)^2} = \dfrac{\cancel{y - z}}{\cancel{x^2}} \cdot \dfrac{\cancel{x}(x - 1)}{(y - z)\cancel{(y - z)}}$

$$= \dfrac{(x - 1)}{x(y - z)}$$

EXAMPLE 26 Multiply $\dfrac{2a + 4}{a^2 - 8a - 9} \cdot \dfrac{2a - 18}{4}$.

Solution: $\dfrac{2a + 4}{a^2 - 8a - 9} \cdot \dfrac{2a - 18}{4} = \dfrac{2(a + 2)}{(a - 9)(a + 1)} \cdot \dfrac{2(a - 9)}{2 \cdot 2}$

$$= \dfrac{\cancel{2}(a + 2)}{\cancel{(a - 9)}(a + 1)} \cdot \dfrac{\cancel{2}\cancel{(a - 9)}}{\cancel{2} \cdot \cancel{2}}$$

$$= \dfrac{a + 2}{a + 1}$$

PROBLEM SET 8-13

Find the product of the following.

1. $\dfrac{a^2 b^2}{a^2 - b^2} \cdot \dfrac{a + b}{4ab}$

2. $\dfrac{\theta^2 \alpha}{\theta \alpha^2} \cdot \dfrac{\theta}{\theta^2 \alpha^2 - \theta \alpha}$

3. $\dfrac{7x^2 + 7}{10x^2 - 10} \cdot \dfrac{25x^2}{35}$

4. $\dfrac{5ab^2}{a + 2b} \cdot \dfrac{1}{5(a - b)}$

5. $\dfrac{15 - 5x}{10a} \cdot \dfrac{a^2 + a}{x - 3}$

6. $\dfrac{8r^2 - 8}{5v^2 - 5} \cdot \dfrac{15w^3}{14r - 7} \cdot \dfrac{7v^2 + 14v + 7}{24w}$

7. $\dfrac{K^2 - 7K + 10}{3K^5 l} \cdot \dfrac{12K^4 l^3}{25 - K^2}$

8. $\dfrac{9c^2 - 6c + 4}{4c^2 - 25} \cdot \dfrac{10 + 19c + 6c^2}{27c^3 + 8}$

8-16
Division of Fractions

To divide two fractions invert the divisor and multiply.

EXAMPLE 27 Divide $\dfrac{18x^2 y^2}{21ab}$ by $\dfrac{12xy^3}{7a^2 b}$.

Solution:

$$\frac{18x^2y^2}{21ab} \div \frac{12xy^3}{7a^2b} = \frac{18x^2y^2}{21ab} \cdot \frac{7a^2b}{12xy^3}$$

$$= \frac{\overset{6x}{\cancel{18x^2y^2}} \cdot \overset{a}{\cancel{7a^2b}}}{\cancel{21ab} \cdot \cancel{12xy^3}}$$

$$= \frac{ax}{2y}$$

EXAMPLE 28   Simplify $\dfrac{2x^2 - 2}{4x} \div \dfrac{x^2 - 3x - 4}{2x - 8}$.

Solution:

$$\frac{2x^2 - 2}{4x} \div \frac{x^2 - 3x - 4}{2x - 8} = \frac{2x^2 - 2}{4x} \cdot \frac{2x - 8}{x^2 - 3x - 4}$$

$$= \frac{\cancel{2}(x - 1)\cancel{(x + 1)}}{\cancel{4}x} \cdot \frac{\cancel{2}\cancel{(x - 4)}}{\cancel{(x - 4)}\cancel{(x + 1)}}$$

$$= \frac{x - 1}{x}$$

PROBLEM SET 8-14

Find the quotient of the following.

1. $\dfrac{4a - 2}{15b} \div \dfrac{6a^2 - 3a}{5b^2}$

2. $\dfrac{25r}{6u^3} \div \dfrac{75r^4}{12u^8}$

3. $\dfrac{32e^2 - 8}{70K^7} \div \dfrac{24e^2 - 6}{35K^3}$

4. $\dfrac{40}{0^2 - 1} \div \dfrac{0^2 - 20 + 1}{100^2}$

5. $\dfrac{x^2 + 2x - 8}{x^2 + x - 12} \div \dfrac{x^2 - x - 2}{x^2 - 3x}$

6. $\dfrac{z^3 - 1}{14a^5c^2} \div \dfrac{1 - z}{7a^6c^9}$

7. $\dfrac{4x^2 + 12x + 9}{9 - 4x^2} \div \dfrac{10x^2 + 27x + 18}{8x^2 - 2x - 15}$

8. $\dfrac{z^2 - 3z}{z^2 - 3z + 9} \div \dfrac{(z + 3)^2(z - 3)^3}{(z^3 + 27)(z^2 - 9)}$

8-17
More Addition and Subtraction of Fractions

At this time we would like to look at addition and subtraction of fractions that involve factoring to find the least common denominator LCD.

EXAMPLE 29 Add $\dfrac{a}{2x} + \dfrac{3}{ax}$.

Solution: To add $\dfrac{a}{2x}$ and $\dfrac{3}{ax}$ it is necessary to find the LCD. To find the LCD multiply all the different factors of both denominators. The first denominator $2x$ has 2 and x as factors. The second denominator ax has a and x as factors. The different factors of both are 2, x, and a. The product of 2, x, and a, which is $2ax$, will be the LCD.

Now change each original fraction to a new fraction having $2ax$ as its denominator. To do this divide each denominator into the LCD and multiply the respective numerator by the result.

For example $2x$ will go into $2ax$, a times. Multiply the numerator a by a which results in a^2. Dividing $2ax$ by ax results in 2; multiplying 3 by 2 results in 6.

$$\frac{a}{2x} + \frac{3}{ax} = \frac{a \cdot a}{2ax} + \frac{3 \cdot 2}{2ax}$$

$$= \frac{a^2}{2ax} + \frac{6}{2ax}$$

Now add the numerators and place the result over $2ax$.

$$= \frac{a^2 + 6}{2ax}$$

EXAMPLE 30 Add $\dfrac{4}{x - 3} + \dfrac{x + 2}{x^2 - 9}$.

Solution: Factor each denominator if possible. Since $(x - 3)$ will not factor, $(x - 3)$ becomes the factor. $(x^2 - 9) = (x - 3)(x + 3)$. Therefore the different factors of both denominators are $(x - 3)$ and $(x + 3)$.

Change each original fraction to a new fraction having $(x - 3)(x + 3)$ as its denominator and add.

$$\frac{4}{x - 3} + \frac{x + 2}{x^2 - 9} = \frac{}{(x - 3)(x + 3)} + \frac{}{(x - 3)(x + 3)}$$

$$= \frac{4(x + 3)}{(x - 3)(x + 3)} + \frac{(x + 2)(1)}{(x - 3)(x + 3)}$$

$$= \frac{4x + 12}{(x - 3)(x + 3)} + \frac{x + 2}{(x - 3)(x + 3)}$$

$$= \frac{4x + 12 + x + 2}{(x - 3)(x + 3)}$$

$$= \frac{5x + 14}{(x - 3)(x + 3)}$$

EXAMPLE 31   Combine $\dfrac{x + 3}{x} - \dfrac{x - 8}{x + 2}$.

Solution: The product of (x) and $(x + 2)$ will be the LCD.

$$\frac{x + 3}{x} - \frac{x - 8}{x + 2} = \frac{}{x(x + 2)} - \frac{}{x(x + 2)}$$

$$= \frac{(x + 3)(x + 2)}{x(x + 2)} - \frac{(x - 8)(x)}{x(x + 2)}$$

$$= \frac{x^2 + 5x + 6}{x(x + 2)} - \frac{x^2 - 8x}{x(x + 2)}$$

$$= \frac{x^2 + 5x + 6 - (x^2 - 8x)}{x(x + 2)}$$

$$= \frac{x^2 + 5x + 6 - x^2 + 8x}{x(x + 2)}$$

$$= \frac{13x + 6}{x(x + 2)}$$

PROBLEM SET 8-15

Perform the indicated operations on the following fractions.

1. $\dfrac{M}{3} + \dfrac{5M}{6} - \dfrac{M}{2}$

2. $\dfrac{1}{21ac} + \dfrac{1}{28ba} + \dfrac{1}{36bc}$

3. $\dfrac{a}{a - w} - \dfrac{w}{a + w} - \dfrac{a^2 + w^2}{a^2 - w^2}$

4. $\dfrac{x}{12yz} + \dfrac{3z}{20xy} + \dfrac{2y}{12xz}$

5. $\dfrac{20}{20e - 50} + \dfrac{e - 10}{8e - 20} - 10$

6. $\dfrac{7\theta^2}{16} - \dfrac{3\theta^2}{4} + \dfrac{\theta^2}{8}$

7. $\dfrac{x^2 + 4}{16 - x^2} + \dfrac{x}{x + 4} + \dfrac{x + 1}{2x - 8}$

8. $\dfrac{3}{6 + I} + \dfrac{2I}{I^2 - 36} + \dfrac{5}{6 - I}$

9. $\dfrac{1}{v - 2} + \dfrac{1}{1 - v} + \dfrac{v}{v^2 + 2v - 3}$

10. $\dfrac{x - y}{2xy} + \dfrac{2z + x}{10xz} + \dfrac{y + 2z}{4yz}$

A *complex fraction* is a fraction which itself has a fraction as its numerator, denominator or both. Some examples of complex fractions are

$$\frac{\frac{1}{2}}{\frac{2}{3}} \qquad \frac{1 + \frac{1}{a}}{a - \frac{1}{3}} \qquad \frac{\frac{x}{a} + \frac{y}{b}}{\frac{a}{x} + \frac{b}{y}}$$

The following examples will show how to simplify complex fractions.

EXAMPLE 32   Simplify $\dfrac{\frac{1}{2}}{\frac{2}{3}}$.

Solution: Change to division.

$$\frac{\frac{1}{2}}{\frac{2}{3}} = \tfrac{1}{2} \div \tfrac{2}{3}$$

Invert the divisor and multiply.

$$= \tfrac{1}{2} \times \tfrac{3}{2}$$
$$= \tfrac{3}{4}$$

EXAMPLE 33   Simplify $\dfrac{\frac{1}{x} + \frac{1}{y}}{\frac{x}{y} - \frac{y}{x}}$.

Solution: When the numerator or denominator or both contain more than one fraction you first must simplify each before inverting.

$$\frac{\frac{1}{x} + \frac{1}{y}}{\frac{x}{y} - \frac{y}{x}} = \frac{\dfrac{y + x}{xy}}{\dfrac{x^2 - y^2}{xy}}$$

$$= \frac{x + y}{xy} \cdot \frac{xy}{x^2 - y^2}$$

$$= \frac{\cancel{x + y}}{\cancel{xy}} \cdot \frac{\cancel{xy}}{(\cancel{x + y})(x - y)}$$

$$= \frac{1}{x - y}$$

Simplify the following.

1. $\dfrac{36xy}{8} \div 3y^3$

2. $\dfrac{36a^5b}{5} \div 4ab^5$

3. $\dfrac{7\phi}{1} \div \dfrac{21\phi^3}{8\phi^4y}$

4. $\left(\dfrac{1}{e^2} + \dfrac{3}{e} - 4\right) \div \left(\dfrac{1}{e^2} + \dfrac{5}{e} + 4\right)$

5. $\left(\lambda - \dfrac{1}{\lambda}\right) \div \left(1 + \dfrac{2}{\lambda} + \dfrac{1}{\lambda^2}\right)$

6. $\left(\dfrac{1}{s - t} + \dfrac{1}{s + t}\right) \div \left(\dfrac{1}{s - t} - \dfrac{1}{s + t}\right)$

7. $\left(\dfrac{3}{B - 4} - \dfrac{16}{B - 3}\right) \div \left(\dfrac{2}{B - 3} - \dfrac{15}{B + 5}\right)$

8. $\left(\dfrac{x + 1}{x - 1} + \dfrac{x - 1}{x + 1}\right) \div \left(\dfrac{x + 1}{x - 1} - \dfrac{x - 1}{x + 1}\right)$

8-19
Solving Fractional Equations

A fractional equation is an equation that contains a fraction. Some examples
of fractional equations are:

$$(1)\ \frac{x}{2} = 7$$

$$(2)\ \frac{x - 3}{4} = \frac{1}{2}$$

$$(3)\ \frac{3}{x} = 4$$

$$(4)\ \frac{x - 1}{4} = \frac{x + 2}{6}$$

$$(5)\ \frac{2x - 1}{x + 3} = \frac{1}{x} + 2$$

Any fractional equation is very easily solved if you will multiply each
individual fraction by the LCD of all the fractions and use all rules for solving
equations.

EXAMPLE 34 Solve $\dfrac{x}{2} = 7$ for x.

Solution: The LCD is 2. Therefore, multiplying each part by 2, we have:

$$(2)\frac{x}{2} = 7(2)$$

$$x = 14$$

EXAMPLE 35 Solve $\dfrac{x-3}{4} = \dfrac{1}{2}$ for x.

Solution: The LCD is 4. Multiplying each fraction by 4 and solving for x we have:

$$\frac{x-3}{4} = \frac{1}{2}$$

$$\frac{x-3}{4}(4) = \frac{1}{2}(4)$$

$$x - 3 = 2$$

$$x = 5$$

EXAMPLE 36 Solve $\dfrac{3}{x} = 4$ for x.

Solution: The LCD is x. Multiplying each part by x and then solving for x, we have:

$$\frac{3}{x} = 4$$

$$\frac{3}{x}(x) = 4(x)$$

$$3 = 4x$$

$$\frac{3}{4} = x$$

EXAMPLE 37 Solve $\dfrac{2x-1}{x+3} + \dfrac{1}{x} = 2$ for x.

Solution: The LCD is $x(x + 3)$. Multiplying each fraction by $x(x + 3)$ and solving for x we have:

$$\frac{2x - 1}{x + 3} + \frac{1}{x} = 2$$

$$\frac{2x - 1}{x + 3}(x)(x + 3) + \frac{1}{x}(x)(x + 3) = 2(x)(x + 3)$$

$$(2x - 1)(x) + 1(x + 3) = 2(x)(x + 3)$$

$$2x^2 - x + x + 3 = 2x^2 + 6x$$

$$2x^2 - 2x^2 - x + x - 6x = -3$$

$$-6x = -3$$

$$x = \tfrac{1}{2}$$

PROBLEM SET 8-17

Solve the following for the letter or symbol involved.

1. $\dfrac{E}{2} + \dfrac{E}{3} = \dfrac{1}{6}$

2. $\dfrac{1}{x} + \dfrac{1}{2} = \dfrac{5}{6} + \dfrac{1}{3}$

3. $\dfrac{r^2}{r^2 - 1} = 1 + \dfrac{1}{r + 1}$

4. $\dfrac{2}{a + 4} = \dfrac{7}{a - 6}$

5. $\dfrac{18}{10x - 6} = \dfrac{10}{6x + 14}$

6. $\dfrac{3\theta + 4}{\theta - 7} + 2 = \dfrac{2\theta + 2}{\theta - 4} + 3$

7. $\dfrac{5}{2I + 1} = \dfrac{1}{2I - 1} + \dfrac{2}{I - 2}$

8. $\dfrac{R + 3}{R^2 - 2R} + \dfrac{4}{R + 2} = \dfrac{5R}{R^2 - 4}$

8-20
Quadratic Equations

An equation which contains a term to the second degree and no higher is called a quadratic equation. The general form of a quadratic equation is given as $ax^2 + bx + c = 0$. Some examples of quadratic equations are:

(a) $x^2 + 2x + 3 = 0$ (b) $2x^2 - x = 0$

(c) $3x^2 = 4$ (d) $2x^2 - x + 7 = 0$

To find the solution of a quadratic equation is to find all values of x which will make a true statement of the quadratic equation. For example, the solution of

$$x^2 - 4x + 3 = 0$$

is 1 and 3, because

$$(1)^2 - 4(1) + 3 = 1 - 4 + 3$$
$$= 4 - 4$$
$$= 0$$
$$(3)^2 - 4(3) + 3 = 9 - 12 + 3$$
$$= 12 - 12$$
$$= 0$$

8-21
Solving Pure Quadratic Equations

The simplest of the quadratic equations is the pure quadratic equation $ax^2 = c$. Some examples of pure quadratic equations are:

$$2x^2 = 3$$
$$2x^2 - 1 = 0$$
$$x^2 + 4 = 0$$

To solve a pure quadratic such as $ax^2 - c = 0$, first isolate the constant c to one side of the equal sign.

$$ax^2 = c$$

Divide both sides by a.

$$x^2 = \frac{c}{a}$$

Take the square root of both sides.

$$x = \pm \sqrt{\frac{c}{a}}$$

Of course you should rationalize $\sqrt{\frac{c}{a}}$.

EXAMPLE 38   Solve $x^2 - 4 = 0$ for x.

Solution:
$$x^2 - 4 = 0$$
$$x^2 = 4$$
$$x = \pm 2$$

EXAMPLE 39   Solve $2x^2 - 3 = 0$ for x.

Solution:
$$2x^2 - 3 = 0$$
$$2x^2 = 3$$
$$x^2 = \frac{3}{2}$$
$$x = \pm\sqrt{\frac{3}{2}}$$
$$x = \pm\frac{\sqrt{6}}{2}$$

EXAMPLE 40   Solve $x^2 + 9 = 0$ for x.

Solution:
$$x^2 + 9 = 0$$
$$x^2 = -9$$
$$x = \pm\sqrt{-9}$$
$$x = \pm3\sqrt{-1}$$
$$x = \pm3j$$

PROBLEM SET 8-18

Solve the following pure quadratic equations.

1. $x^2 - 9 = 0$ 2. $x^2 - 16 = 0$ 3. $2x^2 - 16 = 0$

4. $4x^2 - 1 = 0$ 5. $4x^2 - 25 = 0$ 6. $3x^2 - 27 = 0$

7. $5x^2 - 125 = 0$ 8. $16x^2 - 1 = 0$ 9. $25x^2 - 4 = 0$

8-22
Solving Quadratic Equations by Factoring

Sometimes it is possible to solve a quadratic equation like $ax^2 + bx + c = 0$ by factoring if one side of the equation is equal to zero, and the other side is factorable. For example, if we have an equation like

$$x^2 - 5x - 6 = 0$$

we can factor the left side as

$$(x - 6)(x + 1) = 0$$

This says that the product of two factors is equal to zero. This means that either one factor or the other is equal to zero. Set each factor equal to zero as follows, and solve for x in each case.

$$x - 6 = 0 \qquad x + 1 = 0$$
$$x = 6 \qquad x = -1$$

Therefore, we find that $(6, -1)$ are the solutions for $x^2 - 5x - 6 = 0$. To be certain we have a true statement we should substitute each value for x in $x^2 - 5x - 6 = 0$. Hence,

$$x^2 - 5x - 6 = 0 \qquad\qquad x^2 - 5x - 6 = 0$$
$$(6)^2 - 5(6) - 6 = 0 \qquad (-1)^2 - 5(-1) - 6 = 0$$
$$36 - 30 - 6 = 0 \qquad\qquad 1 + 5 - 6 = 0$$
$$36 - 36 = 0 \qquad\qquad\qquad 0 = 0$$
$$0 = 0$$

To solve quadratic equations by factoring be sure that each term is to one side of the equation and that the other number is zero. Factor the quadratic expression. Set each factor equal to zero and solve for x.

EXAMPLE 41   Solve $x^2 - 10x + 16 = 0$ for x.

Solution: $x^2 - 10x + 16 = 0$
Factor.

$$(x - 8)(x - 2) = 0$$

Set each factor equal to zero and solve for x.

$$x - 8 = 0 \qquad x - 2 = 0$$
$$x = 8 \qquad x = 2$$

EXAMPLE 42   Solve $2x^2 - 3x = 5$ for x.

Solution: Place all terms to one side of the equal sign.
Factor.

$$2x^2 - 3x - 5 = 0$$
$$(2x - 5)(x + 1) = 0$$

Set each factor equal to zero and solve for x.

$$2x - 5 = 0 \qquad x + 1 = 0$$
$$2x = 5 \qquad x = -1$$
$$x = \tfrac{5}{2}$$

In each example the student should check the answers by substituting each value of x into the problem. If the result is a true statement then the solution is correct.

Solving Quadratic Equations by Completing the Square

If a quadratic equation is not factorable it is possible to solve the equation by completing the square. To complete the square means to make a trinomial square out of the quadratic. Consider the following example.

$$x^2 + 6x + 4 = 0$$

If you try to factor $x^2 + 6x + 4$ you will find it impossible. When this is the case move the constant term to the side with the zero, leaving only terms containing x on one side such as:

$$x^2 + 6x \qquad = -4$$

We leave a space where the 4 was on the left side because we will fill this space with the number which will make the left side a perfect square. To find this number divide the coefficient of x by 2.

$$6 \div 2 = 3$$

Square the result.

$$3^2 = 9$$

Place the 9 in the space where the 4 originally was but doing this is actually adding 9 to the left side of the equation $x^2 + 6x = -4$. Therefore it is also necessary to add 9 to the right member.

$$x^2 + 6x + 9 = -4 + 9$$
$$x^2 + 6x + 9 = 5$$

Change $x^2 + 6x + 9$ to $(x + 3)^2$

$$(x + 3)^2 = 5$$

Take the square root of both sides.

$$x + 3 = \pm\sqrt{5}$$

Solving for x.

$$x = -3 \pm\sqrt{5}$$

We should think of this as $x = -3 + \sqrt{5}$ and $x = -3 - \sqrt{5}$, resulting in two answers for x.

In general if we have a quadratic equation such as $ax^2 + bx + c = 0$ and we wish to solve this by completing the square:

(1) Move the constant c to one side.

$$ax^2 + bx = -c$$

(2) Divide both sides by a.

$$x^2 + \frac{b}{a}x = -\frac{c}{a}$$

(3) Divide $\dfrac{b}{a}$ by 2, square the result and add this to both sides.

(4) Find the square root of both sides.

EXAMPLE 43   Solve $x^2 + 8x - 3 = 0$ for x.

Solution: $\qquad\qquad\qquad x^2 + 8x - 3 = 0$

Move the -3 to the right side.

$$x^2 + 8x \qquad = +3$$

Dividing 8 by 2 equals 4, squaring 4 equals 16, and adding 16 to both sides results in

$$x^2 + 8x + 16 = +3 + 16$$
$$x^2 + 8x + 16 = 19$$

Changing $x^2 + 8x + 16$ to $(x + 4)^2$ produces

$$(x + 4)^2 = 19$$

Taking the square root of both sides

$$x + 4 = \pm\sqrt{19}$$

Solving for x,

$$x = -4 \pm \sqrt{19}$$

or

$$x = -4 + \sqrt{19}$$

and

$$x = -4 - \sqrt{19}$$

EXAMPLE 44   Solve $3x^2 - 2x - 5 = 0$.

Solution: $\qquad\qquad\qquad 3x^2 - 2x - 5 = 0$

Move -5 to the right side.

$$3x^2 - 2x \qquad = 5$$

Divide each side by 3.

$$x^2 - \tfrac{2}{3}x \qquad = \tfrac{5}{3}$$

Divide $\tfrac{2}{3}$ by 2, square the result ($\tfrac{1}{3}$).

$$(\tfrac{1}{3})^2 = \tfrac{1}{9}$$

Add $\tfrac{1}{9}$ to both sides.

$$x^2 - \tfrac{2}{3}x + \tfrac{1}{9} = \tfrac{5}{3} + \tfrac{1}{9}$$
$$x^2 - \tfrac{2}{3}x + \tfrac{1}{9} = \tfrac{16}{9}$$

Change $x^2 - \tfrac{2}{3}x + \tfrac{1}{9}$ to $(x - \tfrac{1}{3})^2$.

$$(x - \tfrac{1}{3})^2 = \tfrac{16}{9}$$

Taking the square root of both sides,

$$x - \tfrac{1}{3} = \pm\tfrac{4}{3}$$

Solving for x

$$x = \tfrac{1}{3} \pm \tfrac{4}{3}$$

$$x = \tfrac{5}{3}, -1$$

Look at example 44 again. Can $3x^2 - 2x - 5$ be factored? If so, then it is possible to solve this example by factoring as well as completing the square. This is true of any quadratic equation that is factorable.

PROBLEM SET 8-19

Solve the following quadratic equations by factoring.

1. $5v^2 + 2v - 3 = 0$ 2. $a^2 - a = 12$

3. $\lambda(3\lambda + 11) = 20$ 4. $8k^2 - 40k + 50 = 0$

5. $E^2 = 9E - 18$ 6. $\theta(6\theta + 7) = 3$

7. $9I^2 + 64 = -48I$ 8. $5t(2t - 3) + 35 = (t + 5)3t$

Solve the following equations by the method of completing the square.

1. $a^2 - 10a = 24$ 2. $h^2 = 2h + 24$

3. $2c^2 - 12c + 14 = 0$ 4. $3R^2 - 7R = 6$

5. $14\theta^2 - 22\theta = 12$ 6. $16\lambda^2 - 24\lambda = -5$

7. $E^2 = -(10E + 19)$ 8. $25L^2 - 10L = 8$

8-24

The Quadratic Formula $x = \dfrac{-b \pm \sqrt{b^2 - 4ac}}{2a}$

So far we have solved quadratic equations by factoring and by completing the square. The quadratic formula provides another way to solve quadratic equations. The quadratic formula is the same as completing the square, but makes the solution much simpler than the process of completing the square.

To derive the quadratic formula we start with the general form of a quadratic equation, $ax^2 + bx + c = 0$, and solve for x by completing the square.

$$ax^2 + bx + c = 0$$

Move c to the right side.

$$ax^2 + bx \quad\;\; = -c$$

Divide both sides by a.

$$x^2 + \frac{b}{a}x \quad\;\; = -\frac{c}{a}$$

Divide $\dfrac{b}{a}$ by 2, square the result and add this to both sides.

$$\frac{1}{2}\left(\frac{b}{a}\right) = \frac{b}{2a}$$

$$\left(\frac{b}{2a}\right)^2 = \frac{b^2}{4a^2}$$

$\dfrac{b^2}{4a^2}$ is what we add to both sides.

$$x^2 + \frac{b}{a}x + \frac{b^2}{4a^2} = \frac{b^2}{4a^2} - \frac{c}{a}$$

Find a common denominator for the right side and combine. The common denominator is $4a^2$, thus

$$\frac{b^2}{4a^2} - \frac{c}{a} = \frac{b^2}{4a^2} - \frac{4ac}{4a^2}$$

$$= \frac{b^2 - 4ac}{4a^2}$$

Therefore:

$$x^2 + \frac{b}{a}x + \frac{b^2}{4a^2} = \frac{b^2 - 4ac}{4a^2}$$

$$\left(x + \frac{b}{2a}\right)^2 = \frac{b^2 - 4ac}{4a^2}$$

Taking the square root of both sides

$$x + \frac{b}{2a} = \pm \frac{\sqrt{b^2 - 4ac}}{2a}$$

Solving for x,

$$x = -\frac{b}{2a} \pm \frac{\sqrt{b^2 - 4ac}}{2a}$$

Combining we have

$$x = \frac{-b \pm \sqrt{b^2 - 4ac}}{2a}$$

Using this formula we can now solve quadratic equations by substituting the respective values of a, b, and c into $x = \dfrac{-b \pm \sqrt{b^2 - 4ac}}{2a}$.

EXAMPLE 45   Solve $x^2 - 4x - 6 = 0$ by the quadratic formula.

Solution: $x^2 - 4x - 6 = 0$

$a = 1$ $x = \dfrac{-b \pm \sqrt{b^2 - 4ac}}{2a}$

$b = -4$

$c = -6$ $x = \dfrac{-(-4) \pm \sqrt{(-4)^2 - 4(1)(-6)}}{2(1)}$

$$= \dfrac{4 \pm \sqrt{16 + 24}}{2}$$

$$= \dfrac{4 \pm \sqrt{40}}{2}$$

$$= \dfrac{4 \pm 2\sqrt{10}}{2}$$

$$= \dfrac{4 + 2\sqrt{10}}{2}, \dfrac{4 - 2\sqrt{10}}{2}$$

$$= 2 + \sqrt{10}, 2 - \sqrt{10}$$

PROBLEM SET 8-20

Solve the following equations by using the quadratic formula.

1. $2a^2 - 5a + 3 = 0$ 2. $p^2 + 4p = 60$

3. $10d^2 - 26d + 12 = 0$ 4. $t^2 + 2t - 2 = 0$

5. $24B^2 = 4 - 10B$ 6. $E(24 - 18E) = 106$

7. $14s^2 - 6s + 4 = 0$ 8. $27I^2 = 6I + 8$

9. $2 = 16r(r + 1)$ 10. $20 = 64v - 16v^2$

8-25

Graph of a Function

In Chapter 2, linear functions were discussed. At this time, we turn our attention to functions in which the graph will not be a straight line. The three types of functions which will be discussed are quadratic, power, and exponential.

A function like $y = mx^2 + bx + c$ where m, b, and c are real numbers and $m \neq 0$, is a *quadratic function*. The simplest of quadratic functions is $y = x^2$. To draw the graph of $y = x^2$, assign values to x and solve for y. It is necessary with quadratic, as well as with power and exponential, functions that enough points are found in order to give an accurate representation of the graph. Assigning values to x and solving for y in $y = x^2$ we get the following.

x	-3	-2	-1	0	1	2	3
y	9	4	1	0	1	4	9

If these points are plotted on a graph we have:

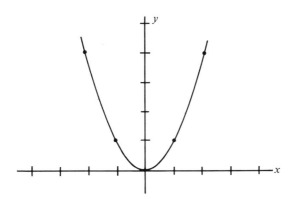

Figure 8-1

EXAMPLE 46   Draw the graph of $y = x^2 + x - 1$.

Solution: We must find some values for x and y which satisfy this function. We assign the following values to x and solve for y.

x	-3	-2	-1	0	1	2	3
y							

Let $x = -3$

$$y = x^2 + x - 1$$
$$= (-3)^2 + (-3) - 1$$
$$= 9 - 3 - 1$$
$$= 5$$

when $x = -3,\ y = 5$

Let $x = -1$

$$y = x^2 + x - 1$$
$$= (-1)^2 + (-1) - 1$$
$$= 1 - 1 - 1$$
$$= -1$$

when $x = -1,\ y = -1$

Let $x = -2$

$$y = x^2 + x - 1$$
$$= (-2)^2 + (-2) - 1$$
$$= 4 - 2 - 1$$
$$= 1$$

when $x = -2,\ y = 1$

Let $x = 0$

$$y = x^2 + x - 1$$
$$= (0)^2 + (0) - 1$$
$$= -1$$

when $x = 0,\ y = -1$

Let $x = 1$ Let $x = 2$

$$y = x^2 + x - 1$$ $$y = x^2 + x - 1$$

$$= (1)^2 + (1) - 1$$ $$= (2)^2 + (2) - 1$$

$$= 1 + 1 - 1$$ $$= 4 + 2 - 1$$

$$= 1$$ $$= 5$$

when $x = 1$, $y = 1$ when $x = 2$, $y = 5$

Let $x = 3$

$$y = x^2 + x - 1$$

$$= (3)^2 + (3) - 1$$

$$= 9 + 3 - 1$$

$$= 11$$

when $x = 3$, $y = 11$

hence,

x	-3	-2	-1	0	1	2	3
y	5	1	-1	-1	1	5	11

If we plot the above points on a graph we get the following.

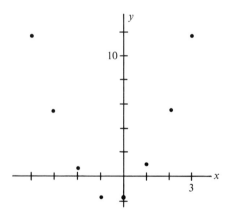

Figure 8-2

Care should be exercised in connecting these points with a smooth curve. It would seem that the curve between the points $(-1, -1)$ and $(0, -1)$ is a straight line, but if values are assigned to x between 0 and -1 the graph of the curve is slightly below a straight line connecting these two points.

The curve for $y = x^2 + x - 1$ is as follows.

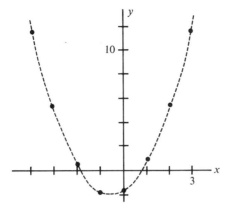

Figure 8-3

A function like $y = ax^n$, for some number n, is called a *power function*. As you can see $y = x^2$ could be called quadratic or power. However, some power functions will be shown where n is some number other than 2.

EXAMPLE 47 Graph the power function $y = x^3$, where $a = 1$ and $n = 3$.

Solution: Assign values for x and solve for y.

Let $x = -2$	Let $x = -1$
$y = x^3$	$y = x^3$
$\quad = (-2)^3$	$\quad = (-1)^3$
$\quad = -8$	$\quad = -1$
when $x = -2,\ y = -8$	when $x = -1,\ y = -1$
Let $x = 0$	Let $x = 1$
$y = x^3$	$y = x^3$
$\quad = 0^3$	$\quad = 1^3$
$\quad = 0$	$\quad = 1$
when $x = 0,\ y = 0$	when $x = 1,\ y = 1$
Let $x = 2$	
$y = x^3$	
$\quad = 2^3$	
$\quad = 8$	
when $x = 2,\ y = 8$	

Therefore we have the following set of values.

x	-2	-1	0	1	2
y	-8	-1	0	1	8

The graph of $y = x^3$ would be as follows.

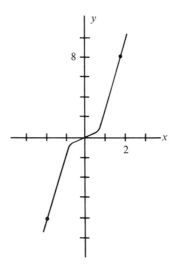

Figure 8-4

An equation of the type $y = mb^x$ where m and x can be any real number and b is some preassigned number, is an *exponential equation*. Let $m = 1$, and $b = 2$, and draw the graph of $y = 2^x$. Again we assign values for x and solve for y.

Let $x = -4$	Let $x = -3$
$y = 2^x$	$y = 2^x$
$= 2^{(-4)}$	$= 2^{-3}$
$= \frac{1}{16}$	$= \frac{1}{8}$
when $x = -4$, $y = \frac{1}{16}$	when $x = -3$, $y = \frac{1}{8}$
Let $x = -2$	Let $x = -1$
$y = 2^x$	$y = 2^x$
$= 2^{-2}$	$= 2^{-1}$
$= \frac{1}{4}$	$= \frac{1}{2}$
when $x = -2$, $y = \frac{1}{4}$	when $x = -1$, $y = \frac{1}{2}$
Let $x = 0$	Let $x = 1$
$y = 2^x$	$y = 2^x$
$= 2^0$	$= 2^1$
$= 1$	$= 2$
when $x = 0$, $y = 1$	when $x = 1$, $y = 2$
Let $x = 2$	Let $x = 3$
$y = 2^x$	$y = 2^x$
$= 2^2$	$= 2^3$
$= 4$	$= 8$
when $x = 2$, $y = 4$	when $x = 3$, $y = 8$

Therefore we have the following table of values.

x	-4	-3	-2	-1	0	1	2	3
y	$\frac{1}{16}$	$\frac{1}{8}$	$\frac{1}{4}$	$\frac{1}{2}$	1	2	4	8

Using these values we obtain the following graph.

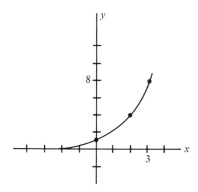

Figure 8-5

EXAMPLE 48   Draw the graph of $y = 2^{-x}$.

Solution: Let x be each of these values, $-4, -3, -2, -1, 0, 1, 2, 3, 4$, to obtain the table:

x	-4	-3	-2	-1	0	1	2	3	4
y	16	8	4	2	1	$\frac{1}{2}$	$\frac{1}{4}$	$\frac{1}{8}$	$\frac{1}{16}$

Hence the graph of $y = 2^{-x}$ would be:

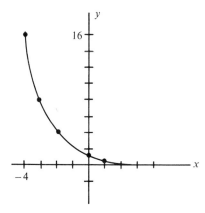

Figure 8-6

Exponential functions are very useful in science-related areas. If ideal conditions exist, the growth of a population, or the growth of a number of bacteria in a culture is exponential. The decay curve of a radioactive element is exponential.

EXAMPLE 49   Let the exponential function $y = 300(3^x)$ represent the number of bacteria in a culture. Find the number of bacteria at the end of 1 hr, 2 hr, 3 hr. What is the number of bacteria initially?

Solution: Let $x = 1$ hr

$$y = 300(3^x)$$
$$= 300(3^1)$$
$$= 900, \text{ bacteria at the end of 1 hr}$$

Let $x = 2$ hr
$$y = 300(3^x)$$
$$= 300(3^2)$$
$$= 300(9)$$
$$= 2700, \text{ bacteria at the end of 2 hr}$$

Let $x = 3$ hr
$$y = 300(3^x)$$
$$= 300(3^3)$$
$$= 300(27)$$
$$= 8100, \text{ bacteria at the end of 3 hr}$$

To find the number of bacteria initially, let $x = 0$ hr and solve for y.

Let $x = 0$ hr
$$y = 300(3^x)$$
$$= 300(3^0)$$
$$= 300(1)$$
$$= 300, \text{ bacteria initially}$$

EXAMPLE 50   The curve representing decay of a radioactive element is exponential in form. If the equation $y = 200(2^{-.05x})$ represents the number of grams present at any time x of a given radioactive element, how much is present to start with, and how much is present after 50 years?

Solution: Let $x = 0$ yrs
$$y = 200(2^{-.05x})$$
$$= 200(2^{-.05 \cdot 0})$$
$$= 200(2^0)$$
$$= 200(1)$$
$$= 200, \text{ grams present at start}$$

$$\text{Let } x = 50 \text{ yrs}$$
$$y = 200(2^{-.05 \cdot 50})$$
$$= 200(2^{-2.50})$$
$$= 200(2^{-5/2})$$
$$= 200 \left(\frac{1}{2^{5/2}} \right)$$
$$= 200(.177)$$
$$= 35.4, \text{ grams present after 50 years}$$

PROBLEM SET 8-21

Draw graphs for each of the following.

1. $y = 10^x$ 2. $y = x^2 + x$ 3. $y = x^4$

4. $y = x^{-1}$ 5. $y = x^2 - 1$ 6. $y = (x + 1)^2$

7. $y = 10^{-x}$ 8. $y = (\frac{1}{2})^x$ 9. $y = x^{1/2}$

10. $y = 2x^2 + 2$ 11. $y = 3^{-x}$ 12. $y = x^{3x-1}$

8-25
The Binomial Theorem

The binomial theorem is a formula used for expanding a binomial to any power. Consider the following.

$$(a + b)^1 = a + b$$
$$(a - b)^2 = a^2 - 2ab + b^2$$
$$(a + b)^3 = a^3 + 2a^2b + 3ab^2 + b^3$$
$$(a - b)^4 = a^4 - 4a^3b + 6a^2b^2 - 4ab^3 + b^4$$
$$(a + b)^5 = a^5 + 5a^4b + 10a^3b^2 + 10a^2b^3 + 5ab^4 + b^5$$

If we examine the preceding equations the following steps are evident which can be used to expand a binomial such as $(a + b)^n$:

1. The first term is a^n, the last term is b^n.
2. The exponent of a decreases by 1 in each term and is not present in the last term.
3. b appears in the second term, and the exponent increases by 1 in each term until it reaches b^n.
4. The coefficient of the first term is 1, and the coefficient of the second term is n.

5. The product of the coefficient of any term and the exponent of *a* in the same term divided by the number of that term, gives the coefficient of the following term.
6. There will be $n + 1$ terms.
7. If the second term of the binomial is positive, the sign of each term is positive.
8. If the second term of the binomial is negative, the signs will be alternately positive and negative.

EXAMPLE 51   Find $(a + b)^7$.

Solution: The first and second terms are:

$$a^7 + 7a^6b$$

Third term:
$$\frac{7 \times 6}{2} a^5b^2 = 21a^5b^2$$

Fourth term:
$$\frac{21 \times 5}{3} a^4b^3 = 35a^4b^3$$

Fifth term:
$$\frac{35 \times 4}{4} a^3b^4 = 35a^3b^4$$

Sixth term:
$$\frac{35 \times 3}{5} a^2b^5 = 21a^2b^5$$

Seventh term:
$$\frac{21 \times 2}{6} ab^6 = 7ab^6$$

Eighth term:
$$\frac{7 \times 1}{7} a^0b^7 = b^7$$

Therefore:

$$(a + b)^7 = a^7 + 7a^6b + 21a^5b^2 + 35a^4b^3 + 35a^3b^4 + 21a^2b^5 + 7ab^6 + b^7$$

EXAMPLE 52   Find $(x + 1)^4$.

Solution:

$$(x + 1)^4 = x^4 + 4x^3(1) + \frac{4(3)}{2} x^2(1)^2 + \frac{4(3)(2)}{2(3)} x^1(1)^3 + \frac{4(3)(2)(1)}{2(3)(4)} x^0(1)^4$$

$$= x^4 + 4x^3 + 6x^2 + 4x + 1$$

EXAMPLE 53   Find $(x - y)^5$.

Solution:

$$(x - y)^5 = x^5 - 5x^4(y) + \frac{5(4)}{2} x^3(y)^2 - \frac{5(4)(3)}{2(3)} x^2(y)^3$$

$$+ \frac{5(4)(3)(2)}{2(3)(4)} x(y)^4 - \frac{5(4)(3)(2)(1)}{(2)(3)(4)(5)} x^0(y)^5$$

$$= x^5 - 5x^4y + 10x^3y^2 - 10x^2y^3 + 5xy^4 - y^5$$

EXAMPLE 54   Find $a - (2b)^4$.

Solution:

$$(a - 2b)^4 = a^4 - 4a^3(2b)^1 + \frac{4(3)}{2} a^2(2b)^2 - \frac{4(3)(2)}{2(3)} a(2b)^3 + \frac{4(3)(2)(1)}{2(3)(4)} a^0(2b)^4$$

$$= a^4 - 4a^3(2b) + 6a^2(4b^2) - 4a(8b^3) + 16b^4$$

$$= a^4 - 8a^3b + 24a^2b^2 - 32ab^3 + 16b^4$$

PROBLEM SET 8-22

Expand the following.

1. $(x + y)^3$ 2. $(y - 1)^4$ 3. $(c + d)^5$

4. $(a + 2)^4$ 5. $(a - 2)^5$ 6. $(a - 1)^4$

7. $(2d + 1)^5$ 8. $(3x - y)^4$ 9. $(x + y)^7$

10. $(c - d)^8$ 11. $(m - n)^9$ 12. $(a + b)^{10}$

9

Differential Calculus

Many science-related problems cannot be mathematically solved without the use of calculus. This is especially true in the case of the technician.

Suppose you wish to bring electricity to your workshop by plugging an extension cord into a 120 volt outlet in your house. The cord has a 12 ampere rating and a resistance of 8 ohms. If you have ten 20 ohm electric heaters available and want to heat your workshop, how many heaters should you plug in at the end of the cord to supply the most heat? This is a problem that cannot be solved by algebraic methods. However, the solution is easily found through the use of calculus. As is the case in all areas of math, we first need to develop the necessary mathematical tools before we attempt to solve the problems.

The fundamental problem of differential calculus, which we are concerned within this chapter, is to determine the rate at which the value of a function $f(x)$ changes with relation to x. Suppose for example, that we increase the load on an electric motor. As the load increases the current required for motor operation also increases. At what rate does the current increase in proportion to the load?

Before attempting to answer questions which involve this fundamental problem of differential calculus we must first consider the idea of the limit of a function, for it is upon this idea that the answer is based.

We state that the limit of a function $f(x)$ as x approaches some constant c is the number L.

$$\lim_{x \to c} f(x) = L$$

This statement means that the value of $f(x)$ can be made to come as near to L as desired, if we take x near to, but not equal to, c. It is not related to the value of $f(x)$ when x equals c.

EXAMPLE 1 $\lim_{x \to 10} 3x = 30$.

Solution: In this expression $f(x) = 3x$, suppose we choose $x = 7$, then $f(7)$ is equal to 21. Now substitute $x = 8$ and $x = 9$. $f(8) = 24$, and $f(9) = 27$. As x approaches 10, $f(x)$ approaches 30.

EXAMPLE 2 $\lim_{x \to 5} (x^2 - 2) = 23$.

Solution: In this expression $f(x) = x^2 - 2$, suppose $x = 3$, then $f(3) = 7$. Now, if $x = 4$, then $f(4) = 14$. If $x = 4.5$, then $f(4.5) = 18.25$. As the value of x approaches 5, the value of $f(x)$ approaches 23.

In the examples given, the limit of the function is fairly obvious. In fact, in these examples the limit of the function can be computed by simply substituting the limiting value of x into the function and calculating the numerical value. This is true as long as the function is continuous at the limiting value of x.

EXAMPLE 3 Find $\lim_{x \to 3} (2x - 1)$.

Solution: Since $f(x) = 2x - 1$, we can substitute the value 3 for x and obtain $f(3) = 2(3) - 1 = 5$. Therefore,

$$\lim_{x \to 3} (2x - 1) = 5$$

All limits are not quite this easy to find. To demonstrate this fact we will try to calculate the limit of another algebraic function.

EXAMPLE 4 Find $\lim_{x \to 2} \dfrac{x^2 + x - 6}{(x - 2)}$.

Solution: In this example $f(x) = \dfrac{x^2 + x - 6}{x - 2}$, if we substitute $x = 2$ into the expression to calculate the limit we obtain $0/0$ for $f(2)$. $0/0$ is undefined, therefore $f(x)$ is undefined and not continuous at $x = 2$. In this case, substituting the limiting value for x into the function does not work. To find the limit, factor $x^2 + x - 6$ into $(x + 3)(x - 2)$. Now $f(x)$ has the form $\dfrac{(x + 3)(x - 2)}{(x - 2)}$. As long as $x \neq 2$ we can write $f(x) = \dfrac{(x + 3)(x - 2)}{(x - 2)}$ or $f(x) = x + 3$. Now as x approaches 2, $f(x)$ approaches 5, and 5 is the limit.

In the above example, the original function was undefined for the limiting value of x. However, we found a new function that was defined for our value of x, and we were able to calculate the limit as in earlier examples.

EXAMPLE 5 Find $\lim\limits_{x \to 5} \dfrac{x^2 - 3x - 10}{x - 5}$.

Solution:
$$\lim_{x \to 5} \frac{x^2 - 3x - 10}{x - 5} = \lim_{x \to 5} \frac{(x + 2)(x - 5)}{(x - 5)}$$

$$= \lim_{x \to 5} (x + 2) \frac{(x - 5)}{(x - 5)}$$

Note that the second factor is 1 for all values of x except $x = 5$. For x equal to any value except 5 we can write:

$$\lim_{x \to 5} \frac{x^2 - 3x - 10}{x - 5} = \lim_{x \to 5} (x + 2) = 7$$

There are a number of theorems on limits that can sometimes be useful in the study of higher mathematics. We state the theorems without proof since we have omitted material necessary for the formal proof.

If $\lim_{x \to a} f(x) = L$ and $\lim_{x \to a} g(x) = M$, then

(1) $\lim\limits_{x \to a} f(x) + g(x) = L + M$

(2) $\lim\limits_{x \to a} f(x) \cdot g(x) = L \cdot M$

(3) $\lim\limits_{x \to a} \dfrac{f(x)}{g(x)} = \dfrac{L}{M}$ if $M \neq 0$

These theorems state that if the limits of $f(x)$ and $g(x)$ exist, then (1) the limit of the sum of two functions is equal to the sum of the limits, (2) the limit of the product of two functions is equal to the product of the limits, and (3) the limit of the quotient of two functions is equal to the quotient of the limits as long as the denominator is not zero.

EXAMPLE 6 Find the $\lim\limits_{x \to 3} [f(x) + g(x)]$

if $\lim\limits_{x \to 3} f(x) = 8$ and   if $\lim\limits_{x \to 3} g(x) = 5$

Solution: $\lim\limits_{x \to 3} [f(x) + g(x)] = 8 + 5 = 13$

EXAMPLE 7   Find $\lim_{x \to 2} [f(x) \cdot g(x)]$ if $f(x) = x^2 - 1$ and $g(x) = x + 2$.

Solution:
$$\lim_{x \to 2} f(x) = \lim_{x \to 2} (x^2 - 1) = 3$$

$$\lim_{x \to 2} g(x) = \lim_{x \to 2} (x + 2) = 4$$

$$\lim_{x \to 2} [f(x) \cdot g(x)] = 3 \cdot 4 = 12$$

Find the limits of the following functions.

1. $\lim_{x \to 2} (x^2 - x + 4)$

2. $\lim_{x \to 3} \dfrac{3x - 4}{x^2 - 3}$

3. $\lim_{x \to 2} \dfrac{x^3 - 8}{x - 2}$

4. $\lim_{x \to 4} \dfrac{x^2 - 2x - 8}{x^2 + 2x - 24}$

5. $\lim_{x \to 0} \dfrac{x^2 - 6x}{2x}$

6. $\lim_{x \to 2} \dfrac{x^2 - x}{x - 1}$

7. $\lim_{x \to -1} \dfrac{x^2 - 1}{x - 1}$

8. $\lim_{x \to 4} \dfrac{x}{x^2 + 2}$

9. $\lim_{x \to 2} \left[\dfrac{3x}{2} + \dfrac{x - 1}{2} \right]$

10. $\lim_{x \to 3} \left[\dfrac{x^2 - 9}{x - 3} \cdot (x + 2) \right]$

11. $\lim_{x \to 4} \left[\dfrac{x^2 - 4x}{x - 4} \div x - 2 \right]$

9-3
Definition of the Derivative

We shall now use the idea of the limit to define the derivative of a function.
Consider the function $y = f(x)$ whose graph is shown in Figure 9-1.

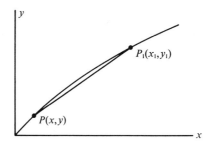

Figure 9-1

Let us select a point $P(x,y)$ of this interval and determine at what rate y is increasing with respect to x or, in other words, find the value of the change in y divided by the change in x at this point.

To answer this question we will select another point $P_1(x_1,y_1)$ such that $x_1 - x = \Delta x$ which is the change in x. The change in y that corresponds to this change in x is $y_1 - y = \Delta y$. Next we divide the change in y by the change in x. This quotient, $\Delta y/\Delta x$ is the average rate of change of y with respect to x for the interval between P and P_1. It is also the slope of the straight line between the two points.

Now think of choosing P_1, closer and closer to P. When this is done the value of Δx becomes smaller and smaller. Now $\Delta y/\Delta x$ is the average rate of change of y with respect to x over a smaller interval. It then appears reasonable that we should define the instantaneous rate at P as the limit of the quotient $\Delta y/\Delta x$ as $\Delta x \to 0$.

This limit, if it exists, is called the derivative of y with respect to x at P. It is usually denoted by one of the symbols, $D_x y$ dy/dx, y', or $f'(x)$.

The value of y at any x can be expressed as $f(x)$ since $y = f(x)$. The value of y at x_1 is expressed as $f(x + \Delta x)$ since $x_1 = x + \Delta x$. From these observations we can write the following expression.

$$\frac{\Delta y}{\Delta x} = \frac{y_1 - y}{x_1 - x} = \frac{f(x + \Delta x) - f(x)}{\Delta x}$$

Now, if we take the limit of this expression as $\Delta x \to 0$ we have the expression for the derivative of the function $f(x)$.

$$\frac{dy}{dx} = \lim_{\Delta x \to 0} \frac{f(x + \Delta x) - f(x)}{\Delta x}$$

Consider the function $f(x) = x^2$ in Figure 9-2.

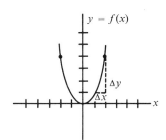

Figure 9-2

We will now find $\displaystyle\lim_{\Delta x \to 0} \frac{f(x + \Delta x) - f(x)}{\Delta x}$ for $f(x) = x^2$.

$$\frac{dy}{dx} = \lim_{\Delta x \to 0} \frac{f(x + \Delta x) - f(x)}{\Delta x} = \lim_{\Delta x \to 0} \frac{x^2 + 2x\,\Delta x + \Delta x^2 - x^2}{\Delta x}$$

$$= \lim_{\Delta x \to 0} 2x + \Delta x$$

$$= 2x$$

This says that the slope, or the derivative, of the function at any point x is equal to $2x$. For example, at $x = 2$ the slope of the function is 4. Figure 9-2 shows the tangent to the function at $x = 2$. The slope of the tangent to the curve is 4 at this point. At $x = 0$ the slope is 0, and at $x = -2$ the slope is -4.

The following examples show you how to find the derivatives of functions using the above equation.

EXAMPLE 8   Find the change in y with respect to x when $f(x) = 2x + 3$.

Solution: We need to find $\lim\limits_{\Delta x \to 0} \dfrac{f(x + \Delta x) - f(x)}{\Delta x}$ when $f(x) = 2x + 3$.

$$\lim_{\Delta x \to 0} \frac{f(x + \Delta x) - f(x)}{\Delta x} = \lim_{\Delta x \to 0} \frac{2(x + \Delta x) + 3 - (2x + 3)}{\Delta x}$$

$$= \lim_{\Delta x \to 0} \frac{2x + 2\Delta x + 3 - 2x - 3}{\Delta x}$$

$$= \lim_{\Delta x \to 0} \frac{2\Delta x}{\Delta x}$$

$$= \lim_{\Delta x \to 0} 2 = 2$$

Thus the change in y with respect to x is 2.

EXAMPLE 9   Find the derivative of y with respect to x when $y = x^2 - 1$.

Solution:

$$\lim_{\Delta x \to 0} \frac{f(x + \Delta x) - f(x)}{\Delta x} = \lim_{\Delta x \to 0} \frac{\{(x + \Delta x)^2 - 1\} - (x^2 - 1)}{\Delta x}$$

$$= \lim_{\Delta x \to 0} \frac{x^2 + 2x\,\Delta x + (\Delta x)^2 - 1 - x^2 + 1}{\Delta x}$$

$$= \lim_{\Delta x \to 0} \frac{2x\,\Delta x + \Delta x^2}{\Delta x}$$

$$= \lim_{\Delta x \to 0} 2x + \Delta x$$

$$= 2x$$

In the first example the derivative was a constant. This means that the value does not depend on x. In the second example the derivative was equal to $2x$. In this case the value of the derivative depends on x. For example, when $x = 3$ the value of the derivative is 2(3) or 6.

EXAMPLE 10 Find the derivative of y with respect to x when $y = \dfrac{x + 2}{x}$.

Solution: $\displaystyle\lim_{\Delta x \to 0} \frac{f(x + \Delta x) - f(x)}{\Delta x}$

$$= \lim_{\Delta x \to 0} \left\{ \frac{\dfrac{x + \Delta x + 2}{x + \Delta x} - \dfrac{x + 2}{x}}{\Delta x} \right\}$$

$$= \lim_{\Delta x \to 0} \left\{ \frac{\dfrac{(x^2 + x\,\Delta x + 2x) - (x^2 + 2x + x\,\Delta x + 2\,\Delta x)}{x(x + \Delta x)}}{\Delta x} \right\}$$

$$= \lim_{\Delta x \to 0} \frac{\dfrac{-2\,\Delta x}{x(x + \Delta x)}}{\Delta x}$$

$$= \lim_{\Delta x \to 0} \frac{-2}{x(x + \Delta x)}$$

$$= \frac{-2}{x^2}$$

One of the most important points to remember is that the derivative of a function is the slope of the function at the point in question. If we find the derivative of a function we have found the slope.

PROBLEM SET 9-2

In Exercises 1–10 find $D_x y$.

1. $y = 3x - 1$

2. $y = 2 - 4x$

3. $y = 3x - x^2$

4. $y = 2x^2 - x + 1$

5. $y = \dfrac{2}{x}$

6. $y = \dfrac{x - 2}{x}$

7. $y = \dfrac{x}{x - 2}$

8. $y = \dfrac{1}{x^2}$

9. $y = x^3$

10. $y = \dfrac{x^2 - 2}{x}$

11. The velocity of a particle is defined as dx/dt where x is the displacement and t is time. Find the velocity of an automobile at the end of $\frac{1}{4}$ mile, if the time required to travel the distance is 10 seconds and the distance is given by the expression $x = 10t^2 + 32t$, where x is in feet and t is in seconds. Find the answer in miles per hour.

Some Derivative Formulas

If it were necessary to use the definition of the derivative in all computations, problem solving would be time consuming. The time required for problem solving is reduced by use of formulas derived from the definition of the derivative. As we introduce the formulas we will derive a few of them so that you may see the procedure involved.

Suppose we want to find the derivative of a constant function such as $y = 3$ or, in general, $y = c$. Using the definition of the derivative we have

$$\frac{dy}{dx} = \lim_{\Delta x \to 0} \frac{f(x + \Delta x) - f(x)}{\Delta x}$$

$$= \lim_{\Delta x \to 0} \frac{3 - 3}{\Delta x}$$

$$= \lim_{\Delta x \to 0} \frac{0}{\Delta x}$$

$$= \lim_{\Delta x \to 0} 0$$

$$= 0$$

Use of the definition of the derivative proves that the derivative of any constant is 0.

Now we will list five derivative formulas which should be memorized.

I. $\dfrac{dc}{dx} = 0$

The derivative of a constant is zero.

II. $\dfrac{dx}{dx} = 1$

The derivative of a variable with respect to itself is 1.

III. $\dfrac{d(u + v)}{dx} = \dfrac{d(u)}{dx} + \dfrac{d(v)}{dx}$

The derivative of the sum of two functions is equal to the sum of their separate derivatives.

IV. $\dfrac{d(c \cdot v)}{dx} = c \dfrac{dv}{dx}$

The derivative of the product of a constant and a function is the constant times the derivative of the function.

V. $\dfrac{d(v^n)}{dx} = nv^{n-1} \dfrac{dv}{dx}$

The derivative of the *n*th power of a function is *n* times the $(n - 1)$th power of the function times the derivative of the function.

Now we shall consider several examples involving the use of the above formulas.

EXAMPLE 11   Find the derivative of $y = x^3 + x^2 + 1$ with respect to *x*.

Solution:
$$\frac{dy}{dx} = \frac{d}{dx}(x^3 + x^2 + 1)$$

Rule III allows us to rewrite this expression as follows.
$$\frac{dy}{dx} = \frac{dx^3}{dx} + \frac{dx^2}{dx} + \frac{d(1)}{dx}$$

Now by Rules V and I we can write
$$\frac{d}{dx} = 3x^2 + 2x + 0 = 3x^2 + 2x$$

This is the derivative of $x^3 + x^2 + 1$.

EXAMPLE 12   Find the derivative of $f(x)$ when $f(x) = 3x^4$.

Solution: First,
$$f'(x) = \frac{d(3x^4)}{dx}$$

By Rule IV we can write
$$f'(x) = 3\frac{dx^4}{dx}$$

and then using Rule V we have
$$f'(x) = 3 \cdot (4x^3) = 12x^3$$

EXAMPLE 13   Find the derivative of $y = (3x - 1)^4$.

Solution: Note that $(3x - 1)$ is a function of *x* raised to a power. Rule V allows us to write
$$\frac{dy}{dx} = \frac{d(3x - 1)^4}{dx} = 4(3x - 1)^3 \cdot \frac{d(3x - 1)}{dx}$$

$$\frac{dy}{dx} = 4(3x - 1)^3 \left(\frac{d(3x)}{dx} - \frac{d1}{dx}\right)$$

$$\frac{dy}{dx} = 4(3x - 1)^3 \left(3\frac{dx}{dx}\right) = 12(3x - 1)^3$$

EXAMPLE 14   Find the derivative of $f(x) = (x^3 + 2x^2)^3 - 3x^2 - 2$.

Solution: First we write

$$f'(x) = \frac{d(x^3 + 2x^2)^3}{dx} - \frac{d(3x^2)}{dx} - \frac{d2}{dx}$$

Rule V allows us to write

$$f'(x) = 3(x^3 + 2x^2)^2 \frac{d(x^3 + 2x^2)}{dx} - \frac{d(3x^2)}{dx} - \frac{d2}{dx}$$

Next, by using Rule III we can write

$$f'(x) = 3(x^3 + 2x^2)^2 \left(\frac{dx^3}{dx} + \frac{d(2x^2)}{dx}\right) - \frac{d(3x^2)}{dx} - \frac{d2}{dx}$$

Now, by Rules I, IV, and V we have

$$f'(x) = 3(x^3 + 2x^2)^2 \left(3x^2 + 2\frac{dx^2}{dx}\right) - 3\frac{dx^2}{dx} - 0$$

$$f'(x) = 3(x^3 + 2x^2)^2(3x^2 + 2(2x)) - 3(2x)$$

$$f'(x) = 3(x^3 + 2x^2)^2(3x^2 + 4x) - 6x$$

Multiplication could be completed and like terms combined, but in this text we will leave the derivative in the above form except when it is necessary to combine terms to find the complete solution to the problem.

EXAMPLE 15   Find the derivative of $f(x) = x^{1/3}$.

Solution: This problem is different from the previous examples because the exponent is a fraction instead of an integer. However, the same rules for differentiation still apply. Rule V allows us to write

$$f'(x) = \frac{dx^{1/3}}{dx} = \frac{1}{3}x^{-2/3}$$

We still simply multiply the function by the exponent and then reduce the exponent by one.

EXAMPLE 16   Find the derivative of $f(x) = x^{-3}$.

Solution: This problem is worked exactly like the previous examples.

$$f'(x) = \frac{d(x^{-3})}{dx} = -3x^{-4}$$

Again, we multiplied by the exponent and then reduced the exponent by one.

Find the derivatives of each of the following expressions.

1. $y = 3$

2. $f(x) = x$

3. $f(x) = 2x$

4. $y = x^3$

5. $y = 2x^4$

6. $y = x^3 + 2x^2 + x$

7. $f(x) = (2x - 1)^2$

8. $f(x) = x^4 + 2x^3 - 1$

9. $f(x) = (3x^2 - 2x + 1)^5$

10. $y = (3x^4 - x^3 + 2x^2)^3 - 3x^2 + 2$

11. $y = x^{-2}$

12. $f(x) = x + 2x^{-1}$

13. $f(x) = x^{1/3}$

14. $f(x) = 1/x^3$

15. $f(x) = x^2 + x^{-1/2}$

16. $f(x) = \dfrac{1}{3x}$

17. A particle moves along the x-axis so that its distance from the origin at the end of t seconds is $x = 3t^2 + t + 1$. Find the position and velocity of the particle at the end of 8 seconds. Remember that the velocity is equal to the derivative of the distance with respect to time.

9-5
Time Rate Problems

We have indicated that the derivative of a function $f(x)$ with respect to x is the rate of change of $f(x)$ with respect to x. Now suppose we have a variable quantity M which varies with time according to some law

$$M = f(t)$$

The value of dM/dt at any instant is the rate of change of M with respect to time.

EXAMPLE 17 Suppose oil is being pumped into a tank such that the number of gallons in the tank at the end of t minutes is given by the equation

$$G = t^2 - t + 1$$

The rate of change of the oil in the tank at any time t is

$$\frac{dG}{dt} = 2t - 1 \text{ gal per minute}$$

This is the rate at which the oil is entering the tank. At the end of 15 minutes the oil would be entering the tank at a rate of

$$\frac{dG}{dt} = 2(15) - 1 = 29 \text{ gal per minute}$$

Consider a particle P moving along a straight line AB in such a way that its distance from A is a function of t.

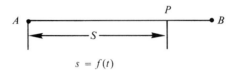

$$s = f(t)$$

Figure 9-3

In a small additional time Δt the particle moves an additional distance Δs. The average velocity over this distance is expressed as

$$v = \frac{\Delta s}{\Delta t}$$

The instantaneous velocity of the particle is defined as the limit of the above quotient as Δt approaches zero.

$$v = \lim_{\Delta t \to 0} \frac{\Delta s}{\Delta t}$$

Since Δs is a function of t

$$v = \lim_{\Delta t \to 0} \frac{\Delta s}{\Delta t} = \frac{ds}{dt}$$

A special case of motion is that of the projectile that is thrown vertically upward with an initial velocity v_0; the distance from the point where it was thrown at the end of t seconds is known to be

$$s = v_0 t - \tfrac{1}{2}gt^2$$

where g is acceleration due to gravity at t is time. Its velocity at any time t in feet per second is

$$\frac{ds}{dt} = \frac{d}{dt}(v_0 t) - \frac{d}{dt}(\tfrac{1}{2}gt^2)$$

Both v_0 and g, which is the acceleration of gravity, are constants.

$$\frac{ds}{dt} = v_0 \frac{dt}{dt} - \tfrac{1}{2}g \frac{dt^2}{dt}$$

$$\frac{ds}{dt} = v_0 - gt$$

EXAMPLE 18   The distance traveled by a missile is given by the expression

$$s = 10t^3 + 3t - 1$$

Find the velocity after 4 seconds.

Solution:
$$v = \frac{ds}{dt}$$

$$\frac{ds}{dt} = 30t^2 + 3$$

When $t = 4$

$$\frac{ds}{dt} = 30(4)^2 + 3$$

$$\frac{ds}{dt} = 480 + 3 = 483 \text{ ft/sec}$$

EXAMPLE 19   What is the velocity of a ball thrown vertically at a velocity of 128 ft/sec after 5 seconds?

Solution:
$$v = \frac{ds}{dt} = v_0 - gt$$

$$v_0 = 128 \text{ ft/sec}$$
$$g = 32 \text{ ft/sec}^2 \text{ (acceleration of gravity)}$$
$$v = 128 \text{ ft/sec} - 32 \text{ ft/sec}^2 \cdot 5 \text{ sec}$$
$$v = 128 \text{ ft/sec} - 160 \text{ ft/sec}$$
$$v = -32 \text{ ft/sec}$$

The negative value for velocity indicates that the ball has reached its highest point and is now falling.

EXAMPLE 20   An electron in an electric field moves according to the distance equation $s = eEt^2/2M_e$. If $e = 1.602 \times 10^{-19}$ coulombs, $M_e = 9.109 \times 10^{-31}$ kilograms, $E = 20$ volts per meter, and t is in seconds, find the velocity of the electron when $t = .2$ microsecond.

Solution:
$$v = \frac{ds}{dt} = \frac{d(eEt^2/2M_e)}{dt}$$

$$v = (eE/2M_e) \frac{dt^2}{dt}$$

$$v = 2(eE/2M_e)t = eEt/M_e$$
$$v = (1.062 \times 10^{-19})(20)(.2 \times 10^{-6})/9.109 \times 10^{-31}$$
$$v = 4.662 \times 10^6 \text{ meter/sec}$$

1. A small lead ball is thrown vertically upward from a point A on the ground with initial velocity 120 ft per second. Find:
   (a) Its velocity at the end of 2 sec; 4 sec.
   (b) Its greatest distance from the ground.
   (c) The total time in the air.

2. The image on the face of a television tube is produced by an electron beam which strikes the face of the tube after being accelerated through an electric field. Determine the velocity of such an electron, if the electric field intensity is 10,000 volts per meter, after .5 microseconds.

3. The distance traveled by an airplane along the runway from the time of touchdown is given by $s = 100t - 2t^2$ ft, t equals the seconds after touchdown. Find:
   (a) Its velocity at touchdown.
   (b) Its acceleration.
   (c) How far it travels before coming to a stop.

4. Find the angular velocity at time $t = 3$ sec of a rotating machine for which the angular displacement at time t seconds is given by $\theta = t^4 - 2t^2$ radians.

5. A small heavy object is dropped from the top of a cliff 600 ft high. With what velocity will it strike the ground below?

6. The work done in moving a body is expressed as $W = 2t - 3/t$ ft-lb over an interval of time t seconds. Find the power of the force creating the motion at $t = 2$ sec. (Hint: power is the rate of change of work with respect to time.)

7. A certain high jumper can jump 6 ft. Regarding him as a "particle" find:
   (a) His vertical speed as he leaves the ground.
   (b) How long he is in the air.

8. The voltage V across a resistance r, carrying a fixed current I, is $V = Ir$. Because of heating, r varies with time according to the formula $r = 2t^3 + 1$. Find a formula for dV/dt. What is the rate of change of the voltage when $t = 2$ seconds? V is in volts, r is in ohms and $I = 4$ amperes.

9. The energy stored in the magnetic field of an inductor is $w = Li^2$ joules. If $L = 6$ henrys and $i = 2 - 3t^2$, what is the rate of change of energy when $t = 6$ seconds.

10. A transmission line hangs in an approximate parabolic shape whose equation is $y = 2 \times 10^{-3}x^2$ where x and y are in feet. x is the distance from the lowest point on the line to the poles. If the poles are 100 ft apart, what is the slope of the line at the point of support. Express your answer in degrees.

Now we will introduce more derivative formulas that should also be memorized.

VI. $\dfrac{d}{dx}(U \cdot V) = U\dfrac{dV}{dx} + V\dfrac{dU}{dx}$

The derivative of the product of two functions is the first function times the derivative of the second plus the second function times the derivative of the first. U and V are both functions of x.

At this point, so that we do not completely ignore the theory involved, we will prove the above expression. Consider the function.

$$f(x) = U \cdot V$$

Where U and V are differentiable functions of x. That is

$$\lim_{\Delta x \to 0} \frac{U(x + \Delta x) - U(x)}{\Delta x} = \frac{dU}{dx}$$

and

$$\lim_{\Delta x \to 0} \frac{V(x + \Delta x) - V(x)}{\Delta x} = \frac{dV}{dx}$$

Now recall that

$$f'(x) = \lim_{\Delta x \to 0} \frac{f(x + \Delta x) - f(x)}{\Delta x}$$

but $f(x) = U(x) \cdot V(x)$ therefore

$$f'(x) = \lim_{\Delta x \to 0} \frac{U(x + \Delta x)V(x + \Delta x) - U(x)V(x)}{\Delta x}$$

Now we will add and subtract $\dfrac{U(x)V(x + \Delta x)}{\Delta x}$.

$f'(x)$

$$= \lim_{\Delta x \to 0} \frac{U(x + \Delta x)V(x + \Delta x) - U(x)V(x) + U(x)V(x + \Delta x) - U(x)V(x + \Delta x)}{\Delta x}$$

$$= \lim_{\Delta x \to 0} \frac{V(x + \Delta x)[U(x + \Delta x) - U(x)] + U(x)[V(x + \Delta x) - V(x)]}{\Delta x}$$

$$= \lim_{\Delta x \to 0} \frac{V(x + \Delta x)[U(x + \Delta x) - U(x)]}{\Delta x} + \lim_{\Delta x \to 0} \frac{U(x)[V(x + \Delta x) - V(x)]}{\Delta x}$$

$$= \lim_{\Delta x \to 0} V(x + \Delta x) \cdot \lim_{\Delta x \to 0} \frac{U(x + \Delta x) - U(x)}{\Delta x} + U(x) \lim_{\Delta x \to 0} \frac{V(x + \Delta x) - V(x)}{\Delta x}$$

$$= V\frac{dU}{dx} + U\frac{dV}{dx}$$

This is the proof of the derivative of the product of two functions. The rules for limits are used extensively in the proof as is the case for many proofs in calculus.

The remaining derivative formulas in the section will be stated without proof.

$$\text{VII.} \quad \frac{d}{dx}\left(\frac{u}{v}\right) = \frac{v\dfrac{du}{dx} - u\dfrac{dv}{dx}}{v^2}$$

The derivative of the quotient of two functions is the denominator times the derivative of the numerator, minus the numerator times the derivative of the denominator, all divided by the square of the denominator.

$$\text{VIII.} \quad \frac{dy}{dx} = \frac{dy}{dv} \cdot \frac{dv}{dx}$$

If y is a function of v, and v is in turn a function of x, then the derivative of y with respect to x equals the product of the derivative of y with respect to v and the derivative of v with respect to x. Rule VIII is usually called the *chain rule*.

EXAMPLE 21 Find the derivative of

$$f(x) = (x^2 + 2)(3x + 1) \qquad \text{when} \quad x = 2$$

Solution: In this problem let $u = x^2 + 2$ and $v = 3x + 1$ and use the product rule to find the derivative.

$$\frac{du}{dx} = 2x \qquad \frac{dv}{dx} = 3$$

$$f'(x) = v\frac{du}{dx} + u\frac{dv}{dx}$$

$$f'(x) = (3x + 1) \cdot 2x + (x^2 + 2) \cdot 3$$

$$f'(x) = 2x(3x + 1) + 3(x^2 + 2)$$

Now when $x = 2$

$$f'(2) = (2 \cdot 2)(3 \cdot 2 + 1) + 3(2^2 + 2)$$

$$f'(2) = 46$$

EXAMPLE 22 Find the derivative of

$$f(x) = \frac{(x^3 + 1)^2}{x^2 + 2} \qquad \text{when} \quad x = 1$$

Solution: Let $u = (x^3 + 1)^2$ and $v = x^2 + 2$. Use the quotient rule.

$$\frac{du}{dx} = 2(x^3 + 1)(3x^2) = 6x^2(x^3 + 1)$$

$$\frac{dv}{dx} = 2x$$

$$f'(x) = \frac{v\dfrac{du}{dx} - u\dfrac{dv}{dx}}{v^2}$$

$$f'(x) = \frac{(x^2 + 2)(6x^2)(x^3 + 1) - (x^3 + 1)^2(2x)}{(x^2 + 2)^2}$$

$$f'(1) = \frac{(1^2 + 2)(6 \cdot 1^2)(1^3 + 1) - (1^3 + 1)^2(2 \cdot 1)}{(1^2 + 2)^2}$$

$$f'(1) = \frac{28}{9}$$

EXAMPLE 23 Find $\dfrac{dy}{dx}$ if $y = 3v^2$ and $v = x^2 - 1$.

Solution: We will use Rule VIII given in this section which is sometimes called the chain rule.

$$\frac{dy}{dv} = 6v$$

$$\frac{dv}{dx} = 2x$$

$$\frac{dy}{dx} = \frac{dy}{dv} \cdot \frac{dv}{dx}$$

$$\frac{dy}{dx} = 6v \cdot 2x = 12vx$$

Since $v = x^2 - 1$ we can write

$$\frac{dy}{dx} = 12x(x^2 - 1)$$

PROBLEM SET 9-5

Find the derivatives of the following expressions and evaluate with the given value of x.

1. $y = (x + 2)(x^2)$, $x = 2$

2. $y = (x^3 + 1)^2(x - 2)$, $x = 1$

3. $f(x) = (x^2 - 1)(x + 1)^2$, $x = 3$

4. $f(x) = (x + 1)(x - 3)^{-2}$, $x = -1$

5. $y = \dfrac{\sqrt{x + 1}}{x - 2}$, $x = -3$

6. $y = \dfrac{x^2}{x - 3}$, $x = 2$

7. $y = \dfrac{x^{1/3}}{\sqrt{x - 1}}$, $x = 3$

8. $f(x) = \dfrac{(x^2 - x + 1)^2}{x^2 + 1}$, $x = 1$

Find $\dfrac{dy}{dx}$ for the following.

9. $y = u^2$
   $u = x + 1$

10. $y = u$
    $u = x^2$

11. $y = z - 2$
    $z = x^2 + 1$

12. $y = (z + 1)^2$
    $z = x$

13. $y = w + x$
    $w = x^2 - 4$

14. $y = v^2 + v$
    $v = x + 1$

15. The side x of a square is increasing at 2 in. per second. At what rate is the area A increasing when each side is 8 in. long? (Note: solve for dA/dt given that $dx/dt = 2$ and $A = x^2$.)

16. A circular plate contracts as it cools. Find the rate of decrease in the area when the radius is 4 ft, if the radius is decreasing at a rate of .1 ft per hour.

17. The side x of an equilateral triangle is increasing at a rate of .5 in. per second. At what rate is the area increasing when each side is 10 in. long.

9-8
Related Rate Problems

A frequently encountered type of rate problem is one in which two variable quantities are related by a function $y = f(x)$. The time rate of change of x is known and the corresponding time rate of change of y is to be found. The chain rule is used to solve this type of problem.

$$\frac{dy}{dt} = \frac{dy}{dx} \cdot \frac{dx}{dt}$$

EXAMPLE 24 The current flowing through a system is changing at the rate of 2 amperes per second. What is the corresponding change in voltage if the resistance is 6 ohms?

The equation relating current and voltage is

$$E = IR$$

where E is the voltage, I is the current, and R is the resistance.

First we determine the rate that is given. In this example it is the rate of change of current.

$$\frac{dI}{dt} = 2 \text{ amperes/sec}$$

Next we determine the rate that is to be calculated. For this problem it is the rate of change of voltage with respect to time. By the chain rule

$$\frac{dE}{dt} = \frac{dE}{dI} \cdot \frac{dI}{dt}$$

Since

$$E = IR$$

and R is a constant

$$\frac{dE}{dI} = R\frac{dI}{dI} = R$$

We can now substitute into the chain rule and find dE/dt.

$$\frac{dE}{dt} = R\frac{dI}{dt} = (6 \text{ ohms}) (2 \text{ amp/sec})$$

$$\frac{dE}{dt} = 12 \text{ volts/sec}$$

EXAMPLE 25 A ladder 25 ft long rests against a wall and the lower end is being pulled away from the wall at 6 ft per minute. At what rate is the top descending at the instant when the lower end is 7 ft from the wall?

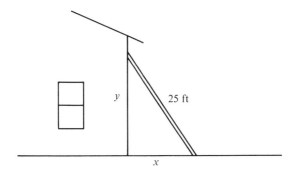

Figure 9-4

Solution: First we note that $y = \sqrt{25^2 - x^2}$ and that $dx/dt = 6$ ft/min. Using the chain rule we have

$$\frac{dy}{dt} = \frac{dy}{dx} \cdot \frac{dx}{dt}$$

where dy/dt is the rate at which the top is descending.

$$\frac{dy}{dx} = \frac{d}{dx}(25^2 - x^2)^{1/2}$$

$$\frac{dy}{dx} = \tfrac{1}{2}(25^2 - x^2)^{-1/2}(-2x)$$

$$\frac{dy}{dx} = -\frac{x}{(25^2 - x^2)^{1/2}}$$

By the chain rule

$$\frac{dy}{dt} = \left[-\frac{x}{(25^2 - x^2)^{1/2}}\right]\frac{dx}{dt}$$

$$\frac{dy}{dt} = -\frac{7}{(625 - 49)^{1/2}} \cdot 6$$

$$\frac{dy}{dt} = -\frac{7}{24} \cdot 6 = -\frac{7}{4} \text{ ft per min}$$

The negative value indicates that the top of the ladder is moving down.

EXAMPLE 26 A cone is 8 in. in diameter and 8 in. deep. Water is poured into it at a rate of 2 cu in. per sec. At what rate is the water level rising at the instant when the depth is 4 in.?

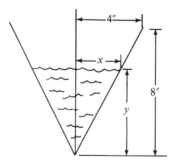

Figure 9-5

Solution: We must first express the volume V of water in terms of y, its depth.

$$V = \tfrac{1}{3}\pi x^2 y$$

Note that $\dfrac{x}{y}$ is proportional to $\dfrac{4}{8}$ since any section of the cone yields a right triangle formed by the center line and the surface of the cone.

$$\frac{x}{y} = \frac{4}{8}$$

$$x = \frac{1}{2}y$$

Substituting into the expression for volume we have:

$$V = \frac{1}{3}\pi \left(\frac{y}{2}\right)^2 y$$

$$V = \pi \frac{y^3}{12}$$

Next we will find dV/dy.

$$\frac{dV}{dy} = \frac{\pi y^2}{4}$$

Since the rate of change of the volume is given

$$\frac{dV}{dt} = 2\,\frac{\text{in.}^3}{\text{sec}}$$

we can apply the chain rule.

$$\frac{dV}{dt} = \frac{dV}{dy}\frac{dy}{dt}$$

$$2 = \frac{\pi y^2}{4} \cdot \frac{dy}{dt}$$

$$\frac{dy}{dt} = \frac{8}{\pi y^2}$$

When $y = 4$ in. we have

$$\frac{dy}{dt} = \frac{8}{\pi(4)^2}$$

$$\frac{dy}{dt} = \frac{1}{2\pi}\ \text{in. per sec}$$

This is the rate at which the water level is rising.

EXAMPLE 27 The power, P, in a circuit varies according to $P = Ri^2$ watts. If $R = 10$ ohms and i varies with time according to $i = 3t^2 + t$ amperes, where t is in seconds, find the rate of change of the power with respect to time when $t = 4$ seconds.

Solution: Using the chain rule we can write the expression

$$\frac{dP}{dt} = \frac{dP}{di} \cdot \frac{di}{dt}$$

$$\frac{dP}{di} = \frac{d}{di}(Ri^2) = 2Ri$$

$$\frac{di}{dt} = \frac{d}{dt}(3t^2 + t) = 6t + 1$$

$$\frac{dP}{dt} = \frac{dP}{di} \cdot \frac{di}{dt} = 2Ri(6t + 1)$$

When $t = 4$, $i = 3(4)^2 + 4 = 40$ amperes.

$$\frac{dP}{dt} = [2(10)(52)][6(4) + 1] = 26,000 \text{ watt/sec}$$

PROBLEM SET 9-6

1. A boat B is 12 miles west of another boat A. B starts east at 8 m.p.h. and at the same time A starts north at 12 m.p.h. At what rate is the distance between them changing at the end of $\frac{1}{2}$ hr?

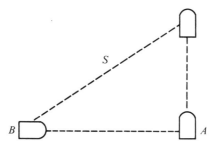

Figure 9-6

(Hint: Let x be the distance that B travels and y be the distance that A travels. This is given $y = \frac{3}{2}x$. The expression for s in terms of x is $s = \sqrt{(12 - x^2) + (\frac{3}{2}x)^2}$. Now we can find $\frac{ds}{dt}$ and use $\frac{ds}{dt} = \frac{dx}{dt} \cdot \frac{ds}{dx}$ to find the required result, since $\frac{dx}{dt}$ is given.)

2. The side of a square is increasing at 2 in. per minute. At what rate is the area increasing when each side is 8 in. long?

3. If the radius of a sphere is 10 in. and is decreasing at the rate of 1 in. per minute, at what rate is the volume decreasing?

4. A baseball diamond is a square 90 ft on each side. A player is running from second to third at a rate of 24 ft per second. At what rate is his distance from first base changing when he reaches a point 30 ft from third?

5. Sand is being poured from a hopper onto a conical pile. The angle of repose of the sand is such that the height of the pile is always equal to one-fourth of its radius. When the pile is 6 ft high, what rate of discharge from the hopper would cause the height of the pile to increase at $\frac{1}{16}$ in. per minute?

6. A boy is flying a kite at a height of 90 ft. If the kite is directly above the boy at $t = 0$ and moves horizontally away from him at a rate of 3 ft per second, how fast is the string being pulled out when the kite is 150 ft from the boy?

7. A spherical snowball is melting at the rate of 10 in.^{3}/min but retains its spherical shape. When the radius is 6 in., how fast is the radius changing?

8. A water trough is 8 ft long and 4 ft deep. Its cross section is a trapezoid 2 ft wide at the bottom and 4 ft wide at the top. At what rate must water be poured in to cause the surface to rise at 5 in. per minute when the depth is 1 ft?

9. The force between two charged particles having fixed charges Q_1 and Q_2 varies with the distance separating them. $F = Q_1Q_2/\pi e s^2$ where F is force, e is a constant, and s is the distance between the particles. If s varies with time according to $s = 3t^{3/2}$, find a formula for dF/dt.

10. The energy stored in the magnetic field of an inductor is $w = Li^2$ joules. If $L = 6$ henrys, and if $i = 2 - 6t^2$, find the rate of change of energy with respect to time when $t = 6$ seconds.

9-9

The Derivatives of the Trigonometric Functions

In this section we will consider only three of the trigonometric functions. We will derive the formula for the derivative of the sine function but will present the derivatives of the cosine and tangent functions without proof. We will again concentrate on using the formulas and see how some of the algebraic formulas can be used when the trigonometric functions are involved.

To find the derivative of the function $y = \sin x$ we must apply the fundamental process of differentiation. Starting at any point P on the curve in Figure 9-7 and letting x increase by an amount Δx we have

$$\Delta y = \sin (x + \Delta x) - \sin x$$

$$\frac{\Delta y}{\Delta x} = \frac{\sin (x + \Delta x) - \sin x}{\Delta x}$$

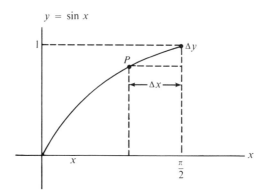

Figure 9-7

The required derivative is the limit of this quotient as $\Delta x \to 0$. To find this limit, we must apply the formula $\sin A - \sin B = 2 \cos \frac{1}{2}(A + B) \sin \frac{1}{2}(A - B)$ to the numerator of the fraction.

$$\frac{\Delta y}{\Delta x} = \frac{2 \cos (x + \frac{1}{2}\Delta x) \sin \frac{1}{2}\Delta x}{\Delta x}$$

$$\frac{\Delta y}{\Delta x} = \cos (x + \frac{1}{2}\Delta x) \frac{\sin \frac{1}{2}\Delta x}{\frac{1}{2}\Delta x}$$

$$\frac{dy}{dx} = \lim_{\Delta x \to 0} \frac{\Delta y}{\Delta x} = \lim_{\Delta x \to 0} \cos (x + \frac{1}{2}\Delta x) \cdot \lim_{\Delta x \to 0} \frac{\sin \frac{1}{2}\Delta x}{\frac{1}{2}\Delta x}$$

We now need to compute

$$\lim_{\Delta x \to 0} \cos (x + \frac{1}{2}\Delta x)$$

and

$$\lim_{\Delta x \to 0} \frac{\sin \frac{1}{2}\Delta x}{\frac{1}{2}\Delta x}$$

The $\lim_{\Delta x \to 0} \cos (x + \frac{1}{2}\Delta x)$ is $\cos x$ which is found by letting $\Delta x = 0$. However, $\lim_{\Delta x \to 0} \sin \frac{1}{2}\Delta x / \frac{1}{2}\Delta x$ is not as easily found. We must find the $\lim_{\theta \to 0} (\sin \theta)/\theta$ where θ is in radians. Let θ, where $0 < \theta < \frac{1}{2}\pi$, be the radian measure of a central angle in a circle of radius $OC = 1$ (Figure 9-8). Then Area of

$$\triangle OBC = \frac{1}{2} OC \cdot BC = \frac{1}{2} BC = \frac{1}{2} \tan \theta \quad \text{since} \quad \tan \theta = \frac{BC}{OC} = BC$$

Area of sector

$$OAC = \frac{1}{2} (OC)^2 \cdot \theta = \frac{1}{2} (1)^2 \theta = \frac{1}{2} \theta$$

Area of

$$\triangle OAC = \frac{1}{2} OC \cdot AD = \frac{1}{2} AD = \frac{1}{2} \sin \theta \quad \text{since} \quad \sin \theta = \frac{AD}{OC} = \frac{AD}{OA} = AD$$

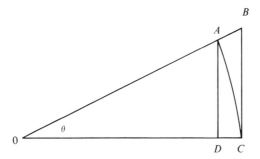

Figure 9-8

Of these three areas the first is the largest and the last is the smallest; therefore

$$\tfrac{1}{2} \tan \theta > \tfrac{1}{2}\theta > \tfrac{1}{2} \sin \theta$$

Dividing by $\tfrac{1}{2} \sin \theta$ (a positive quantity) this becomes

$$\frac{1}{\cos \theta} > \frac{\theta}{\sin \theta} > 1$$

Inverting and reversing the inequality sign we have

$$\cos \theta < \frac{\sin \theta}{\theta} < 1$$

This inequality states that the value of $(\sin \theta)/\theta$ lies between that of $\cos \theta$ and a. The limit of $\cos \theta$ as $\theta \rightarrow 0$ is 1 since the cosine function is continuous and has y value of 1 when $\theta = 0$. Hence $(\sin \theta)/\theta$, since it lies between 1 and a quantit1 approaching 1 as a limit, must also approach 1 when $\theta \rightarrow 0$. That is

$$\lim_{\theta \rightarrow 0} \frac{\sin \theta}{\theta} = 1$$

If $\theta = \tfrac{1}{2}\Delta x$, then $\lim\limits_{\Delta x \rightarrow 0} \dfrac{\sin \tfrac{1}{2}\Delta x}{\tfrac{1}{2}\Delta x} = 1.$

Since $\lim\limits_{\Delta x \rightarrow 0} \cos (x + \tfrac{1}{2}\Delta x) = \cos x$ and $\lim\limits_{\Delta x \rightarrow 0} \dfrac{\sin \tfrac{1}{2}\Delta x}{\tfrac{1}{2}\Delta x} = 1$ we can write

$$\frac{d(\sin x)}{dx} = \cos x$$

In more general terms we can write

$$\frac{d}{dx} \sin v = \cos v \frac{dv}{dx}$$

where v is a function of x. This formula says that the slope of the tangent to the sine curve at any point x is equal to the cos v multiplied by the derivative of v with respect to x.

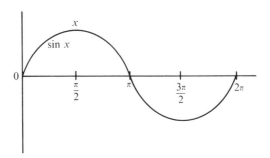

Figure 9-9

Consider the above graph. Suppose you want to find the slope of the tangent to the sine curve when $x = \dfrac{\pi}{2}$.

In this example $v = x$. Our formula now reads as follows.

$$\frac{d \sin x}{dx} = \cos x \frac{dx}{dx} = \cos x$$

When $x = \dfrac{\pi}{2}$

$$\frac{d \sin x}{dx} = \cos \frac{\pi}{2} = 0$$

This says that the slope of the tangent to the sine curve at $x = \dfrac{\pi}{2}$ is 0 and the tangent line is parallel to the x axis.

EXAMPLE 28   Find the derivative of $f(x) = \sin 3x$.

Solution: In this example $v = 3x$.

$$f'(x) = \frac{d \sin 3x}{dx} = \cos 3x \frac{d(3x)}{dx}$$

$$f'(x) = \frac{d \sin 3x}{dx} = 3 \cos 3x$$

EXAMPLE 29   Find the derivative of $y = \sin^2 x$.

Solution: To solve this problem we can use the power rule where $v = \sin x$.

$$\frac{dy}{dx} = nv^{n-1} \frac{dv}{dx}$$

$$\frac{dy}{dx} = 2 \sin x \frac{d \sin x}{dx}$$

$$\frac{dy}{dx} = 2 \sin x \cos x$$

EXAMPLE 30   Find the derivative of $y = \sin^3 (3x^2 + 1)$.

Solution: First we will use

$$\frac{dv^n}{dx} = nv^{n-1} \frac{dv}{dx}$$

with $v = \sin (3x^2 + 1)$

$$\frac{d \sin^3 (3x^2 + 1)}{dx} = 3 \sin^2 (3x^2 + 1) \cdot \frac{d \sin (3x^2 + 1)}{dx}$$

Now we must find $\dfrac{d \sin (3x^2 + 1)}{dx}$.

For this we will use

$$\frac{d \sin v}{dx} = \cos v \frac{dv}{dx}$$

where $v = 3x^2 + 1$.

$$\frac{d \sin (3x^2 + 1)}{dx} = \cos (3x^2 + 1) \cdot \frac{d(3x^2 + 1)}{dx}$$

$$= \cos (3x^2 + 1) \cdot 6x$$

Therefore

$$\frac{d \sin^3 (3x^2 + 1)}{dx} = 3 \sin^2 (3x + 1) \cos (3x^2 + 1) \cdot 6x$$

$$= 18x \sin^2 (3x^2 + 1) \cdot \cos (3x^2 + 1)$$

In the above examples it was necessary to use both algebraic rules and trigonometric rules for finding derivatives. We will now define the rules for finding the derivatives of $\cos x$ and $\tan x$ and then look at some more examples.

The rule for finding the derivative of $\cos v$ is

$$\frac{d \cos v}{dx} = -\sin v \frac{dv}{dx}$$

where v is a function of x.

EXAMPLE 31   Find the derivative of $y = \cos x^2$.

Solution: $$\frac{dy}{dx} = \frac{d \cos x^2}{dx} = -\sin x^2 \frac{dx^2}{dx}$$

$$= -2x \sin x^2$$

The rule for finding the derivative of $\tan v$ is

$$\frac{d \tan v}{dx} = \sec^2 v \frac{dv}{dx}$$

where v is a function of x.

EXAMPLE 32   Find the derivative of $y = \tan (x^2 + x)$.

Solution: $\dfrac{dy}{dx} = \dfrac{d \tan (x^2 + x)}{dx} = \sec^2 (x^2 + x) \dfrac{d(x^2 + x)}{dx}$

$$= (2x + 1) \sec^2 (x^2 + x)$$

EXAMPLE 33   Find the derivative of $y = \tan 2x$ by using the rule for $\sin v$ and $\cos v$.

Solution: First we note that

$$y = \tan 2x = \frac{\sin 2x}{\cos 2x}$$

The above function is a quotient of two functions of x. Therefore, we will apply the quotient rule that we discussed in a previous section.

$$\frac{d}{dx}\left(\frac{u}{v}\right) = \frac{v \dfrac{du}{dx} - u \dfrac{dv}{dx}}{v^2}$$

Let $u = \sin 2x$ and $v = \cos 2x$.

$$\frac{dy}{dx} = \frac{d}{dx}\frac{\sin 2x}{\cos 2x} = \frac{\cos 2x \left(\dfrac{d \sin 2x}{dx}\right) - \sin 2x \left(\dfrac{d \cos 2x}{dx}\right)}{\cos^2 2x}$$

$$\frac{d \sin 2x}{dx} = \cos 2x \left(\frac{d2x}{dx}\right) = 2 \cos 2x$$

$$\frac{d \cos 2x}{dx} = -\sin 2x \left(\frac{d2x}{dx}\right) = -2 \sin 2x$$

$$\frac{dy}{dx} = \frac{\cos 2x[2 \cos 2x] - \sin 2x[-2 \sin 2x]}{\cos^2 2x}$$

$$\frac{dy}{dx} = \frac{2[\cos^2 2x + \sin^2 2x]}{\cos^2 2x}$$

Recall that $\cos^2 2x + \sin^2 2x = 1$.

$$\frac{dy}{dx} = \frac{2 \cdot 1}{\cos^2 2x} = 2 \cdot \frac{1}{\cos^2 2x}$$

Recall that $\sec 2x = \dfrac{1}{\cos 2x}$ therefore $\sec^2 2x = \dfrac{1}{\cos^2 2x}$

$$\frac{dy}{dx} = 2 \cdot \frac{1}{\cos^2 2x} = 2 \sec^2 2x$$

We could have arrived at the result by using the rule for finding the derivative of tan x.

$$\frac{d \tan 2x}{dx} = \sec^2 2x \left(\frac{d2x}{dx}\right)$$

$$= 2 \sec^2 2x$$

EXAMPLE 34 Find the derivative of $y = \cos x \sin x$.

Solution: This function is a product so we must apply the product rule.

$$\frac{d}{dx}(uv) = u\left(\frac{dv}{dx}\right) + v\left(\frac{du}{dx}\right)$$

Let $u = \cos x$ and $v = \sin x$.

$$\frac{d}{dx}(\sin x \cos x) = \sin x \left(\frac{d \cos x}{dx}\right) + \cos x \left(\frac{d \sin x}{dx}\right)$$

$$= (\sin x)(-\sin x) + (\cos x)(\cos x)$$

$$= \cos^2 x - \sin^2 x$$

We should note that finding derivatives of complicated functions is simply a matter of applying the appropriate rules in the proper order. This will be demonstrated again in the following example.

EXAMPLE 35 Find the derivative of $y = \left(\dfrac{\cos 3x}{\sin 2x}\right)^4$.

Solution: The above function is raised to a power so we must first use

$$\frac{dv^n}{dx} = nv^{n-1}\left(\frac{dv}{dx}\right)$$

where $v = \dfrac{\cos 3x}{\sin 3x}$.

$$\frac{dy}{dx} = \frac{d}{dx}\left(\frac{\cos 3x}{\sin 3x}\right)^4 = 4\left(\frac{\cos 3x}{\sin 3x}\right)^3 \frac{d}{dx}\frac{\cos 3x}{\sin 3x}$$

We must now find $\dfrac{d}{dx}\left(\dfrac{\cos 3x}{\sin 3x}\right)$ which is a quotient. The quotient rule is

$$\frac{d}{dx}\left(\frac{u}{v}\right) = \frac{v\dfrac{du}{dx} - u\dfrac{dv}{dx}}{v^2}$$

Let $u = \cos 3x$ and $v = \sin 3x$.

$$\frac{d}{dx}\left(\frac{\cos 3x}{\sin 3x}\right) = \frac{\sin 3x \left(\dfrac{d \cos 3x}{dx}\right) - \cos 3x \left(\dfrac{d \sin 3x}{dx}\right)}{\sin^2 3x}$$

Now we must find $\dfrac{d(\cos 3x)}{dx}$ and $\dfrac{d(\sin 3x)}{dx}$.

$$\frac{d(\cos 3x)}{dx} = -\sin 3x \frac{d(3x)}{dx} = -3 \sin 3x$$

$$\frac{d(\sin 3x)}{dx} = \cos 3x \frac{d(3x)}{dx} = 3 \cos 3x$$

Now we can find $\dfrac{d}{dx}\left(\dfrac{\cos 3x}{\sin 3x}\right)$.

$$\frac{d}{dx}\left(\frac{\cos 3x}{\sin 3x}\right) = \frac{-3 \sin^2 3x - 3 \cos^2 3x}{\sin^2 3x}$$

$$= -3 \left(\frac{\sin^2 3x + \cos^2 3x}{\sin^2 3x}\right)$$

$$= -3 \left(\frac{1}{\sin^2 3x}\right)$$

Now we can find $\dfrac{dy}{dx}$.

$$\frac{dy}{dx} = 4 \left(\frac{\cos 3x}{\sin 3x}\right)^3 \frac{d}{dx}\left(\frac{\cos 3x}{\sin 3x}\right)$$

$$= 4 \left(\frac{\cos 3x}{\sin 3x}\right)^3 \left(\frac{-3}{\sin^2 3x}\right)$$

$$= -12 \frac{\cos^3 3x}{\sin^5 3x}$$

Problems of this type can be quite lengthy, but, if the steps are followed in order, they are relatively easy.

PROBLEM SET 9-7

Find the derivatives of the following expressions.

1. $y = \sin (2x)$ 2. $f(x) = \cos (x)$

3. $y = \tan (x)$ 4. $y = \sin (x^3)$

5. $y = \cos^2 (x)$ 6. $f(x) = \tan (x^2 - 2x)$

7. $f(x) = \tan^3 (x^2)$

8. $f(x) = \sin (x) \tan (x)$

9. $f(x) = \cos (3x^2) \tan (3x^2)$

10. $f(x) = \dfrac{\cos (x)}{\tan (x)}$

11. $y = \dfrac{1}{\cos (x)}$

12. $y = \csc (x)$

13. $y = \cot (3x)$

14. $y = \cos (4x)$

15. $y = \sin (\frac{1}{5}x)$

16. $y = \left(\dfrac{\cos (2x)}{\sin (2x)}\right)^2$

17. $f(x) = \dfrac{1}{\cot (4x)}$

18. $y = \sin^2 x + \cos^2 x$

19. $y = \tan^2 (3x) - \cos x$

20. $f(x) = \sec (3x)$

21. A man starts at a point A and walks 40 ft north; then turns and walks east at 5 ft per second. If a searchlight placed at A follows him, at what rate is it turning at the end of 6 sec?

22. A voltage $v = 2{,}000 \sin 100t$ is impressed across a 20-microfarad capacitor. Find a formula for the resulting current. (Remember $i = dv/dt$ where i is current, c is capacitance, and v is the impressed voltage.)

23. The current in a 20-ohm resistor is $i = 6 \sin 60t$. How fast is the power in the resistor changing at any time? (Remember $P = i^2R$).

9-10

The Derivative of the Logarithmic and Exponential Functions

The formula for finding the derivative of the logarithmic function

$$y = \log_b x$$

can be found by applying the fundamental differentiation process.

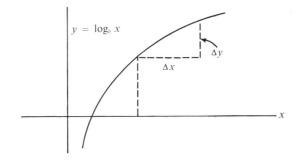

Figure 9-10

Starting at any point P in Figure 9-10 and letting x increase by an amount Δx we have

$$\Delta y = \log_b (x + \Delta x) - \log_b (x)$$

$$= \log_b \left(\frac{x + \Delta x}{x} \right)$$

$$= \log_b \left(1 + \frac{\Delta x}{x} \right)$$

Hence,

$$\frac{\Delta y}{\Delta x} = \frac{1}{\Delta x} \log_b \left(1 + \frac{\Delta x}{x} \right)$$

It is not easy to see what happens to the value of this fraction when $\Delta x \to 0$. However, if we multiply the numerator and denominator by x, we may write the previous expression in the form

$$\frac{\Delta y}{\Delta x} = \frac{1}{x} \frac{x}{\Delta x} \log_b \left(1 + \frac{\Delta x}{x} \right)$$

$$= \frac{1}{x} \log_b \left(1 + \frac{\Delta x}{x} \right)^{x/\Delta x}$$

If now $\Delta x \to 0$, the quantity $\left(1 + \frac{\Delta x}{x} \right)^{x/\Delta x}$ approaches as a limit the number e because it is of the form $(1 + v)^{1/v}$ with v approaching zero. The definition of e is $\lim_{v \to 0} (1 + v)^{1/v}$. We have then

$$\frac{dy}{dx} = \frac{1}{x} \lim_{\Delta x \to 0} \log_b \left(1 + \frac{\Delta x}{x} \right)^{x/\Delta x}$$

$$= \frac{1}{x} \log_b e$$

A more general form of the formula is

$$\frac{d}{dx} \log_b V = \frac{1}{V} \log_b e \frac{dV}{dx}$$

where b is the base, V is a function of x and e is the base for natural logarithms. If $b = e$, then the formula becomes

$$\frac{d}{dx} \log_e V = \frac{1}{V} \log_e e \frac{dV}{dx}$$

$$\frac{d}{dx} \log_e V = \frac{1}{V} \frac{dV}{dx}$$

since $\log_e e$ is equal to one. This is the formula for the derivative of the natural logarithmic function.

EXAMPLE 36 Find the derivative of $y = \log_e (x^3 + 4)$.

Solution: In this example $V = x^3 + 4$.

$$\frac{d}{dx} \log_e (x^3 + 4) = \frac{1}{x^3 + 4} \cdot \frac{d(x^3 + 4)}{dx}$$

$$\frac{d}{dx} \log_e (x^3 + 4) = \frac{1}{x^3 + 4} \cdot 3x^2$$

$$= \frac{3x^2}{x^3 + 4}$$

EXAMPLE 37 Find the derivative of $y = \log_{10} \sin x$.

Solution: In the example $V = \sin x$ and the base is 10.

$$\frac{d}{dx} \log_{10} \sin x = \frac{1}{\sin x} \log_{10} e \frac{d}{dx} \sin x$$

$$= \frac{1}{\sin x} \log_{10} e (\cos x)$$

$$= \frac{\cos x}{\sin x} \log_{10} e$$

$$= \cot x \log_{10} e$$

The derivative of the exponential function is given by the formula

$$\frac{d}{dx} b^V = b^V \log_e b \frac{dV}{dx}$$

where V is a function of x.

As in the case of the logarithm, suppose that $b = e$. The above formula becomes

$$\frac{d}{dx} e^V = e^V \frac{dV}{dx}$$

EXAMPLE 38 Find the derivative of $f(x) = e^{x^2}$.

Solution: In this example, $V = x^2$.

$$\frac{d}{dx} e^{x^2} = e^{x^2} \frac{d}{dx} (x^2)$$

$$= e^{x^2}(2x)$$

$$= 2xe^{x^2}$$

EXAMPLE 39   Find the derivative of $y = 4^{x^2 - 1}$.

Solution: For this example $V = x^2 - 1$.

$$\frac{d}{dx} 4^{x^2 - 1} = 4^{x^2 - 1} \log_e 4 \frac{d}{dx} (x^2 - 1)$$

$$= (4^{x^2 - 1})(\log_e 4)(2x)$$

Differentiate the following functions.

1. $y = \log_e 3x$

2. $y = 4 \log_e x^2$

3. $y = \log_e (\tan^2 \theta)$

4. $y = \log_{10} (4x^3)$

5. $y = x \log_e (x^2)$

6. $y = 4 \log_e^2 (3x)$

7. $y = \log_e (\log_e x)$

8. $y = \log_e \dfrac{x^2}{x^2 + 1}$

9. $y = e^{2x}$

10. $y = 3e^x$

11. $y = e^x \log_e x$

12. $y = e^{(x^2 + 2x)}$

13. $y = e^{-x} \log_e x^2$

14. $y = 3^x$

15. $y = 2^{(x^2 - 3x)}$

16. $y = e^x 4^x$

17. A radioactive substance decomposes at a rate proportional to the amount present. The amount of a radioactive substance present is given by the expression $y = \frac{1}{10} e^{-2t}$ where y is in grams and t is in seconds. What is the rate of change of the substance present when $t = 2$ seconds?

18. The inductance of a coaxial cable is given by $L = .140 \log_{10} (b/a) + .015$ microhenries per foot, where a and b are radii of the inner and outer conductors, respectively. Find the rate of change of the inductance with respect to a.

19. When a metal is dipped into a solution containing ions of that metal an *emf* is produced between the metal and the solution. The value of the *emf* is $V = -A \log_e B/p$ where A and B are constants. Find dV/dp.

20. A current $i = 3e^{-.02t}$ flows through the primary winding of a transformer. If the mutual inductance between the windings is 1.0 henry, what is the induced secondary voltage when $t = 2$ seconds. (Hint: $e = L \, di/dt$ where e is the induced voltage and L is the inductance.)

9-11
Second and Higher Order Derivatives

Suppose we are given an expression for distance s of a particle from some point in terms of time t.

$$s = 3t^3 + 4t^2 + t - 3$$

Recall that we can find the velocity by finding the derivative of s with respect to t.

$$v = \frac{ds}{dt} = 9t^2 + 8t + 1$$

Now acceleration is defined as the rate of change of velocity with respect to time.

$$v = 9t^2 + 8t + 1$$

$$a = \frac{dv}{dt} = 18t + 8$$

$$a = \frac{d}{dt}\left(\frac{ds}{dt}\right) = \frac{d^2s}{dt^2} = 18t + 8$$

The acceleration of a particle is the second derivative of the displacement with respect to time.

Higher order derivatives can be found in exactly the same manner as the second derivative.

EXAMPLE 40 Find the third derivative of $y = 4x^4 - x^3 + x^2 - 2$.

Solution: First we find the first derivative.

$$\frac{dy}{dx} = 16x^3 + 3x^2 + 2x$$

Next we find the second derivative.

$$\frac{d}{dx}\left(\frac{dx}{dx}\right) = \frac{d^2y}{dx^2} = 48x^2 - 6x + 2$$

Finally we find the third derivative.

$$\frac{d}{dx}\left(\frac{d^2y}{dx^2}\right) = \frac{d^3y}{dx^3} = 96x - 6$$

The above procedure is used to find any order of derivative that is desired.

PROBLEM SET 9-9

Find the first, second, and third derivatives for the following functions.

1. $y = x^4 + 3x^3 - x^2 + x - 7$ 2. $y = \sin 3x$

3. $y = 3 \log_e 2x$ 4. $y = (x + 1)^4$

5. $y = (x^2 + 3)^3$ 6. $y = \dfrac{1}{x}$

7. $y = \dfrac{1}{\sin x}$　　　　　　　　8. $y = \cos^2 x$

9. $y = \dfrac{x + 1}{x^2}$

10. The vertical displacement of a rocket is given by the expression $s = 10t^3 + 4t^2 + t$. Find the displacement, velocity, and acceleration when $t = 2$ seconds.

9-12

Maxima and Minima

Suppose that the sum of two positive numbers is 20. Find the numbers if the product obtained by multiplying one of them by the other is a maximum.

This is an example of a problem in which we are asked to find the maximum value of a function that is the product of two numbers. The numbers can be represented by x and $20 - x$ since the sum of the two numbers is 20. The function is

$$f(x) = x(20 - x)$$

Next we will sketch the graph of this function.

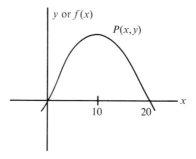

Figure 9-11

The above is a sketch of the graph of the function. We note that the graph has some maximum value and we want to find the corresponding x value.

The slope of a line tangent to the curve at the point p is zero. The procedure for finding the maximum point is to find the derivative of the function, set it equal to zero since the derivative is the slope, and solve for x.

$$f(x) = x(20 - x)$$

$$f(x) = 20x - x^2$$

$$f'(x) = 20 - 2x$$

We now have the derivative of the function. Next we will set it equal to zero and solve for x.

$$20 - 2x = 0$$

$$2x = 20$$

$$x = 10$$

When x is equal to 10 the function has its maximum value. We can try a few values for x to check our result.

$$f(x) = 20x - x^2$$

$$f(8) = 160 - 64 = 96$$

$$f(9) = 180 - 81 = 99$$

$$f(10) = 200 - 100 = 100$$

$$f(11) = 220 - 121 = 99$$

For this problem x is equal to 10 and $20 - x$ is equal to 10. The two numbers that satisfy the problem are both equal to 10.

Now let us consider the problem mentioned at the beginning of this chapter. We wanted to bring electricity into the workshop by using an extension cord. The cord was to be plugged into a 120-volt outlet and it had a 12 ampere rating and a resistance of 8 ohms. We had ten 20 ohm heaters and wanted to know how many to use.

Power, which is equivalent to heat per unit time, is equal to E^2/R where E is the voltage and R is the resistance. We want to find the maximum power that we can obtain with the given condition. We first write the expression for power. The power dissipated by the heaters can be determined by use of basic electrical formulas. Consider Figure 9-12 which represents the twenty heaters in parallel, in series with 8 ohms representing the resistance of the power line and

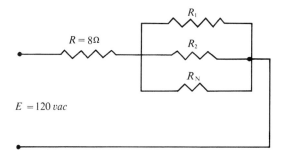

Figure 9-12

fed by 120 volts. The voltage across the heaters is calculated by use of the voltage divider equation:

$$E = 120 \left(\frac{\dfrac{20}{N}}{8 + \dfrac{20}{N}} \right) = 120 \left(\frac{\dfrac{20}{N}}{\dfrac{8N + 20}{N}} \right)$$

$$= \left(\frac{20}{N} \right) \left(\frac{N}{8N + 20} \right) (120) = 120 \left(\frac{20}{8N + 20} \right)$$

Since $P = \dfrac{E^2}{R}$ we can substitute this expression for E:

$$P = \frac{E^2}{R} = \frac{\left[120 \left(\dfrac{20}{8N + 20} \right) \right]^2}{\dfrac{20}{N}} = \frac{20N(120)^2}{(8N + 20)^2}$$

where N is the number of heaters used. Next we need to find the derivative of P with respect to N. By using the quotient rule we find that

$$\frac{dP}{dN} = \frac{(20 + 8N)^2 20(120)^2 - 20N(120)^2 2(20 + 8N)8}{(20 + 8N)^4}$$

If we set $\dfrac{dP}{dN}$ equal to zero, we obtain

$$(20 + 8N)^2 20(120)^2 - 20N(120)^2 2(20 + 8N)8 = 0$$

Now by factoring we have

$$20(120)^2 \{(20 + 8N)^2 - 16N(20 + 8N)\} = 0$$
$$(20 + 8N)^2 - 16N(20 + 8N) = 0$$
$$\{20 + 8N\}\{(20 + 8N) - 16N\} = 0$$
$$\{20 + 8N\}\{20 - 8N\} = 0$$
$$N = \pm \tfrac{20}{8}$$

Taking the positive value and the nearest whole number to $\tfrac{20}{8}$, we obtain the numbers 2 and 3, one of which is the solution. If we substitute both numbers for N into the expression for power we find that 3 gives a slightly larger value for power.

EXAMPLE 41   A farmer has 1,200 ft of fence and wishes to enclose a rectangular plot and divide it into three equal parts by cross fences parallel to the ends. What dimensions would give the largest enclosed area?

Solution: Let W be the width. The length is $\dfrac{1200 - 4W}{2}$ or $600 - 2W$.

We wish the area to be a maximum. Therefore, we will write the expression for the area and differentiate.

$$A = (600 - 2W) \cdot W$$

$$A = 600W - 2W^2$$

$$\frac{dA}{dW} = 600 - 4W$$

Now we set $\dfrac{dA}{dW} = 0$ and solve for W.

$$600 - 4W = 0$$

$$4W = 600$$

$$W = 150$$

The width of the largest area is 150 ft, and the length is equal to $600 - 2W$ or 300 ft.

PROBLEM SET 9-10

1. Divide the number 24 into two parts so that the sum of the squares of the parts will be as small as possible.

2. Find the volume of the largest box that can be made by cutting a square from each corner from a sheet of tin 20 in. long and 12.5 in. wide and turning up the sides.

3. A rectangular box without a lid is to be made to hold 144 cu ft. The length is to be twice the width and the material for the bottom costs 4 times as much per unit area as that for the sides. Find the dimensions for minimum cost.

4. A gutter with a trapezoidal cross section is to be made from a long sheet of tin that is 15 in. wide by turning up one-third of the width on each side. What width across the top will give the greatest capacity?

5. A rectangular swimming pool with square top is to be built with brick sides and bottom. If the volume is to be 500 ft^3 what should the dimensions be in order to minimize the amount of brick used?

6. A building stands 64 ft from a wall 27 ft high. What is the length of the shortest ladder that will reach from the side of the building to the ground on the opposite side of the wall?

7. Show that a cylindrical can (with lid) will require a minimum amount of material for a specified volume V if its diameter and height are equal.

8. A telephone company in a certain city has 10,000 telephones where the charge is $4 per month. It is believed that if the charge is reduced, the number of telephones in use will increase at an estimate rate of 200 additional telephones for each 5 cent reduction. What monthly charge will yield the greatest gross revenue on this basis?

9. A ship B is 40 miles due east of a ship A. B starts due north at 25 m.p.h. and at the same time A starts in the direction north θ east where $\theta =$ arctan $\frac{4}{3}$ at 15 m.p.h. When will they be closest together?

10. The current through a constant resistance R in an AC circuit with constant voltage E is given by $I = E/\sqrt{R^2 + x^2}$. Find the value of x for maximum current in the circuit, and find this maximum current.

9-13
Technical Example

In calculus, or mathematics beyond calculus, it is often necessary to derive an equation which may then be used in seeking the solution to a problem. As an example, Figure 9-11 is a drawing of a slider-crank mechanism. The crankshaft, connecting rod, piston arrangement in your car make up a slider-crank mechanism like this.

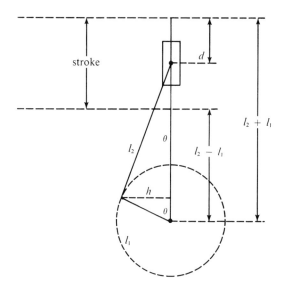

Figure 9-13

Suppose you wish to determine the velocity of the slider (or piston in your car), but have no equation for velocity. We might approach the problem by first deriving an equation for displacement. The derivative of the equation for displacement will then give us an equation for velocity.

The following shows how the equation for displacement is derived. From trigonometry we know that

$$h = l_1 \sin \theta \quad \text{and} \quad h = l_2 \sin \phi$$

Since $h = h$ we can say

$$l_1 \sin \theta = l_2 \sin \phi$$

Solving for $\sin \phi$

$$\sin \phi = \sin \theta \left(\frac{l_1}{l_2}\right)$$

If we let $K = \dfrac{l_2}{l_1}$, then our equation becomes

$$\sin \phi = \sin \theta \left(\frac{1}{K}\right)$$

The displacement of the slider at any instant can be determined by the equation

$$d = (l_2 + l_1) - (l_1 \cos \theta + l_2 \cos \phi)$$

$$= l_2 - l_2 \cos \phi + l_1 - l_1 \cos \theta$$

$$= l_2(1 - \cos \phi) + l_1(1 - \cos \theta)$$

Since this equation contains both angle θ and angle ϕ we need to rewrite it to contain only one or the other. We already have the expression

$$\sin \phi = \frac{\sin \theta}{K}$$

If we square both sides of this we get

$$\sin^2 \phi = \frac{\sin^2 \theta}{K^2}$$

From trigonometry:

$$\cos^2 \phi = 1 - \sin^2 \phi = 1 - \frac{\sin^2 \theta}{K^2}$$

Therefore:

$$\cos \phi = \sqrt{1 - \frac{\sin^2 \theta}{K^2}}$$

The equation for displacement, d, now becomes

$$d = l_1(1 - \cos \theta) + l_2 \left(1 - \sqrt{1 - \frac{\sin^2 \theta}{K^2}}\right)$$

This equation can be simplified by binomial expansion to the form

$$d = l_1(1 - \cos \theta) + \frac{l_1^2 \sin^2 \theta}{2l_2}$$

The derivative of this equation gives us an equation for the velocity of the slider.

$$f'(d) = \frac{d(l_1)}{dt} - \frac{d(l_1 \cos \theta)}{dt} + \frac{d\left(\frac{l_1^2 \sin^2 \theta}{2l_2}\right)}{dt}$$

$$f'(d) = 0 - (l_1)(-\sin \theta)\frac{d\theta}{dt} + \frac{l_1^2}{2l_2}(2 \sin \theta)(\cos \theta)\frac{d\theta}{dt}$$

$$f'(d) = l_1 \sin \theta \frac{d\theta}{dt} + \frac{l_1^2}{2l_2}(2 \sin \theta \cos \theta)\frac{d\theta}{dt}$$

$$f'(d) = l_1 \frac{d\theta}{dt}\left[\sin \theta + \frac{1}{2}\left(\frac{l_1}{l_2}\right)(2 \sin \theta \cos \theta)\right]$$

Substituting ω for $\frac{d\theta}{dt}$ which is angular velocity and $\frac{1}{K}$ for $\frac{l_1}{l_2}$ gives the following

$$f'(d) = l_1\omega\left[\sin \theta + \frac{1}{2}\left(\frac{1}{K}\right)(2 \sin \theta \cos \theta)\right]$$

Again from trigonometry

$$2 \sin \theta \cos \theta = \sin 2\theta$$

Therefore:

$$v = \omega l_1 \sin \theta + \frac{\sin 2\theta}{2K}$$

Integral Calculus

Up to now we have been concerned with obtaining the derivative of a given function. In many applications of calculus we are faced with the inverse problem. We may be given the velocity of a particle as a function of time and asked to find the displacement at a given time.

Suppose we know that the velocity of a particle is given by $v = 3t^2 + 2t + 1$. Now, how can we find the displacement when $t = 2$ seconds if the displacement when $t = 0$ is also zero? Remember that the velocity is equal to the change of distance with respect to time.

$$v = \frac{dx}{dt} = 3t^2 + 2t + 1$$

We can find x as a function of time by trying to find the function of t that will give us $3t^2 + 2t + 1$ when we find the derivative. In other words, we are given the derivative of a function and want to find the function. This is the inverse operation from differentiation.

What function gives us $3t^2 + 2t + 1$ when we find its derivative? Let's look at each term in this function. What will give $3t^2$ when we find its derivative? The way we find a derivative of a *power* function is to multiply by the exponent and then decrease the exponent by one. We will now do the opposite. We will increase the exponent by one and then divide by the new exponent.

If we do this, from $3t^2$ we get t^3. The derivative of t^3 is $3t^2$. Now consider $2t$. We increase the exponent by one and divide by the new exponent and have t^2. What function has the number 1 for its derivative? The answer is t.

One function that will give us $3t^2 + 2t + 1$ for its derivative is $t^3 + t^2 + t$. Note that $t^3 + t^2 + t + 3$ will also give us the same derivative. In fact $t^3 + t^2 + t$ plus any constant has the same derivative. This constant is called the constant of integration and the complete function is called the anti-derivative or the integral of $3t^2 + 2t + 1$.

Now, if $v = 3t^2 + 2t + 1$ and we find the anti-derivative, we have $x = t^3 + t^2 + t + c$. What is the constant? We were given that when $t = 0$, x is also equal to zero. Substituting into $x = t^3 + t^2 + t + c$ we have

$$0 = 0 + 0 + 0 + c$$

$$c = 0$$

If $c = 0$, then $x = t^3 + t^2 + t$. When t is equal to 2 seconds the displacement is equal to 14 ft.

We have completed a problem in which it was necessary to find the anti-derivative of a function. We will now find the anti-derivative of several other functions.

EXAMPLE 1   Find the anti-derivative of $f'(x) = 7x^3$.

Solution: To find the anti-derivative we must increase the exponent by one. We will guess that the anti-derivative of $f'(x) = 7x^3$ is $f(x) = 7x^4$. Now we find the derivative of $f(x) = 7x^4$ and see how it differs from $f'(x) = 7x^3$. If $f(x) = 7x^4$, then $f'(x) = 28x^3$. This function is four times too large so we will try another guess for $f(x)$. Suppose $f(x) = \frac{7}{4}x^4$. Now if we find the derivative of $f(x) = \frac{7}{4}x^4$, we have $f'(x) = 7x^3$ which is what we want the derivative to be. The anti-derivative is $f(x) = \frac{7}{4}x^4 + c$.

The "guessing" method used above is an easy way to find the anti-derivative.

EXAMPLE 2   Find $f(t)$ if $f'(t) = 3t^2$.

Solution: Suppose $f(t) = 3t^3 + c$, then $f'(t) = 9t^2$. $f'(t)$ is three times too large so we let $f(t) = t^3 + c$. If $f(t) = t^3 + c$, then $f'(t) = 3t^2$. We have discovered the right $f(t)$.

EXAMPLE 3   Find $f(x)$ if $f'(x) = (3x + 1)^4$.

Solution: Suppose $f(x) = (3x + 1)^5 + c$, then $f'(x) = 5(3x + 1)^4(3)$ or $f'(x) = 15(3x + 1)$. $f'(x)$ is fifteen times too large so we let $f(x) = \frac{1}{15}(3x + 1)^5 + c$. The derivative of this $f(x)$ is $(3x + 1)^4$.

EXAMPLE 4   Find $f(x)$ if $f'(x) = 3 \cos 2x$.

Solution: Suppose $f(x) = \sin 2x + c$, then $f'(x) = 2 \cos 2x$. $f'(x)$ is $\frac{3}{2}$ of the value we found so $f(x) = \frac{3}{2} \sin 2x + c$.

EXAMPLE 5   Find $f(x)$ if $f'(x) = x^2(3x^3 + 2)^2$.

Solution: We first take the most complicated form and try to decide what form the anti-derivative must have. We will consider $(3x^3 + 2)^2$ and let $f(x) = (3x^3 + 2)^3$ since we have a term raised to a power. $f'(x) = 27x^2(3x^3 + 2)^2$. This function is twenty-seven times too large so we have $f(x) = \frac{1}{27}(3x^3 + 2)^3 + c$.

PROBLEM SET 10-1

Find $f(x)$ in the following problems.

1. $f'(x) = 3x^2$
2. $f'(x) = x^3 + 2x$
3. $f'(x) = x^2 - 3x + 1$
4. $f'(x) = \cos 3x$
5. $f'(x) = 2 \sin x$
6. $f'(x) = x(2x^2 + 1)^3$
7. $f'(x) = \dfrac{x}{(x^2 - 1)}$
8. $f'(x) = (3x + 1)^4$
9. $f'(x) = x(x^2 - 1)$
10. $f'(x) = \dfrac{x^2}{x^3 - 1}$

10-2
Integration Formulas

If it were necessary to always use the "guess" method to find the anti-derivative, the process would be quite time consuming. To avoid this problem we can state several formulas that are used to find the anti-derivative or integral of a function.

The formulas that we will be using are listed below.

(1) $\displaystyle\int (du + dv) = \int du + \int dv$

(2) $\displaystyle\int a\, dv = a \int dv$

(3) $\displaystyle\int v^n\, dv = \frac{v^{n+1}}{n + 1} + c$

(4) $\displaystyle\int \frac{dv}{v} = \log_e v + c$

(5) $\displaystyle\int a^v\, dv = \frac{a^v}{\log a} + c$

(6) $\displaystyle\int e^v \, dv = e^v + c$

(7) $\displaystyle\int \sin v \, dv = -\cos v + c$

(8) $\displaystyle\int \cos v \, dv = \sin v + c$

In each of the above formulas u and v are variables, and a and c are constants.

We will now examine the use of the formulas. First we will consider formula (3).

EXAMPLE 6   Find the anti-derivative or integral of $f(x) = x^3$. The process of finding the integral of a function is denoted by the symbol

$$\int f(x) \, dx$$

which denotes the inverse operation of finding the derivative. The statement

$$\int x^3 \, dx$$

means find the integral of $f(x) = x^3$. Comparing this to formula (3), $v = x$ and $n = 3$. We can now write the answer.

$$\int x^3 \, dx = \frac{x^{3+1}}{3+1} + c = \frac{x^4}{4} + c$$

This is the same answer we would have found using the "guess" method.

EXAMPLE 7   Suppose we have the problem $\int (x^3 + x^2 + 1) \, dx$. Formula (1) says that we can write this expression as $\int x^3 \, dx + \int x^2 \, dx + \int 1 \, dx$. Since l is any variable raised to the zero power we can write the expression as $\int x^3 \, dx + \int x^2 \, dx + \int x^0 \, dx$ and solve as in the previous example. The answer is

$$\frac{x^4}{4} + \frac{x^3}{3} + \frac{x^1}{1} + c$$

EXAMPLE 8   Find $\int \sin x \, dx$. By formula (7) $\int \sin x \, dx = -\cos x + c$.

PROBLEM SET 10-2

Find the following integrals.

1. $\displaystyle\int x^2 \, dx$

2. $\displaystyle\int 3^x \, dx$

3. $\int e^x \, dx$

4. $\int \cos x \, dx$

5. $\int (x^3 + x^2) \, dx$

6. $\int \frac{dx}{x}$

7. $\int \sin x \, dx$

8. $\int 3x^2 \, dx$

9. $\int (x^2 + x^5 + \cos x) \, dx$

10. $\int (3x^2 + 2x^4 + 3e^x) \, dx$

11. We know that the acceleration due to gravity is approximately 32.2 ft/sec². An algebraic expression for this is $g = -32.2$. Find an expression for distance as a function of time. (Hint: $v = -\int 32.2 \, dt$.)

10-3
Use of the Integral Formulas

Consider the formula

$$\int v^n \, dv = \frac{v^{n+1}}{n+1} + c$$

This formula actually applies to any function v and its differential dv.

We need to now examine what we mean by differential. Suppose we have a function $v = x^3$. Now we will find the derivative of v with respect to x.

$$\frac{dv}{dx} = \frac{dx^3}{dx} = 3x^2$$

Now if we write

$$dv = 3x^2 \, dx$$

we have the differential of v which is $3x^2 \, dx$.

EXAMPLE 9 What is the differential of $v = (x^2 + 1)^2$.

$$\frac{dv}{dx} = 2(x^2 + 1) \cdot (2x) = 4x(x^2 + 1)$$

Therefore the differential of v is $4x(x^2 + 1) \, dx$.

Returning to the formula.

$$\int v^n \, dv = \frac{v^{n+1}}{n+1} + c$$

This formula can be used only when we have a function v and its differential.

EXAMPLE 10 Find $\int (x^3 + 1)^2 3x^2 \, dx$.

Solution: Suppose $v = x^3 + 1$, then $\dfrac{dv}{dx} = 3x^2$ and the differential of v is $3x^2\ dx$. If $v = x^3 + 1$ then we have $v^2\ dv$ which can be found using formula (3).

$$\int (x^3 + 1)^2 \cdot 3x^2\ dx = \frac{(x^3 + 1)^3}{3} + c$$

EXAMPLE 11   Find $\int (x^4 + 2)^3 x^3\ dx$.

Solution: The only way we can solve the problem is if we have the form $\int v^n\ dv$. Suppose $v = x^4 + 2$, then $dv = 4x^3\ dx$. We do not have the form $\int v^n\ dv$ but we can use formula (2) to obtain the form. We can express the problem as $\int (x^4 + 2)^3 (\frac{4}{4}) x\ dx$. Now by formula (2) we can write

$$\int (x^4 + 2)^3 x^3\ dx = \tfrac{1}{4} \int (x^4 + 2)^3 4x^3\ dx$$

Now we have the form $\tfrac{1}{4} \int v^n\ dv$ and we can use formula (3).

$$\frac{1}{4} \int (x^4 + 2)^3 4x^3\ dx = \frac{1}{4} \cdot \frac{(x^4 + 2)^4}{4} + c$$

$$= \frac{(x^4 + 2)^4}{16} + c$$

The method used in the above example can be used with all the integral formulas to obtain the proper form providing that the problem needs only a constant multiplier.

EXAMPLE 12   Find $\int \sin (3x)\ dx$.

Solution: We must use the formula

$$\int \sin v\ dv = -\cos v + c$$

Since $v = 3x$, the differential of v is $3\ dx$. If we write $\tfrac{1}{3} \int \sin (3x) \cdot 3\ dx$ we have the proper form.

$$\tfrac{1}{3} \int \sin (3x) \cdot 3\ dx = -\tfrac{1}{3} \cos (3x) + c$$

EXAMPLE 13   Find $\int x^2 e^{x^3}\ dx$.

Solution: We must use the formula $\int e^v\ dv = e^v + c$. If v is x^3 then $dv = 3x^2\ dx$. We can obtain the proper form by writing

$$\int x^2 e^{x^3}\ dx = \tfrac{1}{3} \int e^{x^3} \cdot 3x^2\ dx$$

Now we can use formula (6).

$$\tfrac{1}{3} \int e^{x^3} \cdot 3x^2 \, dx = \tfrac{1}{3} e^{x^3} + c$$

Find the following integrals.

1. $\displaystyle\int (x^2 + 1)^3 x \, dx$

2. $\displaystyle\int \frac{x^2 \, dx}{(x^3 + 2)}$

3. $\displaystyle\int \frac{x^2 \, dx}{(x^3 + 2)^2}$

4. $\displaystyle\int (e^x + 1) e^x \, dx$

5. $\displaystyle\int x \sin (x^2) \, dx$

6. $\displaystyle\int x \cos (x^2 + 1) \, dx$

7. $\displaystyle\int x^2 a^{x^3} \, dx$

8. $\displaystyle\int x^2 \sin (x^3 + 1) \, dx$

9. $\displaystyle\int \sin^2 x \cos x \, dx$

10. $\displaystyle\int e^{\sin x} \cos x \, dx$

11. $\displaystyle\int (x^3 + x^2 + 2)^3 (3x^2 + 2x) \, dx$

12. $\displaystyle\int \frac{(3x^2 + 2x) \, dx}{x^3 + x^2 + 1}$

10-4
Geometric Interpretation of Integration

First we will consider the graph of $f(x) = 3$ shown in Figure 10-1.

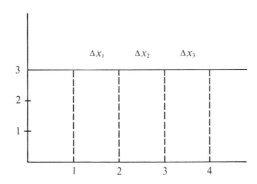

Figure 10-1

Now we will consider the graph of the function between 1 and 4. Suppose we find the area under the curve by adding together the areas of all the rectangles of width $\Delta x = 1$ as shown in Figure 10-1.

$$A = 3\Delta x_1 + 3\Delta x_2 + 3\Delta x_3$$

$$A = 3(2 - 1) + 3(3 - 2) + 3(4 - 3)$$

$$A = 3 \cdot 1 + 3 \cdot 1 + 3 \cdot 1 = 9$$

Now we will find the integral of $f(x) = 3$ and evaluate the integral for $x = 1$ and for $x = 4$.

$$A(x) = \int f(x)\, dx = \int 3\, dx = 3x + c$$

$$A(4) = 3 \cdot 4 + c = 12 + c$$

$$A(1) = 3 \cdot 1 + c = 3 + c$$

Now we will find the difference between $A(4)$ and $A(1)$.

$$A(4) - A(1) = (12 + c) - (3 + c) = 12 + c - 3 - c = 9$$

If we find the integral of a function of x, then evaluate that integral for two values of x and find the difference between the two function values, we will have the area between the graph of the function and the x-axis.

Next we will consider the function $f(x) = x$. The graph of the function is shown in Figure 10-2.

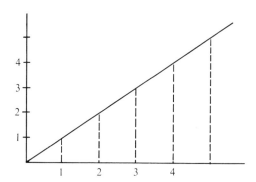

Figure 10-2

We will now find an approximate area under this curve from $x = 2$ to $x = 4$ by adding together the areas of several small strips of width Δx. The approximate area of each strip can be found by multiplying the function value at the end of each Δx by Δx. The sum of all the strip areas will give us the approximate area under the graph for the given interval.

First we will let $\Delta x = \frac{1}{2}$.

$$A = \tfrac{1}{2}(2.5) + \tfrac{1}{2}(3.0) + \tfrac{1}{2}(3.5) + \tfrac{1}{2}(4.0) = 6.5$$

If we find the area under the graph by using the area formula for a triangle, we would find that the area is equal to 6 square units. 6.5 is a close approximation.

Now we will consider $\Delta x = \frac{1}{4}$

$$A = \tfrac{1}{4}(2.25) + \tfrac{1}{4}(2.50) + \tfrac{1}{4}(2.75) + \tfrac{1}{4}(3.00) + \tfrac{1}{4}(3.25) + \tfrac{1}{4}(3.50)$$
$$+ \tfrac{1}{4}(3.75) + \tfrac{1}{4}(4.0) = 6.25$$

The result is even closer to the actual area of 6. If we continue to let Δx get smaller and smaller our computed area will be closer and closer to 6.

Next we will find the integral of $f(x) = x$, evaluate the result at $x = 2$ and at $x = 4$ and then find the difference. The result will be the area under the graph.

$$A(x) = \int x \, dx = \tfrac{1}{2}x^2 + c$$

$$A(2) = \tfrac{1}{2}2^2 + c = 2 + c$$

$$A(4) = \tfrac{1}{2}4^2 + c = \tfrac{16}{2} + c = 8 + c$$

$$A(4) - A(2) = 6$$

Another way of expressing the above operation is

$$A = \int_2^4 x \, dx$$

Here we are saying that the area under the curve between 2 and 4 is the sum of an infinite number of strips that have a length of x and width of dx. Remember, as we let Δx approach zero, we call it dx. We will now perform the operation previously indicated. We first find the anti-derivative.

$$A = \int_2^4 x \, dx$$

$$A = \tfrac{1}{2}x^2 \big|_2^4$$

This can be written

$$A = \tfrac{1}{2}4^2 - \tfrac{1}{2}2^2 = 8 - 2 = 6$$

We found the anti-derivative and then its values at 2 and 4 and, finally, the difference between the values.

PROBLEM SET 10-4

Sketch the graphs and find the approximate areas under the following curves by using the "Δx method."

1. $f(x) = 2x$ between 3 and 8 when $\Delta x = 1$.
2. $f(x) = x^2 + 1$ between 3 and 7 when $\Delta x = 1$.
3. $f(x) = 3x + 1$ between 2 and 5 when $\Delta x = \frac{1}{2}$.

Sketch the graphs and find the areas under the following curves by the use of integration.

4. $f(x) = 2x$ between 3 and 8.

5. $f(x) = x^2 + 1$ between 3 and 7.

6. $f(x) = 3x^3 + 1$ between 2 and 5.

7. $f(x) = x$ between -2 and 2.

8. $f(x) = (x + 2)^3$ between 1 and 4.

9. $f(x) = \sin x$ between 0 and π.

10-5

Applications of the Indefinite Integral

Thus far we have discussed two types of integrals. The definite integral has the form

$$Q = \int_a^b f(x) \, dx$$

which results in a number for an answer. In this problem a and b are constants.

The other type of integral is the indefinite integral. The indefinite integral has the form

$$Q(x) = \int f(x) \, dx$$

where the result is a function of x plus a constant.

We will now look at some applications of the indefinite integral.

EXAMPLE 14 From a point 30 ft above the ground, a small lead ball is thrown vertically upward with an initial velocity of 60 ft per second. Find its velocity and distance from the ground at the end of t sec.

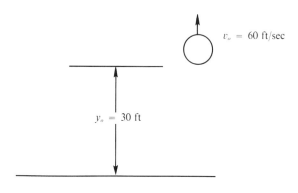

$v_o = 60$ ft/sec

$y_o = 30$ ft

Figure 10-3

Solution: In addition to the displacement with respect to the ground and the initial velocity, we also know that the ball has an acceleration of -32.2 ft/sec^2 because of gravity. Since acceleration is the rate of change of velocity with respect to time we can write

$$\frac{dv}{dt} = -g$$

We can now write the equation in the form

$$dv = -g \, dt$$

Now if we integrate both sides of the equation we have

$$dv = -g \, dt$$

$$\int dv = -g \int dt$$

$$v + c_1 = -gt + c_2$$

$$v = -gt + c$$

We now have an expression for velocity in terms of time. We can solve for the constant c since we know that when $t = 0$, $v_0 = 60$ ft/sec.

$$v = -gt + c$$

$$60 = -g \cdot 0 + c$$

$$c = 60$$

We can now write an expression for velocity at any time t.

$$v = -gt + 60$$

Recall that velocity is the rate of change of distance with respect to time.

$$\frac{dy}{dt} = v = 60 - gt$$

$$dy = (60 - gt) \, dt$$

We will now integrate both sides of the equation just as we did to find the velocity.

$$dy = (60 - gt) \, dt$$

$$\int dy = \int (60 - gt) \, dt$$

$$y = 60t - \tfrac{1}{2}gt^2 + c$$

Since $y = +30$ ft when $t = 0$, $c = 30$; hence, at any time t,

$$y = 30 + 60t - \tfrac{1}{2}gt^2$$

This is the expression for the vertical displacement.

EXAMPLE 15   A 10-microfarad capacitor receives a charge of such that it has a terminal to terminal voltage of 60 volts. Then at an instant that we shall call $t = 0$ we connect the capacitor to a source that sends a current $i = t$ amperes into it. Find the voltage across the capacitor when $t = .1$ seconds.

Solution: First we must recall a few definitions from electronics. The current in a circuit at any instant is

$$i = \frac{dq}{dt}$$

We can rewrite the above expression and integrate both terms.

$$dq = i\, dt$$

$$q = \int i\, dt$$

We also should recall that the charge in a capacitor is

$$q = CV_c \text{ coulombs}$$

where C is the capacitance in farads and v_c represents the voltage between the capacitor terminals. We can combine the above equations to get

$$V_c = \frac{1}{C} \int i\, dt \text{ volts}$$

For our problem, we know that $i = t$. Substituting into the above expression, we have

$$V_c = \frac{1}{C} \int t\, dt$$

$$V_c = \frac{1}{10^{-5}} \int t\, dt = \frac{1}{2 \times 10^{-5}} t^2 + k$$

When $t = 0$, $V_c = 60$. Hence we can solve for k.

$$V_c = \frac{1}{2 \times 10^{-5}} t^2 + k$$

$$60 = \frac{1}{2 \times 10^{-5}} (0)^2 + k$$

$$k = 60$$

Now our equation becomes

$$V_c = \frac{1}{2 \times 10^{-5}} t^2 + 60$$

When $t = .1$ second, we get

$$V_c = 5.0 \times 10^4 \times 10^{-2} + 60 = 560 \text{ volts}$$

1. From the top of a building 50 ft high, a stone is thrown vertically upward with initial velocity of 80 ft per second. Express its velocity and distance above the ground at any time t as a function of t. What is the greatest height reached by the stone?

2. From a point 100 ft above the ground a heavy object is thrown vertically downward with initial velocity 60 ft per second. Derive expressions for its velocity and distance from the ground at any time t. With what velocity does it strike the ground?

3. A block weighing 200 lb is propelled along a rough horizontal surface, starting from rest, by a horizontal force of 40 lb. There is a frictional resistance of 30 lb between the block and the surface. The resistance of the medium is equal to $2v$ where v is the velocity of the block. Derive the expressions for the velocity and displacement of the block in terms of t. (Hint: $F = 40 - 30 - 2v$ and $F = MA$.)

4. The current in a circuit was $i = 3t^2$. How many coulombs were transmitted in 3 seconds. (Hint: $i = dq/dt$ and $q = 0$ when $t = 0$.)

5. If the power in a circuit varied from $t = 0$ to $t = 3$ seconds according to $p = 100(2t^2 + t)$ watts, find the energy expended. (Hint: the relation between power and work or energy is $p = dw/dt$. When $t = 0$ the energy expended is 0.)

6. An 80 microfarad capacitor receives a charge such that it has a terminal to terminal voltage of 100 volts. When $t = 0$ a source is applied that supplies a current $i = t^2$ amperes. Find the voltage across the capacitor when $t = .2$ seconds.

10-6
Applications of the Definite Integral

We saw an example of the definite integral when we computed the area under a curve. In the case of the definite integral we actually evaluate the integral between two limits.

EXAMPLE 16   Evaluate $\displaystyle\int_1^3 x^3\, dx$.

Solution: $\displaystyle\int_1^3 x^3\, dx = \frac{x^4}{4}\Big|_1^3 = \frac{3^4}{4} - \frac{1^4}{4} = \frac{81}{4} - \frac{1}{4}$

$$= 20$$

We will now consider types of problems that can be solved using the definite integral.

The work W done by the force F when applied from position $x = a$ to position $x = b$ is defined to be

$$W = \int_a^b F(x)\, dx$$

EXAMPLE 17   Find the work done in moving a wagon from $x = 0$ ft to $x = 20$ ft if the force is $F = 5x$.

Solution:
$$W = \int_0^{20} 5x\, dx$$

$$W = \frac{5x^2}{2}\bigg|_0^{20}$$

$$W = \frac{5(400)}{2} - \frac{5(0)}{2}$$

$$W = 1000 \text{ ft-lb}$$

Work is also accomplished when a spring is stretched. To stretch a spring a force is required to move the end of the spring through a certain distance. Hooke's law states that an ideal spring exerts a pulling force proportional to the distance that it is stretched. This says that the force is equal to a constant k times a distance x, or $F = kx$.

EXAMPLE 18   What work is required to stretch a spring 6 in. when $K = 2$ lbs per in.

Solution:
$$W = \int_a^b F\, dx$$

$$W = \int_0^6 2x\, dx$$

$$= \frac{2x^2}{2}\bigg|_0^6 = 36 \text{ in. lbs}$$

The volume of geometric solids can be found by using integral calculus. We will next consider an example which involves finding the volume of a cone.

EXAMPLE 19   Find the volume of a right circular cone that has an altitude of 6 in. and base diameter of 6 in.

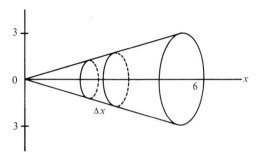

Figure 10-4

Solution: Consider the small sections of the cone each having a thickness of Δx. The approximate volume of each section is the area of the cross section times the thickness Δx. As Δx becomes smaller, the approximate volume approaches the actual volume. If we add all of the volumes together we have an approximate volume of the cone. Recall that we can perform this summation process by using integration and arrive at the exact answer. This was discussed when we found the area under a curve. In other words if we can express the volume as

$$V = \int_0^6 \pi r^2 \, dx$$

where πr^2 is the area of a cross section and dx is the thickness we can integrate and find the sum of all the small volumes represented by $\pi r^2 \, dx$.

From geometry we can write the expression $r = \frac{1}{2}x$ and substitute into the above integral.

$$V = \int_0^6 \pi(\tfrac{1}{2}x)^2 \, dx$$

$$= \int_0^6 \frac{\pi x^2}{4} \, dx$$

$$= \frac{\pi}{4} \int_0^6 x^2 \, dx$$

$$= \frac{\pi}{4} \cdot \frac{x^3}{3} \Big|_0^6$$

$$= \frac{\pi}{4} \left\{ \frac{6^3}{3} - \frac{0^3}{3} \right\} = 18\pi \text{ cu in.}$$

This result can be checked by using the formula $V = \frac{1}{3}bh$ where b is the area of the base and h is the height of the cone.

EXAMPLE 20 The current in a circuit was $i = 4t^3$ amperes. How many coulombs were transmitted in the time period from $t = 0$ seconds to $t = 3$ seconds?

Solution: Recall that

$$q = \int i \, dt$$

We can also express the above integral as a definite integral.

$$q = \int_a^b i \, dt$$

Since $i = 4t^3$ we can write

$$q = \int_0^3 4t^3 \, dt$$

$$= 4 \int_0^3 t^3 \, dt$$

$$= 4 \cdot \frac{t^4}{4} \Big|_0^3 = (3)^4 - (0)^4 = 81 \text{ coulombs}$$

EXAMPLE 21   The velocity of a turtle is given by the expression $v = [2t/(t^2 + 1)]$ in./sec. Find the distance traveled between the times $t = 2$ seconds and $t = 4$ seconds.

Solution: Since $v = dx/dt$ we can write the expression

$$\int dx = \int v \, dt$$

or

$$x = \int v \, dt$$

We can also express this as a definite integral.

$$x = \int_a^b v \, dt$$

Since we know v as a function of t we can write

$$x = \int_2^4 \frac{2t \, dt}{(t^2 + 1)}$$

This expression has the form $\int_a^b du/u$ which integrates to $\log_e u|_a^b$. This formula was presented in an earlier section.

$$x = \int_2^4 \frac{2t \, dt}{(t^2 + 1)} = \log_e (t^2 + 1)|_2^4$$

$$= \log_e (4^2 + 1) - \log_e (2^2 + 1)$$

$$= \log_e 17 - \log_e 5$$

$$= 2.834 - 1.609$$

$$= 1.225 \text{ in.}$$

Needless to say, there are many applications of integrals and all of them can not be presented in the time allowed by a course of this type. To solve many of these problems, much more information than has been presented in this text is required. However, what has been presented should give you an indication of the variety of problems that can be solved by the use of both differential and integral calculus.

<hr>

PROBLEM SET 10-6

1. A certain spring requires a force of 12 lb to stretch it $\frac{1}{4}$ in. Find the work done in stretching it 3 in. beyond its free length.

2. A force $f = 2xe^{x^2}$ is required to move a steel block 2 ft. What is the work done while moving the block?

3. A cable weighing 5 lb per ft is wound on a drum. If one end of the cable was originally 100 ft below the drum, find the work done in winding it up 50 ft. (Hint: the force required is equal to 5 times the length of the cable not yet wound on the drum.)

4. Use the methods of integration to find the volume of a right pyramid whose base is a square with sides 8 in. long and where height is 4 in.

5. Use the methods of integration to find the volume of a frustrum of a right circular cone which has bases with diameters of 4 in. and 12 in. and a height of 8 in.

6. The current in a circuit was $i = 6e^t$ amperes. How many coulombs were transmitted from $t = 2$ seconds to $t = 4$ seconds?

7. The induced current in a circuit is given by the expression

$$i = -\frac{1}{L}\int v\, dt$$

where L is the inductance, v is the induced voltage, and t is time in seconds. If $v = \frac{1}{2}t^3$, what is the current at 2 seconds?

8. The velocity of an automobile is $16t$ ft/sec. What is the distance traveled between $t = 3$ sec and $t = 6$ sec?

11

Curve Fitting

The equations which we have dealt with to this point are called rational equations because they are derived primarily from theoretical considerations. We have learned how to find solutions of rational equations by factoring, completing the square, using the quadratic formula, and various other methods. We have also learned how to differentiate or integrate existing rational equations; however, in the world of work a convenient mathematical equation for a noted phenomena most probably is not available.

Data recorded in the laboratory, in school or on the job, is usually put in tabular form. Often it is desirable to know the equation which best represents, or "fits," the tabular data. Such an equation is called an *empirical equation* and the process of finding it is called curve fitting. Our interest in this chapter is with empirical equations. There are several methods which may be used to determine the equation of the best fitting curve. We shall examine one of the most widely used methods called the *method of average points*.

In general, our first step in determining the equation of the best fitting curve will be to plot the data on ordinary graph paper to determine the type of equation we are dealing with. The types of equations we will be concerned with are the following:

$$y = mx + b \qquad \text{linear}$$
$$y = mx^2 + bx + c \qquad \text{parabolic}$$
$$y = mx^n \qquad \text{power}$$
$$y = mb^{nx} \qquad \text{exponential}$$

In the exponential equation $y = mb^{nx}$, b equals 10, or it may be replaced by an equivalent value $e^{2.303}$ where e is the base of the natural logarithms. After the type equation has been determined, it is necessary to determine the value of the constants such that the empirical equation fits the data satisfactorily.

Several examples, beginning with linear equations, should serve to illustrate how this is accomplished.

11-2
Linear Equations

EXAMPLE 1   A student in an electronics laboratory measured and recorded resistance versus conductor length for No. 28 Nichrome wire. Determine the empirical equation which best fits his tabulated data.

Conductor Length (in.)	Resistance (ohms)
2	.8
4	1.5
6	2.2
8	3.0
10	3.8
12	4.5

Solution: In plotting the data the student noted that the data yielded a linear curve. Knowing this the student could proceed toward a solution recognizing that the general form of the equation for his data was $y = mx + b$. The next step in arriving at a solution is to divide our data into two sets as shown in Figure 11-1, and write linear equations ($y = mx + b$) for each point. In plotting our data we will plot resistance along the vertical axis, by convention the y-axis,

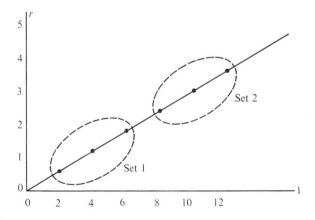

Figure 11-1

and conductor length along the horizontal, or x-axis. Therefore, our first equation is:

$$.8 = 2m + b$$

This says that when $y = .8$, the value of x is 2. Rewriting this equation and the equation for our next two data points yields the following equations for our first set of data points.

$$.8 = 2m + b$$

$$1.5 = 4m + b$$

$$2.2 = 6m + b$$

Set 2 of the data yields the following three equations.

$$3.0 = 8m + b$$

$$3.8 = 10m + b$$

$$4.5 = 12m + b$$

Now if each set of three equations is added we end up with the following two equations in two unknowns which can be solved simultaneously.

$$4.5 = 12m + 3b$$

$$11.3 = 30m + 3b$$

Solving the equations simultaneously yields a value of .378 for m and $-.01$ for b. Therefore the empirical equation for our data is:

$$y = .378x - .01$$

Using this equation, values of y for any value of x may be determined. However, plugging in values of x greater than the largest value recorded in the laboratory is somewhat risky because the curve may not continue to be linear beyond this point. This is known as extrapolation and should be used with caution.

EXAMPLE 2   An electromechanical technician measured the input and output torque on a gear train and recorded the following data.

Input Torque (T_{in})	Output Torque (T_{out})
2 ft lbs	3 ft lbs
4 ft lbs	5.8 ft lbs
6 ft lbs	8.4 ft lbs
8 ft lbs	11.6 ft lbs
10 ft lbs	14.4 ft lbs
12 ft lbs	16.8 ft lbs

Determine the empirical equation for the data.

Solution: Plotting the data yields the linear curve shown in Figure 11.2.

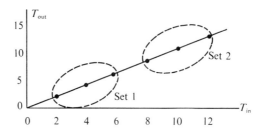

Figure 11-2

Again, the data points are divided into two sets. Linear equations are written for each point.

$$3.0 = m(2) + b$$
$$5.8 = m(4) + b \Big\} \text{Set 1}$$
$$8.4 = m(6) + b$$

$$11.6 = m(8) + b$$
$$14.4 = m(10) + b \Big\} \text{Set 2}$$
$$16.8 = m(12) + b$$

The equations in Set 1 and in Set 2 are added to give a set of two simultaneous equations. These two equations are:

$$17.2 = 12m + 3b$$
$$42.8 = 30m + 3b$$

Solving this set of equations yields the following values for our constants.

$$m = 1.42$$
$$b = .06$$

Therefore the empirical equation for our data is:

$$T_{out} = 1.42T_{in} + .96$$

We can think of many examples of linear relationships between two parameters. Some of these are voltage and current with resistance constant (Ohm's law), force and acceleration (Newton's second law of motion), volume of a gas and absolute temperature (Charles' law), and stress and strain on a body within its elastic limit (Hooke's law).

The examples we have considered previously have fallen in a straight line when plotted. Suppose we consider an example in which the data does not lend itself so conveniently to plotting a linear curve.

EXAMPLE 3 A physics instructor wishes to be able to predict the possibility of students successfully completing his class. To do this he plots the total points versus composite ACT score for former students as shown in Figure 11-3.

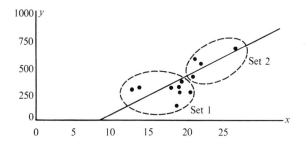

Figure 11-3

The data is shown in tabular form below:

Student	ACT Composite Score (x)	Total Points (y)
A	22	802
B	18	433
C	27	905
D	23	787
E	22	370
F	20	509
G	22	615
H	19	403
I	19	446
J	19	233
K	13	441
L	12	438

Obviously a number of lines, both linear and curved, could be drawn through the plotted points. The instructor believes there is a linear relationship between the variables (composite ACT score and total points). Therefore, two coordinate points are sufficient to identify the best line.

These two points are determined by dividing the data points into two sets and finding the average x and y coordinates.

$$x_a = \frac{12 + 13 + 18 + 19 + 19 + 19 + 20 + 22}{8} = 17.75$$

$$y_a = \frac{441 + 438 + 403 + 446 + 233 + 433 + 509 + 370}{8} = 409.12$$

$$x_b = \frac{22 + 22 + 23 + 27}{4} = 23.50$$

$$y_b = \frac{615 + 802 + 787 + 905}{4} = 777.25$$

The points having coordinates (x_a, y_a) and (x_b, y_b) are located on the graph and a linear curve drawn through them. The equation for the linear curve is determined by solving the set of linear equations.

$$409.12 = m(17.75) + b$$

$$777.25 = m(23.50) + b$$

This gives solutions of:

$$m = 67.15$$

$$b = -800.8$$

Therefore, our empirical equation is:

$$y = 67.15 x - 800.8$$

Knowing the student's composite ACT score, x, the instructor could reasonably predict the grades the student would probably make in physics.

PROBLEM SET 11-1

1. Determine the empirical equation which best fits the following data.

x	0	2	4	6	8	10	12	14
y	0	.3	.57	.84	1.11	1.39	1.66	1.93

2. Determine the empirical equation for the following data.

T	0	1	4	5	7	8
S	30	27	18	15	9	6

3. The following data was recorded by a student in an electronics laboratory. Determine the equation which best fits the data. Using your equation, determine what current will flow in the circuit when 5.2 volts is applied to the circuit.

E(volts)	2	4	6	8	10
I(amps)	.55	1.1	1.6	2.14	2.68

4. A student in a mechanics laboratory performed an experiment to determine the relationship between the load applied to a spring and its length. Determine the empirical equation for his data.

F(lbs)	0	1	2	3	4	5	6
L(in.)	8	8.8	9.7	10.6	11.4	12.2	13

5. According to Charles's law, if pressure remains constant, the volume of a gas decreases almost proportionally to a decrease in temperature. Theoretically, at absolute zero on the Kelvin scale, the gas volume would be zero.

Determine the empirical equation for the following data recorded in a physics laboratory.

$t(°C)$	0	25	50	75	100	125
$v(cm^3$ of He)	100	109	118	127	136	145

Using your equation, determine the Celsius temperature which corresponds to absolute zero.

11-3
Non-Linear Equations

Two points on a straight line are sufficient to allow one to write an empirical equation for a linear curve since the slope at all points must be the same. However, the data must be divided into at least three sets for nonlinear data since the slope is constantly changing.

11-4
Parabolic Equations

The equation $y = mx^2 + bx + c$ is the general form of a parabolic equation. If the parabola passes through the origin its equation is $y = mx^2 + b$. Parabolas are frequently encountered in our physical world. They find application in reflecting telescopes, radar antennas, searchlights, and bridges.

EXAMPLE 4   A student in the physics laboratory is to determine the acceleration due to gravity from the following data which he recorded.

Time (sec)	Distance (in.)
.01	.32
.02	.70
.03	1.10
.04	1.50
.05	1.94
.06	2.48
.07	3.04
.08	3.65
.09	4.23
.10	4.88
.12	6.32
.15	8.68

Solution: In plotting the data the student noted that the resulting curve was parabolic as shown in Figure 11-4. This meant that the general form of the

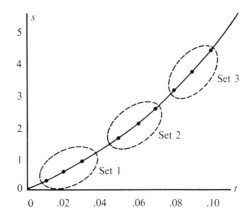

Figure 11-4

equation for the data was $y = mx^2 + bx + c$, or using symbols which are in keeping with the data, $s = mt^2 + bt + c$.

The procedure for determining the empirical equation for a nonlinear curve is to divide the data into three sets, then find the average point for each set.

$$t_1 = \frac{.01 + .02 + .03 + .04}{4} - .025$$

$$s_1 = \frac{.32 + .70 + 1.10 + 1.50}{4} = .905$$

$$t_2 = \frac{.05 + .06 + .07 + .08}{4} = .065$$

$$s_2 = \frac{1.94 + 2.48 + 3.04 + 3.65}{4} = 2.79$$

$$t_3 = \frac{.09 + .10 + .12 + .15}{4} = .115$$

$$s_3 = \frac{4.23 + 4.88 + 6.32 + 8.68}{4} = 6.02$$

Using the three sets of coordinate points, and the general parabolic equation $s = mt^2 + bt + c$, the following three equations may be written.

$$.905 = m(.025)^2 + b(.025) + c$$

$$2.79 = m(.065)^2 + b(.065) + c$$

$$6.02 = m(.115)^2 + b(.115) + c$$

This set of simultaneous equations may be solved using a third-order determinant with the following results for the constants.

$$m = 191$$

$$b = 29.8$$

$$c = .07$$

These constants are then substituted into the equation $s = mt^2 + bt + c$ which yields:

$$s = 191t^2 + 29.8t + .07$$

This is the empirical equation for the data recorded in the laboratory by the student for distance (s) versus time (t). The first derivative of this equation gives an equation for velocity.

$$v = \frac{ds}{dt} = 191(2t) + 29.8$$

$$= 382t + 29.8$$

The derivative of this equation gives an equation for acceleration.

$$a = \frac{dv}{dt} = 382 \text{ in./sec}^2 = 31.83 \text{ ft/sec}^2$$

The acceleration due to gravity is actually 32 ft/sec² so we can see that there is a slight error in our empirical equation.

PROBLEM SET 11-2

1. Determine the empirical equation for the following data.

x	0	1	2	3	4	5
y	3	5	11	21	35	75

2. Determine the empirical equation for the following data.

x	0	1	2	3	4	5
y	-3	-1	3	9	17	27

3. The trajectory of a rocket can be described by a parabolic equation. Determine the empirical equation for the following data. Using your equation, determine how far from its original point the rocket will hit the ground.

x(ft)	1414	2828	4242	5656	7070	8484
y(ft)	1350	2572	3666	4632	5470	6180

4. The support cables on the bridge in the drawing below form a parabolic curve. Determine the equation of the curve from the following data.

x(ft)	0	25	50	75	100	125	150
y(ft)	115	90	65	50	35	26	15

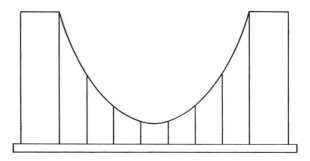

11-5
Power Functions

Suppose we now consider an equation of the type $y = mx^n$ which is nonlinear if n is not equal to one. By taking the logarithm of both sides of this equation we obtain the linear logarithmic equation.

$$\log y = n \log x + \log m$$

If we let $U = \log y$ and $V = \log x$, then our equation can be written as

$$U = nV + \log m$$

where $\log m$ is a constant. We can see that this is a linear equation.

EXAMPLE 5 The current-voltage relationship for a diode is very closely approximated by a power function. An electronics technician records the following data in troubleshooting a logic gate in a computer.

E	.2	.4	.6	.8	1.0	1.2
I	.25	.74	1.35	2.2	3.3	4.4

Plotting the data yields the following curve.

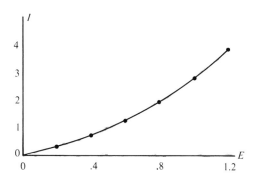

Figure 11-5

Determine the empirical equation for the following data.

Solution: Find the common logarithm for each of the given numbers.

$V = \log E$	$U = \log I$
log .2 $=$ -1.3010 $=$ $-.699$	log .25 $=$ -1.3979 $=$ $-.602$
log .4 $=$ -1.6020 $=$ $-.399$	log .74 $=$ -1.8693 $=$ $-.1307$
log .6 $=$ -1.7781 $=$ $-.222$	log 1.35 $=$ .1303
log .8 $=$ -1.9030 $=$ $-.096$	log 2.2 $=$ .3424
log 1.0 $=$ 0	log 3.3 $=$ .5185
log 1.2 $=$ .0791	log 4.4 $=$ .6434

We must now mention tests by which we can ascertain our original data fits a power function. We can plot the original data on logarithmic-coordinate paper, or we could plot the logarithms of the original data on rectangular-coordinate paper. Both are equivalent, and if the result is a linear curve the original data fit a power function. This is shown in Figure 11-6. Since our data is now linear,

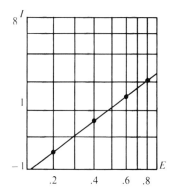

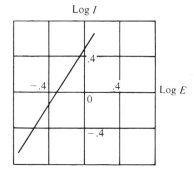

Figure 11-6

we proceed as with a linear function by dividing the data into two sets and finding the average coordinates for each set.

$$Va = \frac{-.699 + (-.399) + (-.222)}{3} = -.440$$

$$Ua = \frac{-.602 + (-.1307) + .1303}{3} = -.201$$

$$Vb = \frac{-.096 + 0 + .0791}{3} = -.0056$$

$$Ub = \frac{+.3424 + .5185 + .6434}{3} = .501$$

The following sets of linear equations can be written for these sets of coordinate points.

$$-.201 = n(-.440) + \log m$$

$$.501 = n(-.0056) + \log m$$

Solving these yields the following.

$$n = 1.61$$

$$\log m = .510$$

The linear equation for our data is

$$U = 1.61V + .510$$

where $\log m = .510$, $m = 3.2$, and $n = 1.61$.
    Therefore our power function is

$$I = 3.2E^{1.61}$$

Some examples of physical relationships which obey the power law are the pressure on a container with temperature constant (Boyle's law), wavelength versus frequency of a sound wave, and the radius of the nucleus of an atom versus the number of neutrons and protons in the nucleus.

PROBLEM SET 11-3

1. Determine the empirical equation for the following data.

x	1	2	3	4	5	6
y	3.3	1.4	.83	.58	.43	.34

2. Determine the empirical equation for the following data.

x	.1	.2	.3	.4	.5	.6
y	22.5	16.5	13.7	12	10.9	10

3. Determine the empirical equation for the following data.

x	10	20	30	40	50	60
y	7.0	1.8	.84	.47	.30	.21

4. The following frequency versus wavelength measurements were made using a wavemeter and a slotted waveguide. Determine the empirical equation in the form $\lambda = mf^n$ for the data.

$f(KHz)$	400	600	800	1000	1200	1400
$\lambda(m)$	750	500	375	300	250	215

Using your equation, determine the wavelength of the upper end of the broadcast band which has a frequency of 1605 KHz.

5. The curve which one obtains by plotting pressure versus volume at a constant temperature is called an isothermal curve and may be described by a power function. The following data represents the pressure-volume relationship in a piston of an internal combustion engine during a portion of the compression cycle. Determine the empirical equation of the form, $P = mV^n$, for the data.

P(psi)	30	25.5	23	20.5	18.5	17
V(in.^{3})	25	30	35	40	45	50

It was at the point where the volume decreased to 15 in.^{3} that the curve became adiabatic, that is, the temperature was no longer constant. From your equation, determine the pressure at that point.

11-6
Exponential Functions

The last type function which we will consider is the exponential function $y = mb^{nx}$. Examples of physical quantities which obey the exponential law are voltage and current in charging or discharging a capacitor, temperature of an object being heated or cooled by putting it in water at a different temperature, and growth or decay of organisms.

EXAMPLE 6 Radium decomposes at a rate proportional to the amount present in an exponential manner. This can be described by the equation $y = m(10)^{nx}$ where b equals 10. If the following data is known, determine the empirical equation which represents the data.

y	100	96	91	59	50
x	0	1	2.6	12	17

Solution: Find the logarithm of the above values of y.

log y	x
log 100 = 2.000	0
log 96 = 1.982	1
log 90 = 1.954	2.6
log 59 = 1.771	12
log 50 = 1.699	17

$v = \log y$	2.0	1.982	1.954	1.771	1.699
x	0	1	2.6	12	17

In order to be certain one's original data fit an exponential curve, the data may be plotted on semi-logarithmic-coordinate paper. The result should be a straight line if log y is plotted along the logarithmic axis and x along the rectangular axis. Divide the data into two sets and find the average coordinates for each set.

$$x_a = \frac{0 + 1 + 2.6}{3} = 1.2$$

$$v_a = \frac{2 + 1.982 + 1.954}{3} = 1.979$$

$$x_b = \frac{12 + 17}{2} = 14.5$$

$$v_b = \frac{1.771 + 1.699}{2} = 1.734$$

From these sets of coordinates the following linear equations may be written.

$$1.979 = n(1.2) + \log m$$

$$1.734 = n(14.5) + \log m$$

Solving this set of equations yields the following.

$$n = -.018$$

$$\log m = 2.00 \quad \text{or} \quad m = 100$$

Therefore, our empirical equation is

$$y = 100(10)^{-.018x}$$

It is often convenient to use the equivalent natural logarithm base, $(e^{2.303})$, in place of the 10. Our equation then becomes

$$y = 100(e^{2.303})^{-.018x}$$

Recognizing that $(e^m)^n = e^{mn}$ we can write our equation as

$$y = 100e^{-.041x}$$

where
$$100 = \text{original quantity of radium}$$
$$y = \text{quantity remaining at time } x$$
$$x = \text{time in centuries.}$$

PROBLEM SET 11-4

1. Determine the empirical equation for the following data.

x	0	1	2	3	4	5	6	7
y	.500	.825	1.36	2.40	3.70	6.09	10.04	16.56

2. Determine the empirical equation for the following data.

x	0	1	2	3	4	5	6	7
y	.200	.775	2.31	7.85	26.41	89.15	295	1000

3. Determine the empirical equation for the following data.

x	0	2	4	6	8	10	12	14
y	.300	.395	.520	.680	.895	1.16	1.52	2.00

4. A student in a physics laboratory performs the following experiment to verify Newton's law of cooling. A thermocouple is attached to a brass sphere and the sphere is heated to 300°F. The sphere is then placed in 70°F water and the following temperature versus time data recorded.

t(sec)	0	25	50	75	100	125	150	175	200
t(°F)	300	260	238	216	201	187	175	164	155

Determine the empirical exponential equation of the form $T = T \cdot e^{-KT}$ for the data. Using your equation, determine the temperature of the sphere after 1000 seconds. (Assume the temperature of the water does not rise.)

5. A certain radioactive substance has a half-life of 1 year which means that for each year that passes the amount of substance remaining is reduced by one-half. Determine the empirical equation for the following data. Using your equation determine the amount of substance remaining after 30 years.

t(years)	0	.5	1	1.5	2	2.5	3
Quantity	10	7.07	5	3.5	2.5	1.8	1.25

Appendix A

UNITS OF MEASURE

The solutions of most scientific and technical problems appear in two parts, a numerical result and a set of units. For example, the solution for the addition of 5 ft and 3 ft is 8 ft with 8 as the numerical result and the units expressed in feet.

There are certain fundamental units of measure by which all physical quantities can be expressed. These include length, time, temperature, and electric charge. A fifth fundamental unit is either force (in engineering) or mass (in physics). All other units of measure are known as *derived units* and are taken from these five fundamental units.

In solving problems, the units are treated with the same laws of algebra that apply to the numbers involved. For example, the area of a rectangle is found by multiplying the numerical values of the length and the width. The units must also be multiplied. The area of a rectangle 4 ft long and 2 ft wide is 8 sq ft. This may be written as 8 ft². The exponent 2 results from multiplying feet times feet.

If the units are correct after solving a problem the results are referred to as *dimensionally correct*. Always check a result to see that it is dimensionally correct.

EXAMPLE 1 What is the speed of an automobile which travels 45 miles in three-quarters of an hour?

Solution: $$\text{speed} = \frac{\text{distance}}{\text{time}}$$

$$\text{speed} = \frac{45 \text{ miles}}{.75 \text{ hours}} = 60 \text{ miles/hour}$$

The correct units for speed or velocity are in distance per unit of time. Therefore the above results are dimensionally correct.

All quantities can be expressed in terms of the fundamental units of measure. However certain quantities which appear very frequently are given special names.

EXAMPLE 2   What force is required to move a mass of one kilogram (Kg) with an acceleration of one meter per second squared (m/s²)?

Solution: Force = (mass)(acceleration)

Force = (1 Kg)(1 m/s²) = 1 Kg · m/s²

The resulting units for force are Kg · m/s². However, force is measured in pounds, dynes, or, in this example, newtons where 1 newton = 1 Kg · m/s². From this example one can see that force has been assigned special units.

There are two systems of units in common use today. These are the British system, widely used in business and industry in English-speaking countries, and the metric system, used in most non-English-speaking countries and in all scientific work. Because technical personnel are exposed to both systems, it is important that one be familiar with both.

In both systems the fundamental units are specified and all others are derived from them. Table 1-1 lists the most common quantities used in scientific or technical applications, the symbol for each, the dimensions, and any special units which apply. The following example illustrates the need for keeping track of the units when solving problems.

EXAMPLE 3   A man walks down a road at 6 m.p.h. His steps are one yard long. On his back he carries a sack of beans. From a hole in the sack one bean per step falls out. If it takes thirty beans to make a bowl of soup and the man walks for 15 minutes how many bowls of soup were lost?

Solution:

$$\frac{6 \text{ miles}}{\text{hour}} \cdot \frac{\text{step}}{\text{yard}} \cdot \frac{\text{bean}}{\text{step}} \cdot \frac{\text{bowl}}{30 \text{ beans}} \cdot \frac{1760 \text{ yards}}{\text{mile}} \cdot \frac{15 \text{ minutes}}{1} \cdot \frac{\text{hour}}{60 \text{ minutes}} = 88 \text{ bowls}$$

All units cancel except "bowl" which is what the problem requires. Hence, the results are dimensionally correct.

Table A-1

Quantity	Symbol	British System Dimensions	British System Special Units	Metric System Dimensions	Metric System Special Units
acceleration	a	ft/sec^2		m/sec^2	
area	A	ft^2		m^2	
energy	E	ft lb		$Kg \cdot m^2/sec^2$	Joule
force	F	lb	pound	$Kg \cdot m/sec^2$	Newton
length	s	ft	foot	m	meter
mass	m	$lb \cdot sec^2/ft$	slug	Kg	kilogram
mass density	d	$slug/ft^3$		Kg/m^3	
momentum	p	$slug \cdot ft/sec$		$Kg \cdot m/sec$	
power	P	$ft \cdot lb/sec$	horsepower	$Kg \cdot m^2/sec^3$	Watt
pressure	p	lb/ft^2		$Kg/m \cdot sec^2$	Newton/meter^2
temperature	T	°F	Fahrenheit	°C	Celsius
time	t	t	second	t	second
torque	τ	$ft \cdot lb$		$Kg \cdot m^2/sec^2$	Newton $\cdot$ meter
velocity	v	ft/sec		m/sec	
volume	V	ft^3		m^3	
work	W	$ft \cdot lb$		$Kg \cdot m^2/sec^2$	Joule
frequency	f	1/sec	Hertz	1/sec	Hertz
period	T	sec		sec	
wave length	λ	ft		m	
capacitance	C			$Q^2 \cdot sec^2/Kg \cdot m^2$	Farad
charge	q			Q	Coulomb
current	I			Q/sec	Ampere
inductance	L			$Kg \cdot m^2/Q^2$	Henry
potential	V			$Kg \cdot m^2/Q \cdot sec^2$	Volt
resistance	R			$Kg \cdot m^2/Q^2 \cdot sec$	Ohm

Appendix B

TABLES

Table B-1 Powers and Roots

No.	Sq	Sq Root	Cube	Cube Root	No.	Sq	Sq Root	Cube	Cube Root
1	1	1.000	1	1.000	51	2,601	7.141	132,651	3.708
2	4	1.414	8	1.260	52	2,704	7.211	140,608	3.733
3	9	1.732	27	1.442	53	2,809	7.280	148,877	3.756
4	16	2.000	64	1.587	54	2,916	7.348	157,464	3.780
5	25	2.236	125	1.710	55	3,025	7.416	166,375	3.803
6	36	2.449	216	1.817	56	3,136	7.483	175,616	3.826
7	49	2.646	343	1.913	57	3,249	7.550	185,193	3.849
8	64	2.828	512	2.000	58	3,364	7.616	195,112	3.871
9	81	3.000	729	2.080	59	3,481	7.681	205,379	3.893
10	100	3.162	1,000	2.154	60	3,600	7.746	216,000	3.915
11	121	3.317	1,331	2.224	61	3,721	7.810	226,981	3.936
12	144	3.464	1,728	2.289	62	3,844	7.874	238,328	3.958
13	169	3.606	2,197	2.351	63	3,969	7.937	250,047	3.979
14	196	3.742	2,744	2.410	64	4,096	8.000	262,144	4.000
15	225	3.873	3,375	2.466	65	4,225	8.062	274,625	4.021
16	256	4.000	4,096	2.520	66	4,356	8.124	287,496	4.041
17	289	4.123	4,913	2.571	67	4,489	8.185	300,763	4.062
18	324	4.243	5,832	2.621	68	4,624	8.246	314,432	4.082
19	361	4.359	6,859	2.668	69	4,761	8.307	328,509	4.102
20	400	4.472	8,000	2.714	70	4,900	8,367	343,000	4.121
21	441	4.583	9,261	2.759	71	5,041	8.426	357,911	4.141
22	484	4.690	10,648	2.802	72	5,184	8.485	373,248	4.160
23	529	4.796	12,167	2.844	73	5,329	8.544	389,017	4.179
24	576	4.899	13,824	2.884	74	5,476	8,602	405,224	4.198
25	625	5.000	15,615	2.924	75	5,625	8.660	421,875	4.217
26	676	5.099	17,576	2.962	76	5,776	8.718	438,976	4.236
27	729	5.196	19,683	3.000	77	5,929	8.775	456,533	4.251
28	784	5.291	21,952	3.037	78	6,084	8.832	474,552	4.273
29	841	5.385	24,389	3.072	79	6,241	8.888	493,039	4.291
30	900	5.477	27,000	3.107	80	6,400	8.944	512,000	4.309
31	961	5.568	29,791	3.141	81	6,561	9.000	531,441	4.327
32	1,024	5.657	32,768	3.175	82	6,724	9.055	551,368	4.344
33	1,089	5.745	35,937	3.208	83	6,889	9.110	571,787	4.362
34	1,156	5.831	39,304	3.240	84	7,056	9.165	592,704	4.380
35	1,225	5.916	42,875	3.271	85	7,225	9.220	614,125	4.397
36	1,296	6.000	46,656	3.302	86	7,396	9.274	636,056	4.414
37	1,369	6.083	50,653	3.332	87	7,569	9.327	658,503	4.431
38	1,444	6.164	54,872	3.362	88	7,744	9.381	681,472	4.448
39	1,521	6.245	59,319	3.391	89	7,921	9.434	704,969	4.465
40	1,600	6.325	64,000	3.420	90	8,100	9.487	729,000	4.481
41	1,681	6.403	68,921	3.448	91	8,281	9.539	753,571	4.498
42	1,764	6.481	74,088	3.476	92	8,464	9.592	778,688	4.514
43	1,849	6.557	79,507	3.503	93	8,649	9.644	804,357	4.531
44	1,936	6.633	85,184	3.530	94	8,836	9.695	830,584	4.547
45	2,025	6.708	91,125	3.557	95	9,025	9.747	857,375	4.563
46	2,116	6.782	97,336	3.583	96	9,216	9.798	884,736	4.579
47	2,209	6.856	103,823	3.609	97	9,409	9.849	912,673	4.595
48	2,304	6.928	110,592	3.634	98	9,604	9.899	941,192	4.610
49	2,401	7.000	117,649	3.659	99	9,801	9.950	970,299	4.626
50	2,500	7.071	125,000	3.685	100	10,000	10.000	1,000,000	4.642

Table B-2 Natural Trigonometric Functions

′	. Sin	Tan	Ctn	Cos	′	′	Sin	Tan	Ctn	Cos	′
0	.00000	.00000	∞	1.0000	60	0	.01745	.01746	57.290	.99985	60
1	.00029	.00029	3437.7	1.0000	59	1	.01774	.01775	56.351	.99984	59
2	.00058	.00058	1718.9	1.0000	58	2	.01803	.01804	55.442	.99984	58
3	.00087	.00087	1145.9	1.0000	57	3	.01832	.01833	54.561	.99983	57
4	.00116	.00116	859.44	1.0000	56	4	.01862	.01862	53.709	.99983	56
5	.00145	.00145	687.55	1.0000	55	5	.01891	.01891	52.882	.99982	55
6	.00175	.00175	572.96	1.0000	54	6	.01920	.01920	52.081	.99982	54
7	.00204	.00204	491.11	1.0000	53	7	.01949	.01949	51.303	.99981	53
8	.00233	.00233	429.72	1.0000	52	8	.01978	.01978	50.549	.99980	52
9	.00262	.00262	381.97	1.0000	51	9	.02007	.02007	49.816	.99980	51
10	.00291	.00291	343.77	1.0000	50	10	.02036	.02036	49.104	.99979	50
11	.00320	.00320	312.52	.99999	49	11	.02065	.02066	48.412	.99979	49
12	.00349	.00349	286.48	.99999	48	12	.02094	.02095	47.740	.99978	48
13	.00378	.00378	264.44	.99999	47	13	.02123	.02124	47.085	.99977	47
14	.00407	.00407	245.55	.99999	46	14	.02152	.02153	46.449	.99977	46
15	.00436	.00436	229.18	.99999	45	15	.02181	.02182	45.829	.99976	45
16	.00465	.00465	214.86	.99999	44	16	.02211	.02211	45.226	.99976	44
17	.00495	.00495	202.22	.99999	43	17	.02240	.02240	44.639	.99975	43
18	.00524	.00524	190.98	.99999	42	18	.02269	.02269	44.066	.99974	42
19	.00553	.00553	180.93	.99998	41	19	.02298	.02298	43.508	.99974	41
20	.00582	.00582	171.89	.99998	40	20	.02327	.02328	42.964	.99973	40
21	.00611	.00611	163.70	.99998	39	21	.02356	.02357	42.433	.99972	39
22	.00640	.00640	156.26	.99998	38	22	.02385	.02386	41.916	.99972	38
23	.00669	.00669	149.47	.99998	37	23	.02414	.02415	41.411	.99971	37
24	.00698	.00698	143.24	.99998	36	24	.02443	.02444	40.917	.99970	36
25	.00727	.00727	137.51	.99997	35	25	.02472	.02473	40.436	.99969	35
26	.00756	.00756	132.22	.99997	34	26	.02501	.02502	39.965	.99969	34
27	.00785	.00785	127.32	.99997	33	27	.02530	.02531	39.506	.99968	33
28	.00814	.00815	122.77	.99997	32	28	.02560	.02560	39.057	.99967	32
29	.00844	.00844	118.54	.99996	31	29	.02589	.02589	38.618	.99966	31
30	.00873	.00873	114.59	.99996	30	30	.02618	.02619	38.188	.99966	30
31	.00902	.00902	110.89	.99996	29	31	.02647	.02648	37.769	.99965	29
32	.00931	.00931	107.43	.99996	28	32	.02676	.02677	37.358	.99964	28
33	.00960	.00960	104.17	.99995	27	33	.02705	.02706	36.956	.99963	27
34	.00989	.00989	101.11	.99995	26	34	.02734	.02735	36.563	.99963	26
35	.01018	.01018	98.218	.99995	25	35	.02763	.02764	36.178	.99962	25
36	.01047	.01047	95.489	.99995	24	36	.02792	.02793	35.801	.99961	24
37	.01076	.01076	92.908	.99994	23	37	.02821	.02822	35.431	.99960	23
38	.01105	.01105	90.463	.99994	22	38	.02850	.02851	35.070	.99959	22
39	.01134	.01135	88.144	.99994	21	39	.02879	.02881	34.715	.99959	21
40	.01164	.01164	85.940	.99993	20	40	.02908	.02910	34.368	.99958	20
41	.01193	.01193	83.844	.99993	19	41	.02938	.02939	34.027	.99957	19
42	.01222	.01222	81.847	.99993	18	42	.02967	.02968	33.694	.99956	18
43	.01251	.01251	79.943	.99992	17	43	.02996	.02997	33.366	.99955	17
44	.01280	.01280	78.126	.99992	16	44	.03025	.03026	33.045	.99954	16
45	.01309	.01309	76.390	.99991	15	45	.03054	.03055	32.730	.99953	15
46	.01338	.01338	74.729	.99991	14	46	.03083	.03084	32.421	.99952	14
47	.01367	.01367	73.139	.99991	13	47	.03112	.03114	32.118	.99952	13
48	.01396	.01396	71.615	.99990	12	48	.03141	.03143	31.821	.99951	12
49	.01425	.01425	70.153	.99990	11	49	.03170	.03172	31.528	.99950	11
50	.01454	.01455	68.750	.99989	10	50	.03199	.03201	31.242	.99949	10
51	.01483	.01484	67.402	.99989	9	51	.03228	.03230	30.960	.99948	9
52	.01513	.01513	66.105	.99989	8	52	.03257	.03259	30.683	.99947	8
53	.01542	.01542	64.858	.99988	7	53	.03286	.03288	30.412	.99946	7
54	.01571	.01571	63.657	.99988	6	54	.03316	.03317	30.145	.99945	6
55	.01600	.01600	62.499	.99987	5	55	.03345	.03346	29.882	.99944	5
56	.01629	.01629	61.383	.99987	4	56	.03374	.03376	29.624	.99943	4
57	.01658	.01658	60.306	.99986	3	57	.03403	.03405	29.371	.99942	3
58	.01687	.01687	59.266	.99986	2	58	.03432	.03434	29.122	.99941	2
59	.01716	.01716	58.261	.99985	1	59	.03461	.03463	28.877	.99940	1
60	.01745	.01746	57.290	.99985	0	60	.03490	.03492	28.636	.99939	0
′	Cos	Ctn	Tan	Sin	′	′	Cos	Ctn	Tan	Sin	′

Table B-2 (*cont.*)

′	Sin	Tan	Ctn	Cos	′	′	Sin	Tan	Ctn	Cos	′
0	.03490	.03492	28.636	.99939	60	0	.05234	.05241	19.081	.99863	60
1	.03519	.03521	28.399	.99938	59	1	.05263	.05270	18.976	.99861	59
2	.03548	.03550	28.166	.99937	58	2	.05292	.05299	18.871	.99860	58
3	.03577	.03579	27.937	.99936	57	3	.05321	.05328	18.768	.99858	57
4	.03606	.03609	27.712	.99935	56	4	.05350	.05357	18.666	.99857	56
5	.03635	.03638	27.490	.99934	55	5	.05379	.05387	18.564	.99855	55
6	.03664	.03667	27.271	.99933	54	6	.05408	.05416	18.464	.99854	54
7	.03693	.03696	27.057	.99932	53	7	.05437	.05445	18.366	.99852	53
8	.03723	.03725	26.845	.99931	52	8	.05466	.05474	18.268	.99851	52
9	.03752	.03754	26.637	.99930	51	9	.05495	.05503	18.171	.99849	51
10	.03781	.03783	26.432	.99929	50	10	.05524	.05533	18.075	.99847	50
11	.03810	.03812	26.230	.99927	49	11	.05553	.05562	17.980	.99846	49
12	.03839	.03842	26.031	.99926	48	12	.05582	.05591	17.886	.99844	48
13	.03868	.03871	25.835	.99925	47	13	.05611	.05620	17.793	.99842	47
14	.03897	.03900	25.642	.99924	46	14	.05640	.05649	17.702	.99841	46
15	.03926	.03929	25.452	.99923	45	15	.05669	.05678	17.611	.99839	45
16	.03955	.03958	25.264	.99922	44	16	.05698	.05708	17.521	.99838	44
17	.03984	.03987	25.080	.99921	43	17	.05727	.05737	17.431	.99836	43
18	.04013	.04016	24.898	.99919	42	18	.05756	.05766	17.343	.99834	42
19	.04042	.04046	24.719	.99918	41	19	.05785	.05795	17.256	.99833	41
20	.04071	.04075	24.542	.99917	40	20	.05814	.05824	17.169	.99831	40
21	.04100	.04104	24.368	.99916	39	21	.05844	.05854	17.084	.99829	39
22	.04129	.04133	24.196	.99915	38	22	.05873	.05883	16.999	.99827	38
23	.04159	.04162	24.026	.99913	37	23	.05902	.05912	16.915	.99826	37
24	.04188	.04191	23.859	.99912	36	24	.05931	.05941	16.832	.99824	36
25	.04217	.04220	23.695	.99911	35	25	.05960	.05970	16.750	.99822	35
26	.04246	.04250	23.532	.99910	34	26	.05999	.05999	16.668	.99821	34
27	.04275	.04279	23.372	.99909	33	27	.06018	.06029	16.587	.99819	33
28	.04304	.04308	23.214	.99907	32	28	.06047	.06058	16.507	.99817	32
29	.04333	.04337	23.058	.99906	31	29	.06076	.06087	16.428	.99815	31
30	.04362	.04366	22.904	.99905	30	30	.06105	.06116	16.350	.99813	30
31	.04391	.04395	22.752	.99904	29	31	.06134	.06145	16.272	.99812	29
32	.04420	.04424	22.602	.99902	28	32	.06163	.06175	16.195	.99810	28
33	.04449	.04454	22.454	.99901	27	33	.06192	.06204	16.119	.99808	27
34	.04478	.04483	22.308	.99900	26	34	.06221	.06233	16.043	.99806	26
35	.04507	.04512	22.164	.99898	25	35	.06250	.06262	15.969	.99804	25
36	.04536	.04541	22.022	.99897	24	36	.06279	.06291	15.895	.99803	24
37	.04565	.04570	21.881	.99896	23	37	.06308	.06321	15.821	.99801	23
38	.04594	.04599	21.743	.99894	22	38	.06337	.06350	15.748	.99799	22
39	.04623	.04628	21.606	.99893	21	39	.06366	.06379	15.676	.99797	21
40	.04653	.04658	21.470	.99892	20	40	.06395	.06408	15.605	.99795	20
41	.04682	.04687	21.337	.99890	19	41	.06424	.06438	15.534	.99793	19
42	.04711	.04716	21.205	.99889	18	42	.06453	.06467	15.464	.99792	18
43	.04740	.04745	21.075	.99888	17	43	.06482	.06496	15.394	.99790	17
44	.04769	.04774	20.946	.99886	16	44	.06511	.06525	15.325	.99788	16
45	.04798	.04803	20.819	.99885	15	45	.06540	.06554	15.257	.99786	15
46	.04827	.04833	20.693	.99883	14	46	.06569	.06584	15.189	.99784	14
47	.04856	.04862	20.569	.99882	13	47	.06598	.06613	15.122	.99782	13
48	.04885	.04891	20.446	.99881	12	48	.06627	.06642	15.056	.99780	12
49	.04914	.04920	20.325	.99879	11	49	.06656	.06671	14.990	.99778	11
50	.04943	.04949	20.206	.99878	10	50	.06685	.06700	14.924	.99776	10
51	.04972	.04978	20.087	.99876	9	51	.06714	.06730	14.860	.99774	9
52	.05001	.05007	19.970	.99875	8	52	.06743	.06759	14.795	.99772	8
53	.05030	.05037	19.855	.99873	7	53	.06773	.06788	14.732	.99770	7
54	.05059	.05066	19.740	.99872	6	54	.06802	.06817	14.669	.99768	6
55	.05088	.05095	19.627	.99870	5	55	.06831	.06847	14.606	.99766	5
56	.05117	.05124	19.516	.99869	4	56	.06860	.06876	14.544	.99764	4
57	.05146	.05153	19.405	.99867	3	57	.06889	.06905	14.482	.99762	3
58	.05175	.05182	19.296	.99866	2	58	.06918	.06934	14.421	.99760	2
59	.05205	.05112	19.188	.99864	1	59	.06947	.06963	14.361	.99758	1
60	.05234	.05241	19.081	.99863	0	60	.06976	.06993	14.301	.99756	0

′	Cos	Ctn	Tan	Sin	′	′	Cos	Ctn	Tan	Sin	′

4° (184°)				**(355°) 175°**		**5° (185°)**				**(354°) 174°**	
′	Sin	Tan	Ctn	Cos	′	′	Sin	Tan	Ctn	Cos	′
0	.06976	.06993	14.301	.99756	60	0	.08716	.08749	11.430	.99619	60
1	.07005	.07022	14.241	.99754	59	1	.08745	.08778	11.392	.99617	59
2	.07034	.07051	14.182	.99752	58	2	.08774	.08807	11.354	.99614	58
3	.07063	.07080	14.124	.99750	57	3	.08803	.08837	11.316	.99612	57
4	.07092	.07110	14.065	.99748	56	4	.08831	.08866	11.279	.99609	56
5	.07121	.07139	14.008	.99746	55	5	.08860	.08895	11.242	.99607	55
6	.07150	.07168	13.951	.99744	54	6	.08889	.08925	11.205	.99604	54
7	.07179	.07197	13.894	.99742	53	7	.08918	.08954	11.168	.99602	53
8	.07208	.07227	13.838	.99740	52	8	.08947	.08983	11.132	.99599	52
9	.07237	.07256	13.782	.99738	51	9	.08976	.09013	11.095	.99596	51
10	.07266	.07285	13.727	.99736	50	10	.09005	.09042	11.059	.99594	50
11	.07295	.07314	13.672	.99734	49	11	.09034	.09071	11.024	.99591	49
12	.07324	.07344	13.617	.99731	48	12	.09063	.09101	10.988	.99588	48
13	.07353	.07373	13.563	.99729	47	13	.09092	.09130	10.953	.99586	47
14	.07382	.07402	13.510	.99727	46	14	.09121	.09159	10.918	.99583	46
15	.07411	.07431	13.457	.99725	45	15	.09150	.09189	10.883	.99580	45
16	.07440	.07461	13.404	.99723	44	16	.09179	.09218	10.848	.99578	44
17	.07469	.07490	13.352	.99721	43	17	.09208	.09247	10.814	.99575	43
18	.07498	.07519	13.300	.99719	42	18	.09237	.09277	10.780	.99572	42
19	.07527	.07548	13.248	.99716	41	19	.09266	.09306	10.746	.99570	41
20	.07556	.07578	13.197	.99714	40	20	.09295	.09335	10.712	.99567	40
21	.07585	.07607	13.146	.99712	39	21	.09324	.09365	10.678	.99564	39
22	.07614	.07636	13.096	.99710	38	22	.09353	.09394	10.645	.99562	38
23	.07643	.07665	13.046	.99708	37	23	.09382	.09423	10.612	.99559	37
24	.07672	.07695	12.996	.99705	36	24	.09411	.09453	10.579	.99556	36
25	.07701	.07724	12.947	.99703	35	25	.09440	.09482	10.546	.99553	35
26	.07730	.07753	12.898	.99701	34	26	.09469	.09511	10.514	.99551	34
27	.07759	.07782	12.850	.99699	33	27	.09498	.09541	10.481	.99548	33
28	.07788	.07812	12.801	.99696	32	28	.09527	.09570	10.449	.99545	32
29	.07817	.07841	12.754	.99694	31	29	.09556	.09600	10.417	.99542	31
30	.07846	.07870	12.706	.99692	30	30	.09585	.09629	10.385	.99540	30
31	.07875	.07899	12.659	.99689	29	31	.09614	.09658	10.354	.99537	29
32	.07904	.07929	12.612	.99687	28	32	.09642	.09688	10.322	.99534	28
33	.07933	.07958	12.566	.99685	27	33	.09671	.09717	10.291	.99531	27
34	.07962	.07987	12.520	.99683	26	34	.09700	.09746	10.260	.99528	26
35	.07991	.08017	12.474	.99680	25	35	.09729	.09776	10.229	.99526	25
36	.08020	.08046	12.429	.99678	24	36	.09758	.09805	10.199	.99523	24
37	.08049	.08075	12.384	.99676	23	37	.09787	.09834	10.168	.99520	23
38	.08078	.08104	12.339	.99673	22	38	.09816	.09864	10.138	.99517	22
39	.08107	.08134	12.295	.99671	21	39	.09845	.09893	10.108	.99514	21
40	.08136	.08163	12.251	.99668	20	40	.09874	.09923	10.078	.99511	20
41	.08165	.08192	12.207	.99666	19	41	.09903	.09952	10.048	.99508	19
42	.08194	.08221	12.163	.99664	18	42	.09932	.09981	10.019	.99506	18
43	.08223	.08251	12.120	.99661	17	43	.09961	.10011	9.9893	.99503	17
44	.08252	.08280	12.077	.99659	16	44	.09990	.10040	9.9601	.99500	16
45	.08281	.08309	12.035	.99657	15	45	.10019	.10069	9.9310	.99497	15
46	.08310	.08339	11.992	.99654	14	46	.10048	.10099	9.9021	.99494	14
47	.08339	.08368	11.950	.99652	13	47	.10077	.10128	9.8734	.99491	13
48	.08368	.08397	11.909	.99649	12	48	.10106	.10158	9.8448	.99488	12
49	.08397	.08427	11.867	.99647	11	49	.10135	.10187	9.8164	.99485	11
50	.08426	.08456	11.826	.99644	10	50	.10164	.10216	9.7882	.99482	10
51	.08455	.08485	11.785	.99642	9	51	.10192	.10246	9.7601	.99479	9
52	.08484	.08514	11.745	.99639	8	52	.10221	.10275	9.7322	.99476	8
53	.08513	.08544	11.705	.99637	7	53	.10250	.10305	9.7044	.99473	7
54	.08542	.08573	11.664	.99635	6	54	.10279	.10334	9.6768	.99470	6
55	.08571	.08602	11.625	.99632	5	55	.10308	.10363	9.6493	.99467	5
56	.08600	.08632	11.585	.99630	4	56	.10337	.10393	9.6220	.99464	4
57	.08629	.08661	11.546	.99627	3	57	.10366	.10422	9.5949	.99461	3
58	.08658	.08690	11.507	.99625	2	58	.10395	.10452	9.5679	.99458	2
59	.08687	.08720	11.468	.99622	1	59	.10424	.10481	9.5411	.99455	1
60	.08716	.08749	11.430	.99619	0	60	.10453	.10510	9.5144	.99452	0
′	Cos	Ctn	Tan	Sin	′	′	Cos	Ctn	Tan	Sin	′

Table B-2 (*cont.*)

′	Sin	Tan	Ctn	Cos	′	′	Sin	Tan	Ctn	Cos	′
0	.10453	.10510	9.5144	.99452	60	0	.12187	.12278	8.1443	.99255	60
1	.10482	.10540	9.4878	.99449	59	1	.12216	.12308	8.1248	.99251	59
2	.10511	.10569	9.4614	.99446	58	2	.12245	.12338	8.1054	.99248	58
3	.10540	.10599	9.4352	.99443	57	3	.12274	.12367	8.0860	.99244	57
4	.10569	.10628	9.4090	.99440	56	4	.12302	.12397	8.0667	.99240	56
5	.10597	.10657	9.3831	.99437	55	5	.12331	.12426	8.0476	.99237	55
6	.10626	.10687	9.3572	.99434	54	6	.12360	.12456	8.0285	.99233	54
7	.10655	.10716	9.3315	.99431	53	7	.12389	.12485	8.0095	.99230	53
8	.10684	.10746	9.3060	.99428	52	8	.12418	.12515	7.9906	.99226	52
9	.10713	.10775	9.2806	.99424	51	9	.12447	.12544	7.9718	.99222	51
10	.10742	.10805	9.2553	.99421	50	10	.12476	.12574	7.9530	.99219	50
11	.10771	.10834	9.2302	.99418	49	11	.12504	.12603	7.9344	.99215	49
12	.10800	.10863	9.2052	.99415	48	12	.12533	.12633	7.9158	.99211	48
13	.10829	.10893	9.1803	.99412	47	13	.12562	.12662	7.8973	.99208	47
14	.10858	.10922	9.1555	.99409	46	14	.12591	.12692	7.8789	.99204	46
15	.10887	.10952	9.1309	.99406	45	15	.12620	.12722	7.8606	.99200	45
16	.10916	.10981	9.1065	.99402	44	16	.12649	.12751	7.8424	.99197	44
17	.10945	.11011	9.0821	.99399	43	17	.12678	.12781	7.8243	.99193	43
18	.10973	.11040	9.0579	.99396	42	18	.12706	.12810	7.8062	.99189	42
19	.11002	.11070	9.0338	.99393	41	19	.12735	.12840	7.7882	.99186	41
20	.11031	.11099	9.0098	.99390	40	20	.12764	.12869	7.7704	.99182	40
21	.11060	.11128	8.9860	.99386	39	21	.12793	.12899	7.7525	.99178	39
22	.11089	.11158	8.9623	.99383	38	22	.12822	.12929	7.7348	.99175	38
23	.11118	.11187	8.9387	.99380	37	23	.12851	.12958	7.7171	.99171	37
24	.11147	.11217	8.9152	.99377	36	24	.12880	.12988	7.6996	.99167	36
25	.11176	.11246	8.8919	.99374	35	25	.12908	.13017	7.6821	.99163	35
26	.11205	.11276	8.8686	.99370	34	26	.12937	.13047	7.6647	.99160	34
27	.11234	.11305	8.8455	.99367	33	27	.12966	.13076	7.6473	.99156	33
28	.11263	.11335	8.8225	.99364	32	28	.12995	.13106	7.6301	.99152	32
29	.11291	.11364	8.7996	.99360	31	29	.13024	.13136	7.6129	.99148	31
30	.11320	.11394	8.7769	.99357	30	30	.13053	.13165	7.5958	.99144	30
31	.11349	.11423	9.7542	.99354	29	31	.13081	.13195	7.5787	.99141	29
32	.11378	.11452	8.7317	.99351	28	32	.13110	.13224	7.5618	.99137	28
33	.11407	.11482	8.7093	.99347	27	33	.13139	.13254	7.5449	.99133	27
34	.11436	.11511	8.6870	.99344	26	34	.13168	.13284	7.5281	.99129	26
35	.11465	.11541	8.6648	.99341	25	35	.13197	.13313	7.5113	.99125	25
36	.11494	.11570	8.6427	.99337	24	36	.13226	.13343	7.4947	.99122	24
37	.11523	.11600	8.6208	.99334	23	37	.13254	.13372	7.4781	.99118	23
38	.11552	.11629	8.5989	.99331	22	38	.13283	.13402	7.4615	.99114	22
39	.11580	.11659	8.5772	.99327	21	39	.13312	.13432	7.4451	.99110	21
40	.11609	.11688	8.5555	.99324	20	40	.13341	.13461	7.4287	.99106	20
41	.11638	.11718	8.5340	.99320	19	41	.13370	.13491	7.4124	.99102	19
42	.11667	.11747	8.5126	.99317	18	42	.13399	.13521	7.3962	.99098	18
43	.11696	.11777	8.4913	.99314	17	43	.13427	.13550	7.3800	.99094	17
44	.11725	.11806	8.4701	.99310	16	44	.13456	.13580	7.3639	.99091	16
45	.11754	.11836	8.4490	.99307	15	45	.13485	.13609	7.3479	.99087	15
46	.11783	.11865	8.4280	.99303	14	46	.13514	.13639	7.3319	.99083	14
47	.11812	.11895	8.4071	.99300	13	47	.13543	.13669	7.3160	.99079	13
48	.11840	.11924	8.3863	.99297	12	48	.13572	.13698	7.3002	.99075	12
49	.11869	.11954	8.3656	.99293	11	49	.13600	.13728	7.2844	.99071	11
50	.11898	.11983	8.3450	.99290	10	50	.13629	.13758	7.2687	.99067	10
51	.11927	.12013	8.3245	.99286	9	51	.13658	.13787	7.2531	.99063	9
52	.11956	.12042	8.3041	.99283	8	52	.13687	.13817	7.2375	.99059	8
53	.11985	.12072	8.2838	.99279	7	53	.13716	.13846	7.2220	.99055	7
54	.12014	.12101	8.2636	.99276	6	54	.13744	.13876	7.2066	.99051	6
55	.12043	.12131	8.2434	.99272	5	55	.13773	.13906	7.1912	.99047	5
56	.12071	.12160	8.2234	.99269	4	56	.13802	.13935	7.1759	.99043	4
57	.12100	.12190	8.2035	.99265	3	57	.13831	.13965	7.1607	.99039	3
58	.12129	.12219	8.1837	.99262	2	58	.13860	.13995	7.1455	.99035	2
59	.12158	.12249	8.1640	.99258	1	59	.13889	.14024	7.1304	.99031	1
60	.12187	.12278	8.1443	.99255	0	60	.13917	.14054	7.1154	.99027	0

′	Cos	Ctn	Tan	Sin	′	′	Cos	Ctn	Tan	Sin	′

Table B-2 (*cont.*)

′	Sin	Tan	Ctn	Cos	′	′	Sin	Tan	Ctn	Cos	′
0	.13917	.14054	7.1154	.99027	60	0	.15643	.15838	6.3138	.98769	60
1	.13946	.14084	7.1004	.99023	59	1	.15672	.15868	6.3019	.98764	59
2	.13975	.14113	7.0855	.99019	58	2	.15701	.15898	6.2901	.98760	58
3	.14004	.14143	7.0706	.99015	57	3	.15730	.15928	6.2783	.98755	57
4	.14033	.14173	7.0558	.99011	56	4	.15758	.15958	6.2666	.98751	56
5	.14061	.14202	7.0410	.99006	55	5	.15787	.15988	6.2549	.98746	55
6	.14090	.14232	7.0264	.99002	54	6	.15816	.16017	6.2432	.98741	54
7	.14119	.14262	7.0117	.98998	53	7	.15845	.16047	6.2316	.98737	53
8	.14148	.14291	6.9972	.98994	52	8	.15873	.16077	6.2200	.98732	52
9	.14177	.14321	6.9827	.98990	51	9	.15902	.16107	6.2085	.98728	51
10	.14205	.14351	6.9682	.98986	50	10	.15931	.16137	6.1970	.98723	50
11	.14234	.14381	6.9538	.98982	49	11	.15959	.16167	6.1856	.98718	49
12	.14263	.14410	6.9395	.98978	48	12	.15988	.16196	6.1742	.98714	48
13	.14292	.14440	6.9252	.98973	47	13	.16017	.16226	6.1628	.98709	47
14	.14320	.14470	6.9110	.98969	46	14	.16046	.16256	6.1515	.98704	46
15	.14349	.14499	6.8969	.98965	45	15	.16074	.16286	6.1402	.98700	45
16	.14378	.14529	6.8828	.98961	44	16	.16103	.16316	6.1290	.98695	44
17	.14407	.14559	6.8687	.98957	43	17	.16132	.16346	6.1178	.98690	43
18	.14436	.14588	6.8548	.98953	42	18	.16160	.16376	6.1066	.98686	42
19	.14464	.14618	6.8408	.98948	41	19	.16189	.16405	6.0955	.98681	41
20	.14493	.14648	6.8269	.98944	40	20	.16218	.16435	6.0844	.98676	40
21	.14522	.14678	6.8131	.98940	39	21	.16246	.16465	6.0734	.98671	39
22	.14551	.14707	6.7994	.98936	38	22	.16275	.16495	6.0624	.98667	38
23	.14580	.14737	6.7856	.98931	37	23	.16304	.16525	6.0514	.98662	37
24	.14608	.14767	6.7720	.98927	36	24	.16333	.16555	6.0405	.98657	36
25	.14637	.14796	6.7584	.98923	35	25	.16361	.16585	6.0296	.98652	35
26	.14666	.14826	6.7448	.98919	34	26	.16390	.16615	6.0188	.98648	34
27	.14695	.14856	6.7313	.98914	33	27	.16419	.16645	6.0080	.98643	33
28	.14723	.14886	6.7179	.98910	32	28	.16447	.16674	5.9972	.98638	32
29	.14752	.14915	6.7045	.98906	31	29	.16476	.16704	5.9865	.98633	31
30	.14781	.14945	6.6912	.98902	30	30	.16505	.16734	5.9758	.98629	30
31	.14810	.14875	6.6779	.98897	29	31	.16533	.16764	5.9651	.98624	29
32	.14838	.15005	6.6646	.98893	28	32	.16562	.16794	5.9545	.98619	28
33	.14867	.15034	6.6514	.98889	27	33	.16591	.16824	5.9439	.98614	27
34	.14896	.15064	6.6383	.98884	26	34	.16620	.16854	5.9333	.98609	26
35	.14925	.15094	6.6252	.98880	25	35	.16648	.16884	5.9228	.98604	25
36	.14954	.15124	6.6122	.98876	24	36	.16677	.16914	5.9124	.98600	24
37	.14982	.15153	6.5992	.98871	23	37	.16706	.16944	5.9019	.98595	23
38	.15011	.15183	6.5863	.98867	22	38	.16734	.16974	5.8915	.98590	22
39	.15040	.15213	6.5734	.98863	21	39	.16763	.17004	5.8811	.98585	21
40	.15069	.15243	6.5606	.98858	20	40	.16792	.17033	5.8708	.98580	20
41	.15097	.15272	6.5478	.98854	19	41	.16820	.17063	5.8605	.98575	19
42	.15126	.15302	6.5350	.98849	18	42	.16849	.17093	5.8502	.98570	18
43	.15155	.15332	6.5223	.98845	17	43	.16878	.17123	5.8400	.98565	17
44	.15184	.15362	6.5097	.98841	16	44	.16906	.17153	5.8298	.98561	16
45	.15212	.15391	6.4971	.98836	15	45	.16935	.17183	5.8197	.98556	15
46	.15241	.15421	6.4846	.98832	14	46	.16964	.17213	5.8095	.98551	14
47	.15270	.15451	6.4721	.98827	13	47	.16992	.17243	5.7994	.98546	13
48	.15299	.15481	6.4596	.98823	12	48	.17021	.17273	5.7894	.98541	12
49	.15327	.15511	6.4472	.98818	11	49	.17050	.17303	5.7794	.98536	11
50	.15356	.15540	6.4348	.98814	10	50	.17078	.17333	5.7694	.98531	10
51	.15385	.15570	6.4225	.98809	9	51	.17107	.17363	5.7594	.98526	9
52	.15414	.15600	6.4103	.98805	8	52	.17136	.17393	5.7495	.98521	8
53	.15442	.15630	6.3980	.98800	7	53	.17164	.17423	5.7396	.98516	7
54	.15471	.15660	6.3859	.98796	6	54	.17193	.17453	5.7297	.98511	6
55	.15500	.15689	6.3737	.98791	5	55	.17222	.17483	5.7199	.98506	5
56	.15529	.15719	6.3617	.98787	4	56	.17250	.17513	5.7101	.98501	4
57	.15557	.15749	6.3496	.98782	3	57	.17279	.17543	5.7004	.98496	3
58	.15586	.15779	6.3376	.98778	2	58	.17308	.17573	5.6906	.98491	2
59	.15615	.15809	6.3257	.98773	1	59	.17336	.17603	5.6809	.98486	1
60	.15643	.15838	6.3138	.98769	0	60	.17365	.17633	5.6713	.98481	0
′	Cos	Ctn	Tan	Sin	′	′	Cos	Ctn	Tan	Sin	′

18° (190°)   (349°) **169°**

′	Sin	Tan	Ctn	Cos	′
0	.17365	.17633	5.6713	.98481	60
1	.17393	.17663	5.6617	.98476	59
2	.17422	.17693	5.6521	.98471	58
3	.17451	.17723	5.6425	.98466	57
4	.17479	.17753	5.6329	.98461	56
5	.17508	.17783	5.6234	.98455	55
6	.17537	.17813	5.6140	.98450	54
7	.17565	.17843	5.6045	.98445	53
8	.17594	.17873	5.5951	.98440	52
9	.17623	.17903	5.5857	.98435	51
10	.17651	.17933	5.5764	.98430	50
11	.17680	.17963	5.5671	.98425	49
12	.17708	.17993	5.5578	.98420	48
13	.17737	.18023	5.5485	.98414	47
14	.17766	.18053	5.5393	.98409	46
15	.17794	.18083	5.5301	.98404	45
16	.17823	.18113	5.5209	.98399	44
17	.17852	.18143	5.5118	.98394	43
18	.17880	.18173	5.5026	.98389	42
19	.17909	.18203	5.4936	.98383	41
20	.17937	.18233	5.4845	.98378	40
21	.17966	.18263	5.4755	.98373	39
22	.17995	.18293	5.4665	.98368	38
23	.18023	.18323	5.4575	.98362	37
24	.18052	.18353	5.4486	.98357	36
25	.18081	.18384	5.4397	.98352	35
26	.18109	.18414	5.4308	.98347	34
27	.18138	.18444	5.4219	.98341	33
28	.18166	.18474	5.4131	.98336	32
29	.18195	.18504	5.4043	.98331	31
30	.18224	.18534	5.3955	.98325	30
31	.18252	.18564	5.3868	.98320	29
32	.18281	.18594	5.3781	.98315	28
33	.18309	.18624	5.3694	.98310	27
34	.18338	.18654	5.3607	.98304	26
35	.18367	.18684	5.3521	.98299	25
36	.18395	.18714	5.3435	.98294	24
37	.18424	.18745	5.3349	.98288	23
38	.18452	.18775	5.3263	.98283	22
39	.18481	.18805	5.3178	.98277	21
40	.18509	.18835	5.3093	.98272	20
41	.18538	.18865	5.3008	.98267	19
42	.18567	.18895	5.2924	.98261	18
43	.18595	.18925	5.2839	.98256	17
44	.18624	.18955	5.2755	.98250	16
45	.18652	.18986	5.2672	.98245	15
46	.18681	.19016	5.2588	.98240	14
47	.18710	.19046	5.2505	.98234	13
48	.18738	.19076	5.2422	.98229	12
49	.18767	.19106	5.2339	.98223	11
50	.18795	.19136	5.2257	.98218	10
51	.18824	.19166	5.2174	.98212	9
52	.18852	.19197	5.2092	.98207	8
53	.18881	.19227	5.2011	.98201	7
54	.18910	.19257	5.1929	.98196	6
55	.18938	.19287	5.1848	.98190	5
56	.18967	.19317	5.1767	.98185	4
57	.18995	.19347	5.1686	.98179	3
58	.19024	.19378	5.1606	.98174	2
59	.19052	.19408	5.1526	.98168	1
60	.19081	.19438	5.1446	.98163	0
′	Cos	Ctn	Tan	Sin	′

100° (280°)   (259°) **79°**

11° (191°)   (348°) **168°**

′	Sin	Tan	Ctn	Cos	′
0	.19081	.19438	5.1446	.98163	60
1	.19109	.19468	5.1366	.98157	59
2	.19138	.19498	5.1286	.98152	58
3	.19167	.19529	5.1207	.98146	57
4	.19195	.19559	5.1128	.98140	56
5	.19224	.19589	5.1049	.98135	55
6	.19252	.19619	5.0970	.98129	54
7	.19281	.19649	5.0892	.98124	53
8	.19309	.19680	5.0814	.98118	52
9	.19338	.19710	5.0736	.98112	51
10	.19366	.19740	5.0658	.98107	50
11	.19395	.19770	5.0581	.98101	49
12	.19423	.19801	5.0504	.98096	48
13	.19452	.19831	5.0427	.98090	47
14	.19481	.19861	5.0350	.98084	46
15	.19509	.19891	5.0273	.98079	45
16	.19538	.19921	5.0197	.98073	44
17	.19566	.19952	5.0121	.98067	43
18	.19595	.19982	5.0045	.98061	42
19	.19623	.20012	4.9969	.98056	41
20	.19652	.20042	4.9894	.98050	40
21	.19680	.20073	4.9819	.98044	39
22	.19709	.20103	4.9744	.98039	38
23	.19737	.20133	4.9669	.98033	37
24	.19766	.20164	4.9594	.98027	36
25	.19794	.20194	4.9520	.98021	35
26	.19823	.20224	4.9446	.98016	34
27	.19851	.20254	4.9372	.98010	33
28	.19880	.20285	4.9298	.98004	32
29	.19908	.20315	4.9225	.97998	31
30	.19937	.20345	4.9152	.97992	30
31	.19965	.20376	4.9078	.97987	29
32	.19994	.20406	4.9006	.97981	28
33	.20022	.20436	4.8933	.97975	27
34	.20051	.20466	4.8860	.97969	26
35	.20079	.20497	4.8788	.97963	25
36	.20108	.20527	4.8716	.97958	24
37	.20136	.20557	4.8644	.97952	23
38	.20165	.20588	4.8573	.97946	22
39	.20193	.20618	4.8501	.97940	21
40	.20222	.20648	4.8430	.97934	20
41	.20250	.20679	4.8359	.97928	19
42	.20279	.20709	4.8288	.97922	18
43	.20307	.20739	4.8218	.97916	17
44	.20336	.20770	4.8147	.97910	16
45	.20364	.20800	4.8077	.97905	15
46	.20393	.20830	4.8007	.97899	14
47	.20421	.20861	4.7937	.97893	13
48	.20450	.20891	4.7867	.97887	12
49	.20478	.20921	4.7798	.97881	11
50	.20507	.20952	4.7729	.97875	10
51	.20535	.20982	4.7659	.97869	9
52	.20563	.21013	4.7591	.97863	8
53	.20592	.21043	4.7522	.97857	7
54	.20620	.21073	4.7453	.97851	6
55	.20649	.21104	4.7385	.97845	5
56	.20677	2.1134	4.7317	.97839	4
57	.20706	.21164	4.7249	.97833	3
58	.20734	.21195	4.7181	.97827	2
59	.20763	.21225	4.7114	.97821	1
60	.20791	.21256	4.7046	.97815	0
′	Cos	Ctn	Tan	Sin	′

101° (281°)   (258°) **78°**

Table B-2 (*cont.*)

′	Sin	Tan	Ctn	Cos	′	′	Sin	Tan	Ctn	Cos	′
0	.20791	.21256	4.7046	.97815	60	0	.22495	.23087	4.3315	.97437	60
1	.20820	.21286	4.6979	.97809	59	1	.22523	.23117	4.3257	.97430	59
2	.20848	.21316	4.6912	.97803	58	2	.22552	.23148	4.3200	.97424	58
3	.20877	.21347	4.6845	.97797	57	3	.22580	.23179	4.3143	.97417	57
4	.20905	.21377	4.6779	.97791	56	4	.22608	.23209	4.3086	.97411	56
5	.20933	.21408	4.6712	.97784	55	5	.22637	.23240	4.3029	.97404	55
6	.20962	.21438	4.6646	.97778	54	6	.22665	.23271	4.2972	.97398	54
7	.20990	.21469	4.6580	.97772	53	7	.22693	.23301	4.2916	.97391	53
8	.21019	.21499	4.6514	.97766	52	8	.22722	.23332	4.2859	.97384	52
9	.21047	.21529	4.6448	.97760	51	9	.22750	.23363	4.2803	.97378	51
10	.21076	.21560	4.6382	.97754	50	10	.22778	.23393	4.2747	.97371	50
11	.21104	.21590	4.6317	.97748	49	11	.22807	.23424	4.2691	.97365	49
12	.21132	.21621	4.6252	.97742	48	12	.22835	.23455	4.2635	.97358	48
13	.21161	.21651	4.6187	.97735	47	13	.22863	.23485	4.2580	.97351	47
14	.21189	.21682	4.6122	.97729	46	14	.22892	.23516	4.2524	.97345	46
15	.21218	.21712	4.6057	.97723	45	15	.22920	.23547	4.2468	.97338	45
16	.21246	.21743	4.5993	.97717	44	16	.22948	.23578	4.2413	.97331	44
17	.21275	.21773	4.5928	.97711	43	17	.22977	.23608	4.2358	.97325	43
18	.21303	.21804	4.5864	.97705	42	18	.23005	.23639	4.2303	.97318	42
19	.21331	.21834	4.5800	.97698	41	19	.23033	.23670	4.2248	.97311	41
20	.21360	.21864	4.5736	.97692	40	20	.23062	.23700	4.2193	.97304	40
21	.21388	.21895	4.5673	.97686	39	21	.23090	.23731	4.2139	.97298	39
22	.21417	.21925	4.5609	.97680	38	22	.23118	.23762	4.2084	.97291	38
23	.21445	.21956	4.5546	.97673	37	23	.23146	.23793	4.2030	.97284	37
24	.21474	.21986	4.5483	.97667	36	24	.23175	.23823	4.1976	.97278	36
25	.21502	.22017	4.5420	.97661	35	25	.23203	.23854	4.1922	.97271	35
26	.21530	.22047	4.5357	.97655	34	26	.23231	.23885	4.1868	.97264	34
27	.21559	.22078	4.5294	.97648	33	27	.23260	.23916	4.1814	.97257	33
28	.21587	.22108	4.5232	.97642	32	28	.23288	.23946	4.1760	.97251	32
29	.21616	.22139	4.5169	.97636	31	29	.23316	.23977	4.1706	.97244	31
30	.21644	.22169	4.5107	.97630	30	30	.23345	.24008	4.1653	.97237	30
31	.21672	.22200	4.5045	.97623	29	31	.23373	.24039	4.1600	.97230	29
32	.21701	.22231	4.4983	.97617	28	32	.23401	.24069	4.1547	.97223	28
33	.21729	.22261	4.4922	.97611	27	33	.23429	.24100	4.1493	.97217	27
34	.21758	.22292	4.4860	.97604	26	34	.23458	.24131	4.1441	.97210	26
35	.21786	.22322	4.4799	.97598	25	35	.23486	.24162	4.1388	.97203	25
36	.21814	.22353	4.4737	.97592	24	36	.23514	.24193	4.1335	.97196	24
37	.21843	.22383	4.4676	.97585	23	37	.23542	.24223	4.1282	.97189	23
38	.21871	.22414	4.4615	.97579	22	38	.23571	.24254	4.1230	.97182	22
39	.21899	.22444	4.4555	.97573	21	39	.23599	.24285	4.1178	.97176	21
40	.21928	.22475	4.4494	.97566	20	40	.23627	.24316	4.1126	.97169	20
41	.21956	.22505	4.4434	.97560	19	41	.23656	.24347	4.1074	.97162	19
42	.21985	.22536	4.4373	.97553	18	42	.23684	.24377	4.1022	.97155	18
43	.22013	.22567	4.4313	.97547	17	43	.23712	.24408	4.0970	.97148	17
44	.22041	.22597	4.4253	.97541	16	44	.23740	.24439	4.0918	.97141	16
45	.22070	.22628	4.4194	.97534	15	45	.23769	.24470	4.0867	.97134	15
46	.22098	.22658	4.4134	.97528	14	46	.23797	.24501	4.0815	.97127	14
47	.22126	.22689	4.4075	.97521	13	47	.23825	.24532	4.0764	.97120	13
48	.22155	.22719	4.4015	.97515	12	48	.23853	.24562	4.0713	.97113	12
49	.22183	.22750	4.3956	.97508	11	49	.23882	.24593	4.0662	.97106	11
50	.22212	.22781	4.3897	.97502	10	50	.23910	.24624	4.0611	.97100	10
51	.22240	.22811	4.3838	.97496	9	51	.23938	.24655	4.0560	.97093	9
52	.22268	.22842	4.3779	.97489	8	52	.23966	.24686	4.0509	.97086	8
53	.22297	.22872	4.3721	.97483	7	53	.23995	.24717	4.0459	.97079	7
54	.22325	.22903	4.3662	.97476	6	54	.24023	.24747	4.0408	.97072	6
55	.22353	.22934	4.3604	.97470	5	55	.24051	.24778	4.0358	.97065	5
56	.22382	.22964	4.3546	.97463	4	56	.24079	.24809	4.0308	.97058	4
57	.22410	.22995	4.3488	.97457	3	57	.24108	.24840	4.0257	.97051	3
58	.22438	.23026	4.3430	.97450	2	58	.24136	.24871	4.0207	.97044	2
59	.22467	.23056	4.3372	.97444	1	59	.24164	.24902	4.0158	.97037	1
60	.22495	.23087	4.3315	.97437	0	60	.24192	.24933	4.0108	.97030	0

′	Cos	Ctn	Tan	Sin	′	′	Cos	Ctn	Tan	Sin	′

Table B-2 (*cont.*)

′	Sin	Tan	Ctn	Cos	′
0	.24192	.24933	4.0108	.97030	60
1	.24220	.24964	4.0058	.97023	59
2	.24249	.24995	4.0009	.97015	58
3	.24277	.25026	3.9959	.97008	57
4	.24305	.25056	3.9910	.97001	56
5	.24333	.25087	3.9861	.96994	55
6	.24362	.25118	3.9812	.96987	54
7	.24390	.25149	3.9763	.96980	53
8	.24418	.25180	3.9714	.96973	52
9	.24446	.25211	3.9665	.96966	51
10	.24474	.25242	3.9617	.96959	50
11	.24503	.25273	3.9568	.96952	49
12	.24531	.25304	3.9520	.96945	48
13	.24559	.25335	3.9471	.96937	47
14	.24587	.25366	3.9423	.96930	46
15	.24615	.25397	3.9375	.96923	45
16	.24644	.25428	3.9327	.96916	44
17	.24672	.25459	3.9279	.96909	43
18	.24700	.25490	3.9232	.96902	42
19	.24728	.25521	3.9184	.96894	41
20	.24756	.25552	3.9136	.96887	40
21	.24784	.25583	3.9089	.96880	39
22	.24813	.25614	3.9042	.96873	38
23	.24841	.25645	3.8995	.96866	37
24	.24869	.25676	3.8947	.96858	36
25	.24897	.25707	3.8900	.96851	35
26	.24925	.25738	3.8854	.96844	34
27	.24954	.25769	3.8807	.96837	33
28	.24982	.25800	3.8760	.96829	32
29	.25010	.25831	3.8714	.96822	31
30	.25038	.25862	3.8667	.96815	30
31	.25066	.25893	3.8621	.96807	29
32	.25094	.25924	3.8575	.96800	28
33	.25122	.25955	3.8523	.96793	27
34	.25151	.25986	3.8482	.96786	26
35	.25179	.26017	3.8436	.96778	25
36	.25207	.26048	3.8391	.96771	24
37	.25235	.26079	3.8345	.96764	23
38	.25263	.26110	3.8299	.96756	22
39	.25291	.26141	3.8254	.96749	21
40	.25320	.26172	3.8208	.96742	20
41	.25348	.26203	3.8163	.96734	19
42	.25376	.26235	3.8118	.96727	18
43	.25404	.26266	3.8073	.96719	17
44	.25432	.26297	3.8028	.96712	16
45	.25460	.26328	3.7983	.96705	15
46	.25488	.26359	3.7938	.96697	14
47	.25516	.26390	3.7893	.96690	13
48	.25545	.26421	3.7848	.96682	12
49	.25573	.26452	3.7804	.96675	11
50	.25601	.26483	3.7760	.96667	10
51	.25629	.26515	3.7715	.96660	9
52	.25657	.26546	3.7671	.96653	8
53	.25685	.26577	3.7627	.96645	7
54	.25713	.26608	3.7583	.96638	6
55	.25741	.26639	3.7539	.96630	5
56	.25769	.26670	3.7495	.96623	4
57	.25798	.26701	3.7451	.96615	3
58	.25826	.26733	3.7408	.96608	2
59	.25854	.26764	3.7364	.96600	1
60	.25882	.26795	3.7321	.96593	0

′	Cos	Ctn	Tan	Sin	′

′	Sin	Tan	Ctn	Cos	′
0	.25882	.26795	3.7321	.96593	60
1	.25910	.26826	3.7277	.96585	59
2	.25938	.26857	3.7234	.96578	58
3	.25966	.26888	3.7191	.96570	57
4	.25994	.26920	3.7148	.96562	56
5	.26022	.26951	3.7105	.96555	55
6	.26050	.26982	3.7062	.96547	54
7	.26079	.27013	3.7019	.96540	53
8	.26107	.27044	3.6976	.96532	52
9	.26135	.27076	3.6933	.96524	51
10	.26163	.27107	3.6891	.96517	50
11	.26191	.27138	3.6848	.96509	49
12	.26219	.27169	3.6806	.96502	48
13	.26247	.27201	3.6764	.96494	47
14	.26275	.27232	3.6722	.96486	46
15	.26303	.27263	3.6680	.96479	45
16	.26331	.27294	3.6638	.96471	44
17	.26359	.27326	3.6596	.96463	43
18	.26387	.27357	3.6554	.96456	42
19	.26415	.27388	3.6512	.96448	41
20	.26443	.27419	3.6470	.96440	40
21	.26471	.27451	3.6429	.96433	39
22	.26500	.27482	3.6387	.96425	38
23	.26528	.27513	3.6346	.96417	37
24	.26556	.27545	3.6305	.96410	36
25	.26584	.27576	3.6264	.96402	35
26	.26612	.27607	3.6222	.96394	34
27	.26640	.27638	3.6181	.96386	33
28	.26668	.27670	3.6140	.96379	32
29	.26696	.27701	3.6100	.96371	31
30	.26724	.27732	3.6059	.96363	30
31	.26752	.27764	3.6018	.96355	29
32	.26780	.27795	3.5978	.96347	28
33	.26808	.27826	3.5937	.96340	27
34	.26836	.27858	3.5897	.96332	26
35	.26864	.27889	3.5856	.96324	25
36	.26892	.27921	3.5816	.96316	24
37	.26920	.27952	3.5776	.96308	23
38	.26948	.27983	3.5736	.96301	22
39	.26976	.28015	3.5696	.96293	21
40	.27004	.28046	3.5656	.96285	20
41	.27032	.28077	3.5616	.96277	19
42	.27060	.28109	3.5576	.96269	18
43	.27088	.28140	3.5536	.96261	17
44	.27116	.28172	3.5497	.96253	16
45	.27144	.28203	3.5457	.96246	15
46	.27172	.28234	3.5418	.96238	14
47	.27200	.28266	3.5379	.96230	13
48	.27228	.28297	3.5339	.96222	12
49	.27256	.28329	3.5300	.96214	11
50	.27284	.28360	3.5261	.96206	10
51	.27312	.28391	3.5222	.96198	9
52	.27340	.28423	3.5183	.96190	8
53	.27368	.28454	3.5144	.96182	7
54	.27396	.28486	3.5105	.96174	6
55	.27424	.28517	3.5067	.96166	5
56	.27452	.28549	3.5028	.96158	4
57	.27480	.28580	3.4989	.96150	3
58	.27508	.28612	3.4951	.96142	2
59	.27536	.28643	3.4912	.96134	1
60	.27564	.28675	3.4874	.96126	0

′	Cos	Ctn	Tan	Sin	′

Table B-2 (cont.)

′	Sin	Tan	Ctn	Cos	′	′	Sin	Tan	Ctn	Cos	′
0	.27564	.28675	3.4874	.96126	60	0	.29237	.30573	3.2709	.95630	60
1	.27592	.28706	3.4836	.96118	59	1	.29265	.30605	3.2675	.95622	59
2	.27620	.28738	3.4798	.96110	58	2	.29293	.30637	3.2641	.95613	58
3	.27648	.28769	3.4760	.96102	57	3	.29321	.30669	3.2607	.95605	57
4	.27676	.28801	3.4722	.96094	56	4	.29348	.30700	3.2573	.95596	56
5	.27704	.28832	3.4684	.96086	55	5	.29376	.30732	3.2539	.95588	55
6	.27731	.28864	3.4646	.96078	54	6	.29404	.30764	3.2506	.95579	54
7	.27759	.28895	3.4608	.96070	53	7	.29432	.30796	3.2472	.95571	53
8	.27787	.28927	3.4570	.96062	52	8	.29460	.30828	3.2438	.95562	52
9	.27815	.28958	3.4533	.96054	51	9	.29487	.30860	3.2405	.95554	51
10	.27843	.28990	3.4495	.96046	50	10	.29515	.30891	3.2371	.95545	50
11	.27871	.29021	3.4458	.96037	49	11	.29543	.30923	3.2338	.95536	49
12	.27899	.29053	3.4420	.96029	48	12	.29571	.30955	3.2305	.95528	48
13	.27927	.29084	3.4383	.96021	47	13	.29599	.30987	3.2272	.95519	47
14	.27955	.29116	3.4346	.96013	46	14	.29626	.31019	3.2238	.95511	46
15	.27983	.29147	3.4308	.96005	45	15	.29654	.31051	3.2205	.95502	45
16	.28011	.29179	3.4271	.95997	44	16	.29682	.31083	3.2172	.95493	44
17	.28039	.29210	3.4234	.95989	43	17	.29710	.31115	3.2139	.95485	43
18	.28067	.29242	3.4197	.95981	42	18	.29737	.31147	3.2106	.95476	42
19	.28095	.29274	3.4160	.95972	41	19	.29765	.31178	3.2073	.95467	41
20	.28123	.29305	3.4124	.95964	40	20	.29793	.31210	3.2041	.95459	40
21	.28150	.29337	3.4087	.95956	39	21	.29821	.31242	3.2008	.95450	39
22	.28178	.29368	3.4050	.95948	38	22	.29849	.31274	3.1975	.95441	38
23	.28206	.29400	3.4014	.95940	37	23	.29876	.31306	3.1943	.95433	37
24	.28234	.29432	3.3977	.95931	36	24	.29904	.31338	3.1910	.95424	36
25	.28262	.29463	3.3941	.95923	35	25	.29932	.31370	3.1878	.95415	35
26	.28290	.29495	3.3904	.95915	34	26	.29960	.31402	3.1845	.95407	34
27	.28318	.29526	3.3868	.95907	33	27	.29987	.31434	3.1813	.95398	33
28	.28346	.29558	3.3832	.95898	32	28	.30015	.31466	3.1780	95389	32
29	.28374	.29590	3.3796	.95890	31	29	.30043	.31498	3.1748	.95380	31
30	.28402	.29621	3.3759	.95882	30	30	.30071	.31530	3.1716	.95372	30
31	.28429	.29653	3.3723	.95874	29	31	.30098	.31562	3.1684	.95363	29
32	.28457	.29685	3.3687	.95865	28	32	.30126	.31594	3.1652	.95354	28
33	.28485	.29716	3.3652	.95857	27	33	.30154	.31626	3.1620	.95345	27
34	.28513	.29748	3.3616	.95849	26	34	.30182	.31658	3.1588	.95337	26
35	.28541	.29780	3.3580	.95841	25	35	.30209	.31690	3.1556	.95328	25
36	.28569	.29811	3.3544	.95832	24	36	.30237	.31722	3.1524	.95319	24
37	.28597	.29843	3.3509	.95824	23	37	.30265	.31754	3.1492	.95310	23
38	.28625	.29875	3.3473	.95816	22	38	.30292	.31786	3.1460	.95301	22
39	.28652	.29906	3.3438	.95807	21	39	.30320	.31818	3.1429	.95293	21
40	.28680	.29938	3.3402	.95799	20	40	.30348	.31850	3.1397	.95284	20
41	.28708	.29970	3.3367	.95791	19	41	.30376	.31882	3.1366	.95275	19
42	.28736	.30001	3.3332	.95782	18	42	.30403	.31914	3.1334	.95266	18
43	.28764	.30033	3.3297	.95774	17	43	.30431	.31946	3.1303	.95257	17
44	.28792	.30065	3.3261	.95766	16	44	.30459	.31978	3.1271	.95248	16
45	.28820	.30097	3.3226	.95757	15	45	.30486	.32010	3.1240	.95240	15
46	.28847	.30128	3.3191	.95749	14	46	.30514	.32042	3.1209	.95231	14
47	.28875	.30160	3.3156	.95740	13	47	.30542	.32074	3.1178	.95222	13
48	.28903	.30192	3.3122	.95732	12	48	.30570	.32106	3.1146	.95213	12
49	.28931	.30224	3.3087	.95724	11	49	.30597	.32139	3.1115	.95204	11
50	.28959	.30255	3.3052	.95715	10	50	.30625	.32171	3.1084	.95195	10
51	.28987	.30287	3.3017	.95707	9	51	.30653	.32203	3.1053	.95186	9
52	.29015	.30319	3.2983	.95698	8	52	.30680	.32235	3.1022	.95177	8
53	.29042	.30351	3.2948	.95690	7	53	.30708	.32267	3.0991	.95168	7
54	.29070	.30382	3.2914	.95681	6	54	.30736	.32299	3.0961	.95159	6
55	.29098	.30414	3.2879	.95673	5	55	.30763	.32331	3.0930	.95150	5
56	.29126	.30446	3.2845	.95664	4	56	.30791	.32363	3.0899	.95142	4
57	.29154	.30478	3.2811	.95656	3	57	.30819	.32396	3.0868	.95133	3
58	.29182	.30509	3.2777	.95647	2	58	.30846	.32428	3.0838	.95124	2
59	.29209	.30541	3.2743	.95639	1	59	.30874	.32460	3.0807	.95115	1
60	.29237	.30573	3.2709	.95630	0	60	.30902	.32492	3.0777	.95106	0

′	Cos	Ctn	Tan	Sin	′	′	Cos	Ctn	Tan	Sin	′

Table B-2 (*cont.*)

′	Sin	Tan	Ctn	Cos	′	′	Sin	Tan	Ctn	Cos	′
0	.30902	.32492	3.0777	.95106	60	0	.32557	.34433	2.9042	.94552	60
1	.30929	.32524	3.0746	.95097	59	1	.32584	.34465	2.9015	.94542	59
2	.30957	.32556	3.0716	.95088	58	2	.32612	.34498	2.8987	.94533	58
3	.30985	.32588	3.0686	.95079	57	3	.32639	.34530	2.8960	.94523	57
4	.31012	.32621	3.0655	.95070	56	4	.32667	.34563	2.8933	.94514	56
5	.31040	.32653	3.0625	.95061	55	5	.32694	.34596	2.8905	.94504	55
6	.31068	.32685	3.0595	.95052	54	6	.32722	.34628	2.8878	.94495	54
7	.31095	.32717	3.0565	.95043	53	7	.32749	.34661	2.8851	.94485	53
8	.31123	.32749	3.0535	.95033	52	8	.32777	.34693	2.8824	.94476	52
9	.31151	.32782	3.0505	.95024	51	9	.32804	.34726	2.8797	.94466	51
10	.31178	.32814	3.0475	.95015	50	10	.32832	.34758	2.8770	.94457	50
11	.31206	.32846	3.0445	.95006	49	11	.32859	.34791	2.8743	.94447	49
12	.31233	.32878	3.0415	.94997	48	12	.32887	.34824	2.8716	.94438	48
13	.31261	.32911	3.0385	.94988	47	13	.32914	.34856	2.8689	.94428	47
14	.31289	.32943	3.0356	.94979	46	14	.32942	.34889	2.8662	.94418	46
15	.31316	.32975	3.0326	.94970	45	15	.32969	.34922	2.8636	.94409	45
16	.31344	.33007	3.0296	.94961	44	16	.32997	.34954	2.8609	.94399	44
17	.31372	.33040	3.0267	.94952	43	17	.33024	.34987	2.8582	.94390	43
18	.31399	.33072	3.0237	.94943	42	18	.33051	.35020	2.8556	.94380	42
19	.31427	.33104	3.0208	.94933	41	19	.33079	.35052	2.8529	.94370	41
20	.31454	.33136	3.0178	.94924	40	20	.33106	.35085	2.8502	.94361	40
21	.31482	.33169	3.0149	.94915	39	21	.33134	.35118	2.8476	.94351	39
22	.31510	.33201	3.0120	.94906	38	22	.33161	.35150	2.8449	.94342	38
23	.31537	.33233	3.0090	.94897	37	23	.33189	.35183	2.8423	.94332	37
24	.31565	.33266	3.0061	.94888	36	24	.33216	.35216	2.8397	.94322	36
25	.31593	.33298	3.0032	.94878	35	25	.33244	.35248	2.8370	.94313	35
26	.31620	.33330	3.0003	.94869	34	26	.33271	.35281	2.8344	.94303	34
27	.31648	.33363	2.9974	.94860	33	27	.33298	.35314	2.8318	.94293	33
28	.31675	.33395	2.9945	.94851	32	28	.33326	.35346	2.8291	.94284	32
29	.31703	.33427	2.9916	.94842	31	29	.33353	.35379	2.8265	.94274	31
30	.31730	.33460	2.9887	.94832	30	30	.33381	.35412	2.8239	.94264	30
31	.31758	.33492	2.9858	.94823	29	31	.33408	.35445	2.8213	.94254	29
32	.31786	.33524	2.9829	.94814	28	32	.33436	.35477	2.8187	.94245	28
33	.31813	.33557	2.9800	.94805	27	33	.33463	.35510	2.8161	.94235	27
34	.31841	.33589	2.9772	.94795	26	34	.33490	.35543	2.8135	.94225	26
35	.31868	.33621	2.9743	.94786	25	35	.33518	.35576	2.8109	.94215	25
36	.31896	.33654	2.9714	.94777	24	36	.33545	.35608	2.8083	.94206	24
37	.31923	.33686	2.9686	.94768	23	37	.33573	.35641	2.8057	.94196	23
38	.31951	.33718	2.9657	.94758	22	38	.33600	.35674	2.8032	.94186	22
39	.31979	.33751	2.9629	.94749	21	39	.33627	.35707	2.8006	.94176	21
40	.32006	.33783	2.9600	.94740	20	40	.33655	.35740	2.7980	.94167	20
41	.32034	.33816	2.9572	.94730	19	41	.33682	.35772	2.7955	.94157	19
42	.32061	.33848	2.9544	.94721	18	42	.33710	.35805	2.7929	.94147	18
43	.32089	.33881	2.9515	.94712	17	43	.33737	.35838	2.7903	.94137	17
44	.32116	.33913	2.9487	.94702	16	44	.33764	.35871	2.7878	.94127	16
45	.32144	.33945	2.9459	.94693	15	45	.33792	.35904	2.7852	.94118	15
46	.32171	.33978	2.9431	.94684	14	46	.33819	.35937	2.7827	.94108	14
47	.32199	.34010	2.9403	.94674	13	47	.33846	.35969	2.7801	.94098	13
48	.32227	.34043	2.9375	.94665	12	48	.33874	.36002	2.7776	.94088	12
49	.32254	.34075	2.9347	.94656	11	49	.33901	.36035	2.7751	.94078	11
50	.32282	.34108	2.9319	.94646	10	50	.33929	.36068	2.7725	.94068	10
51	.32309	.34140	2.9291	.94637	9	51	.33956	.36101	2.7700	.94058	9
52	.32337	.34173	2.9263	.94627	8	52	.33983	.36134	2.7675	.94049	8
53	.32364	.34205	2.9235	.94618	7	53	.34011	.36167	2.7650	.94039	7
54	.32392	.34238	2.9208	.94609	6	54	.34038	.36199	2.7625	.94029	6
55	.32419	.34270	2.9180	.94599	5	55	.34065	.36232	2.7600	.94019	5
56	.32447	.34303	2.9152	.94590	4	56	.34093	.36265	2.7575	.94009	4
57	.32474	.34335	2.9125	.94580	3	57	.34120	.36298	2.7550	.93999	3
58	.32502	.34368	2.9097	.94571	2	58	.34147	.36331	2.7525	.93989	2
59	.32529	.34400	2.9070	.94561	1	59	.34175	.36364	2.7500	.93979	1
60	.32557	.34433	2.9042	.94552	0	60	.34202	.36397	2.7475	.93969	0
′	Cos	Ctn	Tan	Sin	′	′	Cos	Ctn	Tan	Sin	′

20° (200°)				(339°) **159°**		21° (201°)				(338°) **158°**
′	Sin	Tan	Ctn	Cos	′	′	Sin	Tan	Ctn	Cos	′
0	.34202	.36397	2.7475	.93969	60	0	.35837	.38386	2.6051	.93358	60
1	.34229	.36430	2.7450	.93959	59	1	.35864	.38420	2.6028	.93348	59
2	.34257	.36463	2.7425	.93949	58	2	.35891	.38453	2.6006	.93337	58
3	.34284	.36496	2.7400	.93939	57	3	.35918	.38487	2.5983	.93327	57
4	.34311	.36529	2.7376	.93929	56	4	.35945	.38520	2.5961	.93316	56
5	.34339	.36562	2.7351	.93919	55	5	.35973	.38553	2.5938	.93306	55
6	.34366	.36595	2.7326	.93909	54	6	.36000	.38587	2.5916	.93295	54
7	.34393	.36628	2.7302	.93899	53	7	.36027	.38620	2.5893	.93285	53
8	.34421	.36661	2.7277	.93889	52	8	.36054	.38654	2.5871	.93274	52
9	.34448	.36694	2.7253	.93879	51	9	.36081	.38687	2.5848	.93264	51
10	.34475	.36727	2.7228	.93869	50	10	.36108	.38721	2.5826	.93253	50
11	.34503	.36760	2.7204	.93859	49	11	.36135	.38754	2.5804	.93243	49
12	.34530	.36793	2.7179	.93849	48	12	.36162	.38787	2.5782	.93232	48
13	.34557	.36826	2.7155	.93839	47	13	.36190	.38821	2.5759	.93222	47
14	.34584	.36859	2.7130	.93829	46	14	.36217	.38854	2.5737	.93211	46
15	.34612	.36892	2.7106	.93819	45	15	.36244	.38888	2.5715	.93201	45
16	.34639	.36925	2.7082	.93809	44	16	.36271	.38921	2.5693	.93190	44
17	.34666	.36958	2.7058	.93799	43	17	.36298	.38955	2.5671	.93180	43
18	.34694	.36991	2.7034	.93789	42	18	.36325	.38988	2.5649	.93169	42
19	.34721	.37024	2.7009	.93779	41	19	.36352	.39022	2.5627	.93159	41
20	.34748	.37057	2.6985	.93769	40	20	.36379	.39055	2.5605	.93148	40
21	.34775	.37090	2.6961	.93759	39	21	.36406	.39089	2.5583	.93137	39
22	.34803	.37123	2.6937	.93748	38	22	.36434	.39122	2.5561	.93127	38
23	.34830	.37157	2.6913	.93738	37	23	.36461	.39156	2.5539	.93116	37
24	.34857	.37190	3.6889	.93728	36	24	.36488	.39190	2.5517	.93106	36
25	.34884	.37223	2.6865	.93718	35	25	.36515	.39223	2.5495	.93095	35
26	.34912	.37256	2.6841	.93708	34	26	.36542	.39257	2.5473	.93084	34
27	.34939	.37289	2.6818	.93698	33	27	.36569	.39290	2.5452	.93074	33
28	.34966	.37322	2.6794	.93688	32	28	.36596	.39324	2.5430	.93063	32
29	.34993	.37355	2.6770	.93677	31	29	.36623	.39357	2.5408	.93052	31
30	.35021	.37388	2.6746	.93667	30	30	.36650	.39391	2.5386	.93042	30
31	.35048	.37422	2.6723	.93657	29	31	.36677	.39425	2.5365	.93031	29
32	.35075	.37455	2.6699	.93647	28	32	.36704	.39458	2.5343	.93020	28
33	.35102	.37488	2.6675	.93637	27	33	.36731	.39492	2.5322	.93010	27
34	.35130	.37521	2.6652	.93626	26	34	.36758	.39526	2.5300	.92999	26
35	.35157	.37554	2.6628	.93616	25	35	.36785	.39559	2.5279	.92988	25
36	.35184	.37588	2.6605	.93606	24	36	.36812	.39593	2.5257	.92978	24
37	.35211	.37621	2.6581	.93596	23	37	.36839	.39626	2.5236	.92967	23
38	.35239	.37654	2.6558	.93585	22	38	.36867	.39660	2.5214	.92956	22
39	.35266	.37687	2.6534	.93575	21	39	.36894	.39694	2.5193	.92945	21
40	.35293	.37720	2.6511	.93565	20	40	.36921	.39727	2.5172	.92935	20
41	.35320	.37754	2.6488	.93555	19	41	.36948	.39761	2.5150	.92924	19
42	.35347	.37787	2.6464	.93544	18	42	.36975	.39795	2.5129	.92913	18
43	.35375	.37820	2.6441	.93534	17	43	.37002	.39829	2.5108	.92902	17
44	.35402	.37853	2.6418	.93524	16	44	.37029	.39862	2.5086	.92892	16
45	.35429	.37887	2.6395	.93514	15	45	.37056	.39896	2.5065	.92881	15
46	.35456	.37920	2.6371	.93503	14	46	.37083	.39930	2.5044	.92870	14
47	.35484	.37953	2.6348	.93493	13	47	.37110	.39963	2.5023	.92859	13
48	.35511	.37986	2.6325	.93483	12	48	.37137	.39997	2.5002	.92849	12
49	.35538	.38020	2.6302	.93472	11	49	.37164	.40031	2.4981	.92838	11
50	.35565	.38053	2.6279	.93462	10	50	.37191	.40065	2.4960	.92827	10
51	.35592	.38086	2.6256	.93452	9	51	.37218	.40098	2.4939	.92816	9
52	.35619	.38120	2.6233	.93441	8	52	.37245	.40132	2.4918	.92805	8
53	.35647	.38153	2.6210	.93431	7	53	.37272	.40166	2.4897	.92794	7
54	.35674	.38186	2.6187	.93420	6	54	.37299	.40200	2.4876	.92784	6
55	.35701	.38220	2.6165	.93410	5	55	.37326	.40234	2.4855	.92773	5
56	.35728	.38253	2.6142	.93400	4	56	.37353	.40267	2.4834	.92762	4
57	.35755	.38286	2.6119	.93389	3	57	.37380	.40301	2.4813	.92751	3
58	.35782	.38320	2.6096	.93379	2	58	.37407	.40335	2.4792	.92740	2
59	.35810	.38353	2.6074	.93368	1	59	.37434	.40369	2.4772	.92729	1
60	.35837	.38386	2.6051	.93358	0	60	.37461	.40403	2.4751	.92718	0
′	Cos	Ctn	Tan	Sin	′	′	Cos	Ctn	Tan	Sin	′

22° (202°)   (337°) **157°**   **23°** (203°)   (336°) **156°**

′	Sin	Tan	Ctn	Cos	′	′	Sin	Tan	Ctn	Cos	′
0	.37461	.40403	2.4751	.92718	60	0	.39073	.42447	2.3559	.92050	60
1	.37488	.40436	2.4730	.92707	59	1	.39100	.42482	2.3539	.92039	59
2	.37515	.40470	2.4709	.92697	58	2	.39127	.42516	2.3520	.92028	58
3	.37542	.40504	2.4689	.92686	57	3	.39153	.42551	2.3501	.92016	57
4	.37569	.40538	2.4668	.92675	56	4	.39180	.42585	2.3483	.92005	56
5	.37595	.40572	2.4648	.92664	55	5	.39207	.42619	2.3464	.91994	55
6	.37622	.40606	2.4627	.92653	54	6	.39234	.42654	2.3445	.91982	54
7	.37649	.40640	2.4606	.92642	53	7	.39260	.42688	2.3426	.91971	53
8	.37676	.40674	2.4586	.92631	52	8	.39287	.42722	2.3407	.91959	52
9	.37703	.40707	2.4566	.92620	51	9	.39314	.42757	2.3388	.91948	51
10	.37730	.40741	2.4545	.92609	50	10	.39341	.42791	2.3369	.91936	50
11	.37757	.40775	2.4525	.92598	49	11	.39367	.42826	2.3351	.91925	49
12	.37784	.40809	2.4504	.92587	48	12	.29394	.42860	2.3332	.91914	48
13	.37811	.40843	2.4484	.92576	47	13	.39421	.42894	2.3313	.91902	47
14	.37838	.40877	2.4464	.92565	46	14	.39448	.42929	2.3294	.91891	46
15	.37865	.40911	2.4443	.92554	45	15	.39474	.42963	2.3276	.91879	45
16	.37892	.40945	2.4423	.92543	44	16	.39501	.42998	2.3257	.91868	44
17	.37919	.40979	2.4403	.92532	43	17	.39528	.43032	2.3238	.91856	43
18	.37946	.41013	2.4383	.92521	42	18	.39555	.43067	2.3220	.91845	42
19	.37973	.41047	2.4362	.92510	41	19	.39581	.43101	2.3201	.91833	41
20	.37999	.41081	2.4342	.92499	40	20	.39608	.43136	2.3183	.91822	40
21	.38026	.41115	2.4322	.92488	39	21	.39635	.43170	2.3164	.91810	39
22	.38053	.41149	2.4302	.92477	38	22	.39661	.43205	2.3146	.91799	38
23	.38080	.41183	2.4282	.92466	37	23	.39688	.43239	2.3127	.91787	37
24	.38107	.41217	2.4262	.92455	36	24	.39715	.43274	2.3109	.91775	36
25	.38134	.41251	2.4242	.92444	35	25	.39741	.43308	2.3090	.91764	35
26	.38161	.41285	2.4222	.92432	34	26	.39768	.43343	2.3072	.91752	34
27	.38188	.41319	2.4202	.92421	33	27	.39795	.43378	2.3053	.91741	33
28	.38215	.41353	2.4182	.92410	32	28	.39822	.43412	2.3035	.91729	32
29	.38241	.41387	2.4162	.92399	31	29	.39848	.43447	2.3017	.91718	31
30	.38268	.41421	2.4142	.92388	30	30	.39875	.43481	2.2998	.91706	30
31	.38295	.41455	2.4122	.92377	29	31	.39902	.43516	2.2980	.91694	29
32	.38322	.41490	2.4102	.92366	28	32	.39928	.43550	2.2962	.91683	28
33	.38349	.41524	2.4083	.92355	27	33	.39955	.43585	2.2944	.91671	27
34	.38376	.41558	2.4063	.92343	26	34	.39982	.43620	2.2925	.91660	26
35	.38403	.41592	2.4043	.92332	25	35	.40008	.43654	2.2907	.91648	25
36	.38430	.41626	2.4023	.92321	24	36	.40035	.43689	2.2889	.91636	24
37	.38456	.41660	2.4004	.92310	23	37	.40062	.43724	2.2871	.91625	23
38	.38483	.41694	2.3984	.92299	22	38	.40088	.43758	2.2853	.91613	22
39	.38510	.41728	2.3964	.92287	21	39	.40115	.43793	2.2835	.91601	21
40	.38537	.41763	2.3945	.92276	20	40	.40141	.43828	2.2817	.91590	20
41	.38564	.41797	2.3925	.92265	19	41	.40168	.43862	2.2799	.91578	19
42	.38591	.41831	2.3906	.92254	18	42	.40195	.43897	2.2781	.91566	18
43	.38617	.41865	2.3886	.92243	17	43	.40221	.43932	2.2763	.91555	17
44	.38644	.41899	2.3867	.92231	16	44	.40248	.43966	2.2745	.91543	16
45	.38671	.41933	2.3847	.92220	15	45	.40275	.44001	2.2727	.91531	15
46	.38698	.41968	2.3828	.92209	14	46	.40301	.44036	2.2709	.91519	14
47	.38725	.42002	2.3808	.92198	13	47	.40328	.44071	2.2691	.91508	13
48	.38752	.42036	2.3789	.92186	12	48	.40355	.44105	2.2673	.91496	12
49	.38778	.42070	2.3770	.92175	11	49	.40381	.44140	2.2655	.91484	11
50	.38805	.42105	2.3750	.92164	10	50	.40408	.44175	2.2637	.91472	10
51	.38832	.42139	2.3731	.92152	9	51	.40434	.44210	2.2620	.91461	9
52	.38859	.42173	2.3712	.92141	8	52	.40461	.44244	2.2602	.91449	8
53	.38886	.42207	2.3693	.92130	7	53	.40488	.44279	2.2584	.91437	7
54	.38912	.42242	2.3673	.92119	6	54	.40514	.44314	2.2566	.91425	6
55	.38939	.42276	2.3654	.92107	5	55	.40541	.44349	2.2549	.91414	5
56	.38966	.42310	2.3635	.92096	4	56	.40567	.44384	2.2531	.91402	4
57	.38993	.42345	2.3616	.92085	3	57	.40594	.44418	2.2513	.91390	3
58	.39020	.42379	2.3597	.92073	2	58	.40621	.44453	2.2496	.91378	2
59	.39046	.42413	2.3578	.92062	1	59	.40647	.44488	2.2478	.91366	1
60	.39073	.42447	2.3559	.92050	0	60	.40674	.44523	2.2460	.91355	0

′	Cos	Ctn	Tan	Sin	′	′	Cos	Ctn	Tan	Sin	′

112° (292°)   (247°) **67°**   **113°** (293°)   (246°) **66°**

Table B-2 (*cont.*)

′	Sin	Tan	Ctn	Cos	′	′	Sin	Tan	Ctn	Cos	′
0	.40674	.44523	2.2460	.91355	60	0	.42262	.46631	2.1445	.90631	60
1	.40700	.44558	2.2443	.91343	59	1	.42288	.46666	2.1429	.90618	59
2	.40727	.44593	2.2425	.91331	58	2	.42315	.46702	2.1413	.90606	58
3	.40753	.44627	2.2408	.91319	57	3	.42341	.46737	2.1396	.90594	57
4	.40780	.44662	2.2390	.91307	56	4	.42367	.46772	2.1380	.90582	56
5	.40806	.44697	2.2373	.91295	55	5	.42394	.46808	2.1364	.90569	55
6	.40833	.44732	2.2355	.91283	54	6	.42420	.46843	2.1348	.90557	54
7	.40860	.44767	2.2338	.91272	53	7	.42446	.46879	2.1332	.90545	53
8	.40886	.44802	2.2320	.91260	52	8	.42473	.46914	2.1315	.90532	52
9	.40913	.44837	2.2303	.91248	51	9	.42499	.46950	2.1299	.90520	51
10	.40939	.44872	2.2286	.91236	50	10	.42525	.46985	2.1283	.90507	50
11	.40966	.44907	2.2268	.91224	49	11	.42552	.47021	2.1267	.90495	49
12	.40992	.44942	2.2251	.91212	48	12	.42578	.47056	2.1251	.90483	48
13	.41019	.44977	2.2234	.91200	47	13	.42604	.47092	2.1235	.90470	47
14	.41045	.45012	2.2216	.91188	46	14	.42631	.47128	2.1219	.90458	46
15	.41072	.45047	2.2199	.91176	45	15	.42657	.47163	2.1203	.90446	45
16	.41098	.45082	2.2182	.91164	44	16	.42683	.47199	2.1187	.90433	44
17	.41125	.45117	2.2165	.91152	43	17	.42709	.47234	2.1171	.90421	43
18	.41151	.45152	2.2148	.91140	42	18	.42736	.47270	2.1155	.90408	42
19	.41178	.45187	2.2130	.91128	41	19	.42762	.47305	2.1139	.90396	41
20	.41204	.45222	2.2113	.91116	40	20	.42788	.47341	2.1123	.90383	40
21	.41231	.45257	2.2096	.91104	39	21	.42815	.47377	2.1107	.90371	39
22	.41257	.45292	2.2079	.91092	38	22	.42841	.47412	2.1092	.90358	38
23	.41284	.45327	2.2062	.91080	37	23	.42867	.47448	2.1076	.90346	37
24	.41310	.45362	2.2045	.91068	36	24	.42894	.47483	2.1060	.90334	36
25	.41337	.45397	2.2028	.91056	35	25	.42920	.47519	2.1044	.90321	35
26	.41363	.45432	2.2011	.91044	34	26	.42946	.47555	2.1028	.90309	34
27	.41390	.45467	2.1994	.91032	33	27	.42972	.47590	2.1013	.90296	33
28	.41416	.45502	2.1977	.91020	32	28	.42999	.47626	2.0997	.90284	32
29	.41443	.45538	2.1960	.91008	31	29	.43025	.47662	2.0981	.90271	31
30	.41469	.45573	2.1943	.90996	30	30	.43051	.47698	2.0965	.90259	30
31	.41496	.45608	2.1926	.90984	29	31	.43077	.47733	2.0950	.90246	29
32	.41522	.45643	2.1909	.90972	28	32	.43104	.47769	2.0934	.90233	28
33	.41549	.45678	2.1892	.90960	27	33	.43130	.47805	2.0918	.90221	27
34	.41575	.45713	2.1876	.90948	26	34	.43156	.47840	2.0903	.90208	26
35	.41602	.45748	2.1859	.90936	25	35	.43182	.47876	2.0887	.90196	25
36	.41628	.45784	2.1842	.90924	24	36	.43209	.47912	2.0872	.90183	24
37	.41655	.45819	2.1825	.90911	23	37	.43235	.47948	2.0856	.90171	23
38	.41681	.45854	2.1808	.90899	22	38	.43261	.47984	2.0840	.90158	22
39	.41707	.45889	2.1792	.90887	21	39	.43287	.48019	2.0825	.90146	21
40	.41734	.45924	2.1775	.90875	20	40	.43313	.48055	2.0809	.90133	20
41	.41760	.45960	2.1758	.90863	19	41	.43340	.48091	2.0794	.90120	19
42	.41787	.45995	2.1742	.90851	18	42	.43366	.48127	2.0778	.90108	18
43	.41813	.46030	2.1725	.90839	17	43	.43392	.48163	2.0763	.90095	17
44	.41840	.46065	2.1708	.90826	16	44	.43418	.48198	2.0748	.90082	16
45	.41866	.46101	2.1692	.90814	15	45	.43445	.48234	2.0732	.90070	15
46	.41892	.46136	2.1675	.90802	14	46	.43471	.48270	2.0717	.90057	14
47	.41919	.46171	2.1659	.90790	13	47	.43497	.48306	2.0701	.90045	13
48	.41945	.46206	2.1642	.90778	12	48	.43523	.48342	2.0686	.90032	12
49	.41972	.46242	2.1625	.90766	11	49	.43549	.48378	2.0671	.90019	11
50	.41998	.46277	2.1609	.90753	10	50	.43575	.48414	2.0655	.90007	10
51	.42024	.46312	2.1592	.90741	9	51	.43602	.48450	2.0640	.89994	9
52	.42051	.46348	2.1576	.90729	8	52	.43628	.48486	2.0625	.89981	8
53	.42077	.46383	2.1560	.90717	7	53	.43654	.48521	2.0609	.89968	7
54	.42104	.46418	2.1543	.90704	6	54	.43680	.48557	2.0594	.89956	6
55	.42130	.46454	2.1527	.90692	5	55	.43706	.48593	2.0579	.89943	5
56	.42156	.46489	2.1510	.90680	4	56	.43733	.48629	2.0564	.89930	4
57	.42183	.46525	2.1494	.90668	3	57	.43759	.48665	2.0549	.89918	3
58	.42209	.46560	2.1478	.90655	2	58	.43785	.48701	2.0533	.89905	2
59	.42235	.46595	2.1461	.90643	1	59	.43811	.48737	2.0518	.89892	1
60	.42262	.46631	2.1445	.90631	0	60	.43837	.48773	2.0503	.89879	0
′	Cos	Ctn	Tan	Sin	′	′	Cos	Ctn	Tan	Sin	′

Table B-2 (*cont.*)

′	Sin	Tan	Ctn	Cos	′	′	Sin	Tan	Ctn	Cos	′
0	.43837	.48773	2.0503	.89879	60	0	.45399	.50953	1.9626	.89101	60
1	.43863	.48809	2.0488	.89867	59	1	.45425	.50989	1.9612	.89087	59
2	.43889	.48845	2.0473	.89854	58	2	.45451	.51026	1.9598	.89074	58
3	.43916	.48881	2.0458	.89841	57	3	.45477	.51063	1.9584	.89061	57
4	.43942	.48917	2.0443	.89828	56	4	.45503	.51099	1.9570	.89048	56
5	.43968	.48953	2.0428	.89816	55	5	.45529	.51136	1.9556	.89035	55
6	.43994	.48989	2.0413	.89803	54	6	.45554	.51173	1.9542	.89021	54
7	.44020	.49026	2.0398	.89790	53	7	.45580	.51209	1.9528	.89008	53
8	.44046	.49062	2.0383	.89777	52	8	.45606	.51246	1.9514	.88995	52
9	.44072	.49098	2.0368	.89764	51	9	.45632	.51283	1.9500	.88981	51
10	.44098	.49134	2.0353	.89752	50	10	.45658	.51319	1.9486	.88968	50
11	.44124	.49170	2.0338	.89739	49	11	.45684	.51356	1.9472	.88955	49
12	.44151	.49206	2.0323	.89726	48	12	.45710	.51393	1.9458	.88942	48
13	.44177	.49242	2.0308	.89713	47	13	.45736	.51430	1.9444	.88928	47
14	.44203	.49278	2.0293	.89700	46	14	.45762	.51467	1.9430	.88915	46
15	.44229	.49315	2.0278	.89687	45	15	.45787	.51503	1.9416	.88902	45
16	.44255	.49351	2.0263	.89674	44	16	.45813	.51540	1.9402	.88888	44
17	.44281	.49387	2.0248	.89662	43	17	.45839	.51577	1.9388	.88875	43
18	.44307	.49423	2.0233	.89649	42	18	.45865	.51614	1.9375	.88862	42
19	.44333	.49459	2.0219	.89636	41	19	.45891	.51651	1.9361	.88848	41
20	.44359	.49495	2.0204	.89623	40	20	.45917	.51688	1.9347	.88835	40
21	.44385	.49532	2.0189	.89610	39	21	.45942	.51724	1.9333	.88822	39
22	.44411	.49568	2.0174	.89597	38	22	.45968	.51761	1.9319	.88808	38
23	.44437	.49604	2.0160	.89584	37	23	.45994	.51798	1.9306	.88795	37
24	.44464	.49640	2.0145	.89571	36	24	.46020	.51835	1.9292	.88782	36
25	.44490	.49677	2.0130	.89558	35	25	.46046	.51872	1.9278	.88768	35
26	.44516	.49713	2.0115	.89545	34	26	.46072	.51909	1.9265	.88755	34
27	.44542	.49749	2.0101	.89532	33	27	.46097	.51946	1.9251	.88741	33
28	.44568	.49786	2.0086	.89519	32	28	.46123	.51983	1.9237	.88728	32
29	.44594	.49822	2.0072	.89506	31	29	.46149	.52020	1.9223	.88715	31
30	.44620	.49858	2.0057	.89493	30	30	.46175	.52057	1.9210	.88701	30
31	.44646	.49894	2.0042	.89480	29	31	.46201	.52094	1.9196	.88688	29
32	.44672	.49931	2.0028	.89467	28	32	.46226	.52131	1.9183	.88674	28
33	.44698	.49967	2.0013	.89454	27	33	.46252	.52168	1.9169	.88661	27
34	.44724	.50004	1.9999	.89441	26	34	.46278	.52205	1.9155	.88647	26
35	.44750	.50040	1.9984	.89428	25	35	.46304	.52242	1.9142	.88634	25
36	.44776	.50076	1.9970	.89415	24	36	.46330	.52279	1.9128	.88620	24
37	.44802	.50113	1.9955	.89402	23	37	.46355	.52316	1.9115	.88607	23
38	.44828	.50149	1.9941	.89389	22	38	.46381	.52353	1.9101	.88593	22
39	.44854	.50185	1.9926	.89376	21	39	.46407	.52390	1.9088	.88580	21
40	.44880	.50222	1.9912	.89363	20	40	.46433	.52427	1.9074	.88566	20
41	.44906	.50258	1.9897	.89350	19	41	.46458	.52464	1.9061	.88553	19
42	.44932	.50295	1.9883	.89337	18	42	.46484	.52501	1.9047	.88539	18
43	.44958	.50331	1.9868	.89324	17	43	.46510	.52538	1.9034	.88526	17
44	.44984	.50368	1.9854	.89311	16	44	.46536	.52575	1.9020	.88512	16
45	.45010	.50404	1.9840	.89298	15	45	.46561	.52613	1.9007	.88499	15
46	.45036	.50441	1.9825	.89285	14	46	.46587	.52650	1.8993	.88485	14
47	.45062	.50477	1.9811	.89272	13	47	.46613	.52687	1.8980	.88472	13
48	.45088	.50514	1.9797	.89259	12	48	.46639	.52724	1.8967	.88458	12
49	.45114	.50550	1.9782	.89245	11	49	.46664	.52761	1.8953	.88445	11
50	.45140	.50587	1.9768	.89232	10	50	.46690	.52798	1.8940	.88431	10
51	.45166	.50623	1.9754	.89219	9	51	.46716	.52836	1.8927	.88417	9
52	.45192	.50660	1.9740	.89206	8	51	.46742	.52873	1.8913	.88404	8
53	.45218	.50696	1.9725	.89193	7	53	.46767	.52910	1.8900	.88390	7
54	.45243	.50733	1.9711	.89180	6	54	.46793	.52947	1.8887	.88377	6
55	.45269	.50769	1.9697	.89167	5	55	.46819	.52985	1.8873	.88363	5
56	.45295	.50806	1.9683	.89153	4	56	.46844	.53022	1.8860	.88349	4
57	.45321	.50843	1.9669	.89140	3	57	.46870	.53059	1.8847	.88336	3
58	.45347	.50879	1.9654	.89127	2	58	.46896	.53096	1.8834	.88322	2
59	.45373	.50916	1.9640	.89114	1	59	.46921	.53134	1.8820	.88308	1
60	.45399	.50953	1.9626	.89101	0	60	.46947	.53171	1.8807	.88295	0
′	Cos	Ctn	Tan	Sin	′	′	Cos	Ctn	Tan	Sin	′

Table B-2 (*cont.*)

′	Sin	Tan	Ctn	Cos	′	′	Sin	Tan	Ctn	Cos	′
0	.46947	.53171	1.8807	.88295	60	0	.48481	.55431	1.8040	.87462	60
1	.46973	.53208	1.8794	.88281	59	1	.48506	.55469	1.8028	.87448	59
2	.46999	.53246	1.8781	.88267	58	2	.48532	.55507	1.8016	.87434	58
3	.47024	.53283	1.8768	.88254	57	3	.48557	.55545	1.8003	.87420	57
4	.47050	.53320	1.8755	.88240	56	4	.48583	.55583	1.7991	.87406	56
5	.47076	.53358	1.8741	.88226	55	5	.48608	.55621	1.7979	.87391	55
6	.47101	.53395	1.8728	.88213	54	6	.48634	.55659	1.7966	.87377	54
7	.47127	.53432	1.8715	.88199	53	7	.48659	.55697	1.7954	.87363	53
8	.47153	.53470	1.8702	.88185	52	8	.48684	.55736	1.7942	.87349	52
9	.47178	.53507	1.8689	.88172	51	9	.48710	.55774	1.7930	.87335	51
10	.47204	.53545	1.8676	.88158	50	10	.48735	.55812	1.7917	.87321	50
11	.47229	.53582	1.8663	.88144	49	11	.48761	.55850	1.7905	.87306	49
12	.47255	.53620	1.8650	.88130	48	12	.48786	.55888	1.7893	.87292	48
13	.47281	.53657	1.8637	.88117	47	13	.48811	.55926	1.7881	.87278	47
14	.47306	.53694	1.8624	.88103	46	14	.48837	.55964	1.7868	.87264	46
15	.47332	.53732	1.8611	.88089	45	15	.48862	.56003	1.7856	.87250	45
16	.47358	.53769	1.8598	.88075	44	16	.48888	.56041	1.7844	.87235	44
17	.47383	.53807	1.8585	.88062	43	17	.48913	.56079	1.7832	.87221	43
18	.47409	.53844	1.8572	.88048	42	18	.48938	.56117	1.7820	.87207	42
19	.47434	.53882	1.8559	.88034	41	19	.48964	.56156	1.7808	.87193	41
20	.47460	.53920	1.8546	.88020	40	20	.48989	.56194	1.7796	.87178	40
21	.47486	.53957	1.8533	.88006	39	21	.49014	.56232	1.7783	.87164	39
22	.47511	.53995	1.8520	.87993	38	22	.49040	.56270	1.7771	.87150	38
23	.47537	.54032	1.8507	.87979	37	23	.49065	.56309	1.7759	.87136	37
24	.47562	.54070	1.8495	.87965	36	24	.49090	.56347	1.7747	.87121	36
25	.47588	.54107	1.8482	.87951	35	25	.49116	.56385	1.7735	.87107	35
26	.47614	.54145	1.8469	.87937	34	26	.49141	.56424	1.7723	.87093	34
27	.47639	.54183	1.8456	.87923	33	27	.49166	.56462	1.7711	.87079	33
28	.47665	.54220	1.8443	.87909	32	28	.49192	.56501	1.7699	.87064	32
29	.47690	.54258	1.8430	.87896	31	29	.49217	.56539	1.7687	.87050	31
30	.47716	.54296	1.8418	.87882	30	30	.49242	.56577	1.7675	.87036	30
31	.47741	.54333	1.8405	.87868	29	31	.49268	.56616	1.7663	.87021	29
32	.47767	.54371	1.8392	.87854	28	32	.49293	.56654	1.7651	.87007	28
33	.47793	.54409	1.8379	.87840	27	33	.49318	.56693	1.7639	.86993	27
34	.47818	.54446	1.8367	.87826	26	34	.49344	.56731	1.7627	.86978	26
35	.47844	.54484	1.8354	.87812	25	35	.49369	.56769	1.7615	.86964	25
36	.47869	.54522	1.8341	.87798	24	36	.49394	.56808	1.7603	.86949	24
37	.47895	.54560	1.8329	.87784	23	37	.49419	.56846	1.7591	.86935	23
38	.47920	.54597	1.8316	.87770	22	38	.49445	.56885	1.7579	.86921	22
39	.47946	.54635	1.8303	.87756	21	39	.49470	.56923	1.7567	.86906	21
40	.47971	.54673	1.8291	.87743	20	40	.49495	.56962	1.7556	.86892	20
41	.47997	.54711	1.8278	.87729	19	41	.49521	.57000	1.7544	.86878	19
42	.48022	.54748	1.8265	.87715	18	42	.49546	.57039	1.7532	.86863	18
43	.48048	.54786	1.8253	.87701	17	43	.49571	.57078	1.7520	.86849	17
44	.48073	.54824	1.8240	.87687	16	44	.49596	.57116	1.7508	.86834	16
45	.48099	.54862	1.8228	.87673	15	45	.49622	.57155	1.7496	.86820	15
46	.48124	.54900	1.8215	.87659	14	46	.49647	.57193	1.7485	.86805	14
47	.48150	.54938	1.8202	.87645	13	47	.49672	.57232	1.7473	.86791	13
48	.48175	.54975	1.8190	.87631	12	48	.49697	.57271	1.7461	.86777	12
49	.48201	.55013	1.8177	.87617	11	49	.49723	.57309	1.7449	.86762	11
50	.48226	.55051	1.8165	.87603	10	50	.49748	.57348	1.7437	.86748	10
51	.48252	.55089	1.8152	.87589	9	51	.49773	.57386	1.7426	.86733	9
52	.48277	.55127	1.8140	.87575	8	52	.49798	.57425	1.7414	.86719	8
53	.48303	.55165	1.8127	.87561	7	53	.49824	.57464	1.7402	.86704	7
54	.48328	.55203	1.8115	.87546	6	54	.49849	.57508	1.7391	.86690	6
55	.48354	.55241	1.8103	.87532	5	55	.49874	.57541	1.7379	.86675	5
56	.48379	.55279	1.8090	.87518	4	56	.49899	.57580	1.7367	.86661	4
57	.48405	.55317	1.8078	.87504	3	57	.49924	.57619	1.7355	.86646	3
58	.48430	.55355	1.8065	.87490	2	58	.49950	.57657	1.7344	.86632	2
59	.48456	.55393	1.8053	.87476	1	59	.49975	.57696	1.7332	.86617	1
60	.48481	.55431	1.8040	.87462	0	60	.50000	.57735	1.7321	.86603	0

′	Cos	Ctn	Tan	Sin	′	′	Cos	Ctn	Tan	Sin	′

Table B-2 (*cont.*)

′	Sin	Tan	Ctn	Cos	′	′	Sin	Tan	Ctn	Cos	′
0	.50000	.57735	1.7321	.86603	60	0	.51504	.60086	1.6643	.85717	60
1	.50025	.57774	1.7309	.86588	59	1	.51529	.60126	1.6632	.85702	59
2	.50050	.57813	1.7297	.86573	58	2	.51554	.60165	1.6621	.85687	58
3	.50076	.57851	1.7286	.86559	57	3	.51579	.60205	1.6610	.85672	57
4	.50101	.57890	1.7274	.86544	56	4	.51604	.60245	1.6599	.85657	56
5	.50126	.57929	1.7262	.86530	55	5	.51628	.60284	1.6588	.85642	55
6	.50151	.57968	1.7251	.86515	54	6	.51653	.60324	1.6577	.85627	54
7	.50176	.58007	1.7239	.86501	53	7	.51678	.60364	1.6566	.85612	53
8	.50201	.58046	1.7228	.86486	52	8	.51703	.60403	1.6555	.85597	52
9	.50227	.58085	1.7216	.86471	51	9	.51728	.60443	1.6545	.85582	51
10	.50252	.58124	1.7205	.86457	50	10	.51753	.60483	1.6534	.85567	50
11	.50277	.58162	1.7193	.86442	49	11	.51778	.60522	1.6523	.85551	49
12	.50302	.48201	1.7182	.86427	48	12	.51803	.60562	1.6512	.85536	48
13	.50327	.58240	1.7170	.86413	47	13	.51828	.60602	1.6501	.85521	47
14	.50352	.58279	1.7159	.86398	46	14	.51852	.60642	1.6490	.85506	46
15	.50377	.58318	1.7147	.86384	45	15	.51877	.60681	1.6479	.85491	45
16	.50403	.58357	1.7136	.86369	44	16	.51902	.60721	1.6469	.85476	44
17	.50428	.58396	1.7124	.86354	43	17	.51927	.60761	1.6458	.85461	43
18	.50453	.58435	1.7113	.86340	42	18	.51952	.60801	1.6447	.85446	42
19	.50478	.58474	1.7102	.86325	41	19	.51977	.60841	1.6436	.85431	41
20	.50503	.58513	1.7090	.86310	40	20	.52002	.60881	1.6426	.85416	40
21	.50528	.58552	1.7079	.86295	39	21	.52026	.60921	1.6415	.85401	39
22	.50553	.58591	1.7067	.86281	38	22	.52051	.60960	1.6404	.85385	38
23	.50578	.58631	1.7056	.86266	37	23	.52076	.61000	1.6393	.85370	37
24	.50603	.58670	1.7045	86251	36	24	.52101	.61040	1.6383	.85355	36
25	.50628	.58709	1.7033	.86237	35	25	.52126	.61080	1.6372	.85340	35
26	.50654	.58748	1.7022	.86222	34	26	.52151	.61120	1.6361	.85325	34
27	.50679	58787	1.7011	86207	33	27	.52175	.61160	1.6351	.85310	33
28	.50704	.58826	1.6999	.86192	32	28	.52200	.61200	1.6340	.85294	32
29	.50729	.58865	1.6988	.86178	31	29	.52225	.61240	1.6329	.85279	31
30	.50754	.58905	1.6977	.86163	30	30	.52250	.61280	1.6319	.85264	30
31	.50779	.58944	1.6965	.86148	29	31	.52275	.61320	1.6308	.85249	29
32	.50804	.58983	1.6954	.86133	28	32	.52299	.61360	1.6297	.85234	28
33	.50829	.59022	1.6943	.86119	27	33	.52324	.61400	1.6287	.85218	27
34	.50854	.59061	1.6932	.86104	26	34	.52349	.61440	1.6276	.85203	26
35	.50879	.59101	1.6920	.86089	25	35	.52374	.61480	1.6265	.85188	25
36	.50904	.59140	1.6909	.86074	24	36	.52399	.61520	1.6255	.85173	24
37	.50929	.59179	1.6898	.86059	23	37	.52423	.61561	1.6244	.85157	23
38	.50954	.59218	1.6887	.86045	22	38	.52448	.61601	1.6234	.85142	22
39	.50979	.59258	1.6875	.86030	21	39	.52473	.61641	1.6223	.85127	21
40	.51004	.59297	1.6864	.86015	20	40	.52498	.61681	1.6212	.85112	20
41	.51029	.59336	1.6853	.86000	19	41	.52522	.61721	1.6202	.85096	19
42	.51054	.59376	1.6842	.85985	18	42	.52547	.61761	1.6191	.85081	18
43	.51079	.59415	1.6831	.85970	17	43	.52572	.61801	1.6181	.85066	17
44	.51104	.59454	1.6820	.85956	16	44	.52597	.61842	1.6170	.85051	16
45	.51129	.59494	1.6808	.85941	15	45	.52621	.61882	1.6160	.85035	15
46	.51154	.59533	1.6797	.85926	14	46	.52646	.61922	1.6149	.85020	14
47	.51179	.59573	1.6786	.85911	13	47	.52671	.61962	1.6139	.85005	13
48	.51204	.59612	1.6775	.85896	12	48	.52696	.62003	1.6128	.84989	12
49	.51229	.59651	1.6764	.85881	11	49	.52720	.62043	1.6118	.84974	11
50	.51254	.59691	1.6753	.85866	10	50	.52745	.62083	1.6107	.84959	10
51	.51279	.59730	1.6742	.85851	9	51	.52770	.62124	1.6097	.84943	9
52	.51304	.59770	1.6731	.85836	8	52	.52794	.62164	1.6087	.84928	8
53	.51329	.59809	1.6720	.85821	7	53	.52819	.62204	1.6076	.84913	7
54	.51354	.59849	1.6709	.85806	6	54	.52844	.62245	1.6066	.84897	6
55	.51379	.59888	1.6698	.85792	5	55	.52869	.62285	1.6055	.84882	5
56	.51404	.59928	1.6687	.85777	4	56	.52893	.62325	1.6045	.84866	4
57	.51429	.59967	1.6676	.85762	3	57	.52918	.62366	1.6034	.84851	3
58	.51454	.60007	1.6665	.85747	2	58	.52943	.62406	1.6024	.84836	2
59	.51479	.60046	1.6654	.85732	1	59	.52967	.62446	1.6014	.84820	1
60	.51504	.60086	1.6643	.85717	0	60	.52992	.62487	1.6003	.84805	0
′	Cos	Ctn	Tan	Sin	′	′	Cos	Ctn	Tan	Sin	′

Table B-2 (*cont.*)

′	Sin	Tan	Ctn	Cos	′	′	Sin	Tan	Ctn	Cos	′
0	.52992	.62487	1.6003	.84805	60	0	.54464	.64941	1.5399	.83867	60
1	.53017	.62527	1.5993	.84789	59	1	.54488	.64982	1.5389	.83851	59
2	.53041	.62568	1.5983	.84774	58	2	.54513	.65024	1.5379	.83835	58
3	.53066	.62608	1.5972	.84759	57	3	.54537	.65065	1.5369	.83819	57
4	.53091	.62649	1.5962	.84743	56	4	.54561	.65106	1.5359	.83804	56
5	.53115	.62689	1.5952	.84728	55	5	.54586	.65148	1.5350	.83788	55
6	.53140	.62730	1.5941	.84712	54	6	.54610	.65189	1.5340	.83772	54
7	.53164	.62770	1.5931	.84697	53	7	.54635	.65231	1.5330	.83756	53
8	.53189	.62811	1.5921	.84681	52	8	.54659	.65272	1.5320	.83740	52
9	.53214	.62852	1.5911	.84666	51	9	.54683	.65314	1.5311	.83724	51
10	.53238	.62892	1.5900	.84650	50	10	.54708	6.5355	1.5301	.83708	50
11	.53263	.62933	1.5890	.84635	49	11	.54732	.65397	1.5291	.83692	49
12	.53288	.62973	1.5880	.84619	48	12	.54756	.65438	1.5282	.83676	48
13	.53312	.63014	1.5869	.84604	47	13	.54781	.65480	1.5272	.83660	47
14	.53337	.63055	1.5859	.84588	46	14	.54805	.65521	1.5262	.83645	46
15	.53361	.63095	1.5849	.84573	45	15	.54829	.65563	1.5253	.83629	45
16	.53386	.63136	1.5839	.84557	44	16	.54854	.65604	1.5243	.83613	44
17	.53411	.63177	1.5829	.84542	43	17	.54878	.65646	1.5233	.83597	43
18	.53435	.63217	1.5818	.84526	42	18	.54902	.65688	1.5224	.83581	42
19	.53460	.63258	1.5808	.84511	41	19	.54927	.65729	1.5214	.83565	41
20	.53484	.63299	1.5798	.84495	40	20	.54951	.65771	1.5204	.83549	40
21	.53509	.63340	1.5788	.84480	39	21	.54975	.65813	1.5195	.83533	39
22	.53534	.63380	1.5778	.84464	38	22	.54999	.65854	1.5185	.83517	38
23	.53558	.63421	1.5768	.84448	37	23	.55024	.65896	1.5175	.83501	37
24	.53583	.63462	1.5757	.84433	36	24	.55048	.65938	1.5166	.83485	36
25	.53607	.63503	1.5747	.84417	35	25	.55072	.65980	1.5156	.83469	35
26	.53632	.63544	1.5737	.84402	34	26	.55097	.66021	1.5147	.83453	34
27	.53656	.63584	1.5727	.84386	33	27	.55121	.66063	1.5137	.83437	33
28	.53681	.63625	1.5717	.84370	32	28	.55145	.66105	1.5127	.83421	32
29	.53705	.63666	1.5707	.84355	31	29	.55169	.66147	1.5118	.83405	31
30	.53730	.63707	1.5697	.84339	30	30	.55194	.66189	1.5108	.83389	30
31	.53754	.63748	1.5687	.84324	29	31	.55218	.66230	1.5099	.83373	29
32	.53779	.63789	1.5677	.84308	28	32	.55242	.66272	1.5089	.83356	28
33	.53804	.63830	1.5667	.84292	27	33	.55266	.66314	1.5080	.83340	27
34	.53828	.63871	1.5657	.84277	26	34	.55291	.66356	1.5070	.83324	26
35	.53853	.63912	1.5647	.84261	25	35	.55315	.66398	1.5061	.83308	25
36	.53877	.63953	1.5637	.84245	24	36	.55339	.66440	1.5051	.83292	24
37	.53902	.63994	1.5627	.84230	23	37	.55363	.66482	1.5042	.83276	23
38	.53926	.64035	1.5617	.84214	22	38	.55388	.66524	1.5032	.83260	22
39	.53951	.64076	1.5607	.84198	21	39	.55412	.66566	1.5023	.83244	21
40	.53975	.64117	1.5597	.84182	20	40	.55436	.66608	1.5013	.83228	20
41	.54000	.64158	1.5587	.84167	19	41	.55460	.66650	1.5004	.83212	19
42	.54024	.64199	1.5577	.84151	18	42	.55484	.66692	1.4994	.83195	18
43	.54049	.64240	1.5567	.84135	17	43	.55509	.66734	1.4985	.83179	17
44	.54073	.64281	1.5557	.84120	16	44	.55533	.66776	1.4975	.83163	16
45	.54097	.64322	1.5547	.84104	15	45	.55557	.66818	1.4966	.83147	15
46	.54122	.64363	1.5537	.84088	14	46	.55581	.66860	1.4957	.83131	14
47	.54146	.64404	1.5527	.84072	13	47	.55605	.66902	1.4947	.83115	13
48	.54171	.64446	1.5517	.84057	12	48	.55630	.66944	1.4938	.83098	12
49	.54195	.64487	1.5507	.84041	11	49	.55654	.66986	1.4928	.83082	11
50	.54220	.64528	1.5497	.84025	10	50	.55678	.67028	1.4919	.83066	10
51	.54244	.64569	1.5487	.84009	9	51	.55702	.67071	1.4910	.83050	9
52	.54269	.64610	1.5477	.83994	8	52	.55726	.67113	1.4900	.83034	8
53	.54293	.64652	1.5468	.83978	7	53	.55750	.67155	1.4891	.83017	7
54	.54317	.64693	1.5458	.83962	6	54	.55775	.67197	1.4882	.83001	6
55	.54342	.64734	1.5448	.83946	5	55	.55799	.67239	1.4872	.82985	5
56	.54366	.64775	1.5438	.83930	4	56	.55823	.67282	1.4863	.82969	4
57	.54391	.64817	1.5428	.83915	3	57	.55847	.67324	1.4854	.82953	3
58	.54415	.64858	1.5418	.83899	2	58	.55871	.67366	1.4844	.82936	2
59	.54440	.64899	1.5408	.83883	1	59	.55895	.67409	1.4835	.82920	1
60	.54464	.64941	1.5399	.83867	0	60	.55919	.67451	1.4826	.82904	0

′	Cos	Ctn	Tan	Sin	′	′	Cos	Ctn	Tan	Sin	′

Table B-2 (*cont.*)

′	Sin	Tan	Ctn	Cos	′	′	Sin	Tan	Ctn	Cos	′
0	.55919	.67451	1.4826	.82904	60	0	.57358	.70021	1.4281	.81915	60
1	.55943	.67493	1.4816	.82887	59	1	.57381	.70064	1.4273	.81899	59
2	.55968	.67536	1.4807	.82871	58	2	.57405	.70107	1.4264	.81882	58
3	.55992	.67578	1.4798	.82855	57	3	.57429	.70151	1.4255	.81865	57
4	.56016	.67620	1.4788	.82839	56	4	.57453	.70194	1.4246	.81848	56
5	.56040	.67663	1.4779	.82822	55	5	.57477	.70238	1.4237	.81832	55
6	.56064	.67705	1.4770	.82806	54	6	.57501	.70281	1.4229	.81815	54
7	.56088	.67748	1.4761	.82790	53	7	.57524	.70325	1.4220	.81798	53
8	.56112	.67790	1.4751	.82773	52	8	.57548	.70368	1.4211	.81782	52
9	.56136	.67832	1.4742	.82757	51	9	.57572	.70412	1.4202	.81765	51
10	.56160	.67875	1.4733	.82741	50	10	.57596	.70455	1.4193	.81748	50
11	.56184	.67917	1.4724	.82724	49	11	.57619	.70499	1.4185	.81731	49
12	.56208	.67960	1.4715	.82708	48	12	.57643	.70542	1.4176	.81714	48
13	.56232	.68002	1.4705	.82692	47	13	.57667	.70586	1.4267	.81698	47
14	.56256	.68045	1.4696	.82675	46	14	.57691	.70629	1.4158	.81681	46
15	.56280	.68088	1.4687	.82659	45	15	.57715	.70673	1.4150	.81664	45
16	.56305	.68130	1.4678	.82643	44	16	.57738	.70717	1.4141	.81647	44
17	.56329	.68173	1.4669	.82626	43	17	.57762	.70760	1.4132	.81631	43
18	.56353	.68215	1.4659	.82610	42	18	.57786	.70804	1.4124	.81614	42
19	.56377	.68258	1.4650	.82593	41	19	.57810	.70848	1.4115	.81597	41
20	.56401	.68301	1.4641	.82577	40	20	.57833	.70891	1.4106	.81580	40
21	.56425	.68343	1.4632	.82561	39	21	.57857	.70935	1.4097	.81563	39
22	.56449	.68386	1.4623	.82544	38	22	.57881	.70979	1.4089	.81546	38
23	.56473	.68429	1.4614	.82528	37	23	.57904	.71023	1.4080	.81530	37
24	.56497	.68471	1.4605	.82511	36	24	.57928	.71066	1.4071	.81513	36
25	.56521	.68514	1.4596	.82495	35	25	.57952	.71110	1.4063	.81496	35
26	.56545	.68557	1.4586	.82478	34	26	.57976	.71154	1.4054	.81479	34
27	.56569	.68600	1.4577	.82462	33	27	.57999	.71198	1.4045	.81462	33
28	.56593	.68642	1.4568	.82446	32	28	.58023	.71242	1.4037	.81445	32
29	.56617	.68685	1.4559	.82429	31	29	.58047	.71285	1.4028	.81428	31
30	.56641	.68728	1.4550	.82413	30	30	.58070	.71329	1.4019	.81412	30
31	.56665	.68771	1.4541	.82396	29	31	.58094	.71373	1.4011	.81395	29
32	.56689	.68814	1.4532	.82380	28	32	.58118	.71417	1.4002	.81378	28
33	.56713	.68857	1.4523	.82363	27	33	.58141	.71461	1.3994	.81361	27
34	.56736	.68900	1.4514	.82347	26	34	.58165	.71505	1.3985	.81344	26
35	.56760	.68942	1.4505	.82330	25	35	.58189	.71549	1.3976	.81327	25
36	.56784	.68985	1.4496	.82314	24	36	.58212	.71593	1.3968	.81310	24
37	.56808	.69028	1.4487	.82297	23	37	.58236	.71637	1.3959	.81293	23
38	.56832	.69071	1.4478	.82281	22	38	.58260	.71681	1.3951	.81276	22
39	.56856	.69114	1.4469	.82264	21	39	.58283	.71725	1.3942	.81259	21
40	.56880	.69157	1.4460	.82248	20	40	.58307	.71769	1.3934	.81242	20
41	.56904	.69200	1.4451	.82231	19	41	.58330	.71813	1.3925	.81225	19
42	.56928	.69243	1.4442	.82214	18	42	.58354	.71857	1.3916	.81208	18
43	.56952	.69286	1.4433	.82198	17	43	.58378	.71901	1.3908	.81191	17
44	.56976	.69329	1.4424	.82181	16	44	.58401	.71946	1.3899	.81174	16
45	.57000	.69372	1.4415	.82165	15	45	.58425	.71990	1.3891	.81157	15
46	.57024	.69416	1.4406	.82148	14	46	.58449	.72034	1.3882	.81140	14
47	.57047	.69459	1.4397	.82132	13	47	.58472	.72078	1.3874	.81123	13
48	.57071	.69502	1.4388	.82115	12	48	.58496	.72122	1.3865	.81106	12
49	.57095	.69545	1.4379	.82098	11	49	.58519	.72167	1.3857	.81089	11
50	.57119	.69588	1.4370	.82082	10	50	.58543	.72211	1.3848	.81072	10
51	.57143	.69631	1.4361	.82065	9	51	.58567	.72255	1.3840	.81055	9
52	.57167	.69675	1.4352	.82048	8	52	.58590	.72299	1.3831	.81038	8
53	.57191	.69718	1.4344	.82032	7	53	.58614	.72344	1.3823	.81021	7
54	.57215	.69761	1.4335	.82015	6	54	.58637	.72388	1.3814	.81004	6
55	.57238	.69804	1.4326	.81999	5	55	.58661	.72432	1.3806	.80987	5
56	.47262	.69847	1.4317	.81982	4	56	.58684	.72477	1.3798	.80970	4
57	.57286	.69891	1.4308	.81965	3	57	.58708	.72521	1.3789	.80953	3
58	.57310	.69934	1.4299	.81949	2	58	.58731	.72565	1.3781	.80936	2
59	.57334	.69977	1.4290	.81932	1	59	.58755	.72610	1.3772	.80919	1
60	.57358	.70021	1.4281	.81915	0	60	.58779	.72654	1.3764	.80902	0

′	Cos	Ctn	Tan	Sin	′	′	Cos	Ctn	Tan	Sin	′

Table B-2 (*cont.*)

′	Sin	Tan	Ctn	Cos	′	′	Sin	Tan	Ctn	Cos	′
0	.58779	.72654	1.3764	.80902	60	0	.60182	.75355	1.3270	.79864	60
1	.58802	.72699	1.3755	.80885	59	1	.60205	.75401	1.3262	.79846	59
2	.58826	.72743	1.3747	.80867	58	2	.60228	.75447	1.3254	.79829	58
3	.58849	.72788	1.3739	.80850	57	3	.60251	.75492	1.3246	.79811	57
4	.58873	.72832	1.3730	.80833	56	4	.60274	.75538	1.3238	.79793	56
5	.58896	.72877	1.3722	.80816	55	5	.60298	.75584	1.3230	.79776	55
6	.58920	.72921	1.3713	.80799	54	6	.60321	.75629	1.3222	.79758	54
7	.58943	.72966	1.3705	.80782	53	7	.60344	.75675	1.3214	.79741	53
8	.58967	.73010	1.3697	.80765	52	8	.60367	.75721	1.3206	.79723	52
9	.58990	.73055	1.3688	.80748	51	9	.60390	.75767	1.3198	.79706	51
10	.59014	.73100	1.3680	.80730	50	10	.60414	.75812	1.3190	.79688	50
11	.59037	.73144	1.3672	.80713	49	11	.60437	.75858	1.3182	.79671	49
12	.59061	.73189	1.3663	.80696	48	12	.60460	.75904	1.3175	.79653	48
13	.59084	.73234	1.3655	.80679	47	13	.60483	.75950	1.3167	.79635	47
14	.59108	.73278	1.3647	.80662	46	14	.60506	.75996	1.3159	.79618	46
15	.59131	.73323	1.3638	.80644	45	15	.60529	.76042	1.3151	.79600	45
16	.59154	.73368	1.3630	.80627	44	16	.60553	.76088	1.3143	.79583	44
17	.59178	.73413	1.3622	.80610	43	17	.60576	.76134	1.3135	.79565	43
18	.59201	.73457	1.3613	.80593	42	18	.60599	.76180	1.3127	.79547	42
19	.59225	.73502	1.3605	.80576	41	19	.60622	.76226	1.3119	.79530	41
20	.59248	.73547	1.3597	.80558	40	20	.60645	.76272	1.3111	.79512	40
21	.59272	.73592	1.3588	.80541	39	21	.60668	.76318	1.3103	.79494	39
22	.59295	.73637	1.3580	.80524	38	22	.60691	.76364	1.3095	.79477	38
23	.59318	.73681	1.3572	.80507	37	23	.60714	.76410	1.3087	.79459	37
24	.59342	.73726	1.3564	.80489	36	24	.60738	.76456	1.3079	.79441	36
25	.59365	.73771	1.3555	.80472	35	25	.60761	.76502	1.3072	.79424	35
26	.59389	.73816	1.3547	.80455	34	26	.60784	.76548	1.3064	.79406	34
27	.59412	.73861	1.3539	.80438	33	27	.60807	.76594	1.3056	.79388	33
28	.59436	.73906	1.3531	.80420	32	28	.60830	.76640	1.3048	.79371	32
29	.59459	.73951	1.3522	.80403	31	29	.60853	.76686	1.3040	.79353	31
30	.59482	.73996	1.3514	.80386	30	30	.60876	.76733	1.3032	.79335	30
31	.59506	.74041	1.3506	.80368	29	31	.60899	.76779	1.3024	.79318	29
32	.59529	.74086	1.3498	.80351	28	32	.60922	.76825	1.3017	.79300	28
33	.59552	.74131	1.3490	.80334	27	33	.60945	.76871	1.3009	.79282	27
34	.59576	.74176	1.3481	.80316	26	34	.60968	.76918	1.3001	.79264	26
35	.59599	.74221	1.3473	.80299	25	35	.60991	.76964	1.2993	.79247	25
36	.59622	.74267	1.3465	.80282	24	36	.61015	.77010	1.2985	.79229	24
37	.59646	.74312	1.3457	.80264	23	37	.61038	.77057	1.2977	.79211	23
38	.59669	.74357	1.3449	.80247	22	38	.61061	.77103	1.2970	.79193	22
39	.59693	.74402	1.3440	.80230	21	39	.61084	.77149	1.2962	.79176	21
40	.59716	.74447	1.3432	.80212	20	40	.61107	.77196	1.2954	.79158	20
41	.59739	.74492	1.3424	.80195	19	41	.61130	.77242	1.2946	.79140	19
42	.59763	.74538	1.3416	.80178	18	42	.61153	.77289	1.2938	.79122	18
43	.59786	.74583	1.3408	.80160	17	43	.61176	.77335	1.2931	.79105	17
44	.59809	.74628	1.3400	.80143	16	44	.61199	.77382	1.2923	.79087	16
45	.59832	.74674	1.3392	.80125	15	45	.61222	.77428	1.2915	.79069	15
46	.59856	.74719	1.3384	.80108	14	46	.61245	.77475	1.2907	.79051	14
47	.59879	.74764	1.3375	.80091	13	47	.61268	.77521	1.2900	.79033	13
48	.59902	.74810	1.3367	.80073	12	48	.61291	.77568	1.2892	.79016	12
49	.59926	.74855	1.3359	.80056	11	49	.61314	.77615	1.2884	.78998	11
50	.59949	.74900	1.3351	.80038	10	50	.61337	.77661	1.2876	.78980	10
51	.59972	.74946	1.3343	.80021	9	51	.61360	.77708	1.2869	.78962	9
52	.59995	.74991	1.3335	.80003	8	52	.61383	.77754	1.2861	.78944	8
53	.60019	.75037	1.3327	.79986	7	53	.61406	.77801	1.2853	.78926	7
54	.60042	.75082	1.3319	.79968	6	54	.61429	.77848	1.2846	.78908	6
55	.60065	.75128	1.3311	.79951	5	55	.61451	.77895	1.2838	.78891	5
56	.60089	.75173	1.3303	.79934	4	56	.61474	.77941	1.2830	.78873	4
57	.60112	.75219	1.3295	.79916	3	57	.61497	.77988	1.2822	.78855	3
58	.60135	.75264	1.3287	.79899	2	58	.61520	.78035	1.2815	.78837	2
59	.60158	.75310	1.3278	.79881	1	59	.61543	.78082	1.2807	.78819	1
60	.60182	.75355	1.3270	.79864	0	60	.61566	.78129	1.2799	.78801	0

′	Cos	Ctn	Tan	Sin	′	′	Cos	Ctn	Tan	Sin	′

Table B-2 (*cont.*)

′	Sin	Tan	Ctn	Cos	′	′	Sin	Tan	Ctn	Cos	′
0	.61566	.78129	1.2799	.78801	60	0	.62932	.80978	1.2349	.77715	60
1	.61589	.78175	1.2792	.78783	59	1	.62955	.81027	1.2342	.77696	59
2	.61612	.78222	1.2784	.78765	58	2	.62977	.81075	1.2334	.77678	58
3	.61635	.78269	1.2776	.78747	57	3	.63000	.81123	1.2327	.77660	57
4	.61658	.78316	1.2769	.78729	56	4	.63022	.81171	1.2320	.77641	56
5	.61681	.78363	1.2761	.78711	55	5	.63045	.81220	1.2312	.77623	55
6	.61704	.78410	1.2753	.78694	54	6	.63068	.81268	1.2305	.77605	54
7	.61726	.78457	1.2746	.78676	53	7	.63090	.81316	1.2298	.77586	53
8	.61749	.78504	1.2738	.78658	52	8	.63113	.81364	1.2290	.77568	52
9	.61772	.78551	1.2731	.78640	51	9	.63135	.81413	1.2283	.77550	51
10	.61795	.78598	1.2723	.78622	50	10	.63158	.81461	1.2276	.77531	50
11	.61818	.78645	1.2715	.78604	49	11	.63180	.81510	1.2268	.77513	49
12	.61841	.78692	1.2708	.78586	48	12	.63203	.81558	1.2261	.77494	48
13	.61864	.78739	1.2700	.78568	47	13	.63225	.81606	1.2254	.77476	47
14	.61887	.78786	1.2693	.78550	46	14	.63248	.81655	1.2247	.77458	46
15	.61909	.78834	1.2685	.78532	45	15	.63271	.81703	1.2239	.77439	45
16	.61932	.78881	1.2677	.78514	44	16	.63293	.81752	1.2232	.77421	44
17	.61955	.78928	1.2670	.78496	43	17	.63316	.81800	1.2225	.77402	43
18	.61978	.78975	1.2662	.78478	42	18	.63338	.81849	1.2218	.77384	42
19	.62001	.79022	1.2655	.78460	41	19	.63361	.81898	1.2210	.77366	41
20	.62024	.79070	1.2647	.78442	40	20	.63383	.81946	1.2203	.77347	40
21	.62046	.79117	1.2640	.78424	39	21	.63406	.81995	1.2196	.77329	39
22	.62069	.79164	1.2632	.78405	38	22	.63428	.82044	1.2189	.77310	38
23	.62092	.79212	1.2624	.78387	37	23	.63451	.82092	1.2181	.77292	37
24	.62115	.79259	1.2617	.78369	36	24	.63473	.82141	1.2174	.77273	36
25	.62138	.79306	1.2609	.78351	35	25	.63496	.82190	1.2167	.77255	35
26	.62160	.79354	1.2602	.78333	34	26	.63518	.82238	1.2160	.77236	34
27	.62183	.79401	1.2594	.78315	33	27	.63540	.82287	1.2153	.77218	33
28	.62206	.79449	1.2587	.78297	32	28	.63563	.82336	1.2145	.77199	32
29	.62229	.79496	1.2579	.78279	31	29	.63585	.82385	1.2138	.77181	31
30	.62251	.79544	1.2572	.78261	30	30	.63608	.82434	1.2131	.77162	30
31	.62274	.79591	1.2564	.78243	29	31	.63630	.82483	1.2124	.77144	29
32	.62297	.79639	1.2557	.78225	28	32	.63653	.82531	1.2117	.77125	28
33	.62320	.79686	1.2549	.78206	27	33	.63675	.82580	1.2109	.77107	27
34	.62342	.79734	1.2542	.78188	26	34	.63698	.82629	1.2102	.77088	26
35	.62365	.79781	1.2534	.78170	25	35	.63720	.82678	1.2095	.77070	25
36	.62388	.79829	1.2527	.78152	24	36	.63742	.82727	1.2088	.77051	24
37	.62411	.79877	1.2519	.78134	23	37	.63765	.82776	1.2081	.77033	23
38	.62433	.79924	1.2512	.78116	22	38	.63787	.82825	1.2074	.77014	22
39	.62456	.79972	1.2504	.78098	21	39	.63810	.82874	1.2066	.76996	21
40	.62479	.80020	1.2497	.78079	20	40	.63832	.82923	1.2059	.76977	20
41	.62502	.80067	1.2489	.78061	19	41	.63854	.82972	1.2052	.76959	19
42	.62524	.80115	1.2482	.78043	18	42	.63877	.83022	1.2045	.76940	18
43	.62547	.80163	1.2475	.78025	17	43	.63899	.83071	1.2038	.76921	17
44	.62570	.80211	1.2467	.78007	16	44	.63922	.83120	1.2031	.76903	16
45	.62592	.80258	1.2460	.77988	15	45	.63944	.83169	1.2024	.76884	15
46	.62615	.80306	1.2452	.77970	14	46	.63966	.83218	1.2017	.76866	14
47	.62638	.80354	1.2445	.77952	13	47	.63989	.83268	1.2009	.76847	13
48	.62660	.80402	1.2437	.77934	12	48	.64011	.83317	1.2002	.76828	12
49	.62683	.80450	1.2430	.77916	11	49	.64033	.83366	1.1995	.76810	11
50	.62706	.80498	1.2423	.77897	10	50	.64056	.83415	1.1988	.76791	10
51	.62728	.80546	1.2415	.77879	9	51	.64078	.83465	1.1981	.76772	9
52	.62751	.80594	1.2408	.77861	8	52	.64100	.83514	1.1974	.76754	8
53	.62774	.80642	1.2401	.77843	7	53	.64123	.83564	1.1967	.76735	7
54	.62796	.80690	1.2393	.77824	6	54	.64145	.83613	1.1960	.76717	6
55	.62819	.80738	1.2386	.77806	5	55	.64167	.83662	1.1953	.76698	5
56	.62842	.80786	1.2378	.77788	4	56	.64190	.83712	1.1946	.76679	4
57	.62864	.80834	1.2371	.77769	3	57	.64212	.83761	1.1939	.76661	3
58	.62887	.80882	1.2364	.77751	2	58	.64234	.83811	1.1932	.76642	2
59	.62909	.80930	1.2356	.77733	1	59	.64256	.83860	1.1925	.76623	1
60	.62932	.80978	1.2349	.77715	0	60	.64279	.83910	1.1918	.76604	0
′	Cos	Ctn	Tan	Sin	′	′	Cos	Ctn	Tan	Sin	′

Table B-2 (cont.)

′	Sin	Tan	Ctn	Cos	′	′	Sin	Tan	Ctn	Cos	′
0	.64279	.83910	1.1918	.76604	60	0	.65606	.86929	1.1504	.75471	60
1	.64301	.83960	1.1910	.76586	59	1	.65628	.86980	1.1497	.75452	59
2	.64323	.84009	1.1903	.76567	58	2	.65650	.87031	1.1490	.75433	58
3	.64346	.84059	1.1896	.76548	57	3	.65672	.87082	1.1483	.75414	57
4	.64368	.84108	1.1889	.76530	56	4	.65694	.87133	1.1477	.75395	56
5	.64390	.84158	1.1882	.76511	55	5	.65716	.87184	1.1470	.75375	55
6	.64412	.84208	1.1875	.76492	54	6	.65738	.87236	1.1463	.75356	54
7	.64435	.84258	1.1868	.76473	53	7	.65759	.87287	1.1456	.75337	53
8	.64457	.84307	1.1861	.76455	52	8	.65781	.87338	1.1450	.75318	52
9	.64479	.84357	1.1854	.76436	51	9	.65803	.87389	1.1443	.75299	51
10	.64501	.84407	1.1847	.76417	50	10	.65825	.87441	1.1436	.75280	50
11	.64524	.84457	1.1840	.76398	49	11	.65847	.87492	1.1430	.75261	49
12	.64546	.84507	1.1833	.76380	48	12	.65869	.87543	1.1423	.75241	48
13	.64568	.84556	1.1826	.76361	47	13	.65891	.87595	1.1416	.75222	47
14	.64590	.84606	1.1819	.76342	46	14	.65913	.87646	1.1410	.75203	46
15	.64612	.84656	1.1812	.76323	45	15	.65935	.87698	1.1403	.75184	45
16	.64635	.84706	1.1806	.76304	44	16	.65956	.87749	1.1396	.75165	44
17	.64657	.84756	1.1799	.76286	43	17	.65978	.87801	1.1389	.75146	43
18	.64679	.84806	1.1792	.76267	42	18	.66000	.87852	1.1383	.75126	42
19	.64701	.84856	1.1785	.76248	41	19	.66022	.87904	1.1376	.75107	41
20	.64723	.84906	1.1778	.76229	40	20	.66044	.87955	1.1369	.75088	40
21	.64746	.84956	1.1771	.76210	39	21	.66066	.88007	1.1363	.75069	39
22	.64768	.85006	1.1764	.76192	38	22	.66088	.88059	1.1356	.75050	38
23	.64790	.85057	1.1757	.76173	37	23	.66109	.88110	1.1349	.75030	37
24	.64812	.85107	1.1750	.76154	36	24	.66131	.88162	1.1343	.75011	36
25	.64834	.85157	1.1743	.76135	35	25	.66153	.88214	1.1336	.74992	35
26	.64856	.85207	1.1736	.76116	34	26	.66175	.88265	1.1329	.74973	34
27	.64878	.85257	1.1729	.76097	33	27	.66197	.88317	1.1323	.74953	33
28	.64901	.85308	1.1722	.76078	32	28	.66218	.88369	1.1316	.74934	32
29	.64923	.85358	1.1715	.76059	31	29	.66240	.88421	1.1310	.74915	31
30	.64945	.95408	1.1708	.76041	30	30	.66262	.88473	1.1303	.74896	30
31	.64967	.85458	1.1702	.76022	29	31	.66284	.88524	1.1296	.74876	29
32	.64989	.85509	1.1695	.76003	28	32	.66306	.88576	1.1290	.74857	28
33	.65011	.85559	1.1688	.75984	27	33	.66327	.88628	1.1283	.74838	27
34	.65033	.85609	1.1681	.75965	26	34	.66349	.88680	1.1276	.74818	26
35	.65055	.85660	1.1674	.75946	25	35	.66371	.88732	1.1270	.74799	25
36	.65077	.85710	1.1667	.75927	24	36	.66393	.88784	1.1263	.74780	24
37	.65100	.85761	1.1660	.75908	23	37	.66414	.88836	1.1257	.74760	23
38	.65122	.85811	1.1653	.75889	22	38	.66436	.88888	1.1250	.74741	22
39	.65144	.85862	1.1647	.75870	21	39	.66458	.88940	1.1243	.74722	21
40	.65166	.85912	1.1640	.75851	20	40	.66480	.88992	1.1237	.74703	20
41	.65188	.85963	1.1633	.75832	19	41	.66501	.89045	1.1230	.74683	19
42	.65210	.86014	1.1626	.75813	18	42	.66523	.89097	1.1224	.74664	18
43	.65232	.86064	1.1619	.75794	17	42	.66545	.89149	1.1217	.74644	17
44	.65254	.86115	1.1612	.75775	16	44	.66566	.89201	1.1211	.74625	16
45	.65276	.86166	1.1606	.75756	15	45	.66588	.89253	1.1204	.74606	15
46	.65298	.86216	1.1599	.75738	14	46	.66610	.89306	1.1197	.74586	14
47	.65320	.86267	1.1592	.75719	13	47	.66632	.89358	1.1191	.74567	13
48	.65342	.86318	1.1585	.75700	12	48	.66653	.89410	1.1184	.74548	12
49	.65364	.86368	1.1578	.75680	11	49	.66675	.89463	1.1178	.74528	11
50	.65386	.86419	1.1571	.75661	10	50	.66697	.89515	1.1171	.74509	10
51	.65408	.86470	1.1565	.75642	9	51	.66718	.89567	1.1165	.74489	9
52	.65430	.86521	1.1558	.75623	8	52	.66740	.89620	1.1158	.74470	8
53	.65452	.86572	1.1551	.75604	7	53	.66762	.89672	1.1152	.74451	7
54	.65474	.86623	1.1544	.75585	6	54	.66783	.89725	1.1145	.74431	6
55	.65496	.86674	1.1538	.75566	5	55	.66805	.89777	1.1139	.74412	5
56	.65518	.86725	1.1531	.75547	4	56	.66827	.89830	1.1132	.74392	4
57	.65540	.86776	1.1524	.75528	3	57	.66848	.89883	1.1126	.74373	3
58	.65562	.86827	1.1517	.75509	2	58	.66870	.89935	1.1119	.74353	2
59	.65584	.86878	1.1510	.75490	1	59	.66891	.89988	1.1113	.74334	1
60	.65606	.86929	1.1504	.75471	0	60	.66913	.90040	1.1106	.74314	0
′	Cos	Ctn	Tan	Sin	′	′	Cos	Ctn	Tan	Sin	′

Table B-2 (*cont.*)

42° (222°)				(317°) 137°	43° (223°)				(316°) 136°
′	Sin	Tan	Ctn	Cos	′	Sin	Tan	Ctn	Cos
0	.66913	.90040	1.1106	.74314	60	.68200	.93252	1.0724	.73135
1	.66935	.90093	1.1100	.74295	59	.68221	.83306	1.0717	.73116
2	.66956	.90146	1.1093	.74276	58	.68242	.93360	1.0711	.73096
3	.66978	.90199	1.1087	.74256	57	.68264	.93415	1.0705	.73076
4	.66999	.90251	1.1080	.74237	56	.68285	.93469	1.0699	.73056
5	.67021	.90304	1.1074	.74217	55	.68306	.93524	1.0692	.73036
6	.67043	.90357	1.1067	.74198	54	.68327	.93578	1.0686	.73016
7	.67064	.90410	1.1061	.74178	53	.68349	.93633	1.0680	.72996
8	.67086	.90463	1.1054	.74159	52	.68370	.93688	1.0674	.72976
9	.67107	.90516	1.1048	.74139	51	.68391	.93742	1.0668	.72957
10	.67129	.90569	1.1041	.74120	50	.68412	.93797	1.0661	.72937
11	.67151	.90621	1.1035	.74100	49	.68434	.93852	1.0655	.72917
12	.67172	.90674	1.1028	.74080	48	.68455	.93906	1.0649	.72897
13	.67194	.90727	1.1022	.74061	47	.68476	.93961	1.0643	.72877
14	.67215	.90781	1.1016	.74041	46	.68497	.94016	1.0637	.72857
15	.67237	.90834	1.1009	.74022	45	.68518	.94071	1.0630	.72837
16	.67258	.90887	1.1003	.74002	44	.68539	.94125	1.0624	.72817
17	.67280	.90940	1.0996	.73983	43	.68561	.94180	1.0618	.72797
18	.67301	.90993	1.0990	.73963	42	.68582	.94235	1.0612	.72777
19	.67323	.91046	1.0983	.73944	41	.68603	.94290	1.0606	.72757
20	.67344	.91099	1.0977	.73924	40	.68624	.94345	1.0599	.72737
21	.67366	.91153	1.0971	.73904	39	.68645	.94400	1.0593	.72717
22	.67387	.91206	1.0964	.73885	38	.68666	.94455	1.0587	.72697
23	.67409	.91259	1.0958	.73865	37	.68688	.94510	1.0581	.72677
24	.67430	.91313	1.0951	.73846	36	.68709	.94565	1.0575	.72657
25	.67452	.91366	1.0945	.73826	35	.68730	.94620	1.0569	.72637
26	.67473	.91419	1.0939	.73806	34	.68751	.94676	1.0562	.72617
27	.67495	.91473	1.0932	.73787	33	.68772	.94731	1.0556	.72597
28	.67516	.91526	1.0926	.73767	32	.68793	.94786	1.0550	.72577
29	.67538	.91580	1.0919	.73747	31	.68814	.94841	1.0544	.72557
30	.67559	.91633	1.0913	.73728	30	.68835	.94896	1.0538	.72537
31	.67580	.91687	1.0907	.73708	29	.68857	.94952	1.0532	.72517
32	.67602	.91740	1.0900	.73688	28	.68878	.95007	1.0526	.72497
33	.67623	.91794	1.0894	.73669	27	.68899	.95062	1.0519	.72477
34	.67645	.91847	1.0888	.73649	26	.68920	.95118	1.0513	.72457
35	.67666	.91901	1.0881	.73629	25	.68941	.95173	1.0507	.72437
36	.67688	.91955	1.0875	.73610	24	.68962	.95229	1.0501	.72417
37	.67709	.92008	1.0869	.73590	23	.68983	.95284	1.0495	.72397
38	.67730	.92062	1.0862	.73570	22	.69004	.95340	1.0489	.72377
39	.67752	.92116	1.0856	.73551	21	.69025	.95395	1.0483	.72357
40	.67773	.92170	1.0850	.73531	20	.69046	.95451	1.0477	.72337
41	.67795	.92224	1.0843	.73511	19	.69067	.95506	1.0470	.72317
42	.67816	.92277	1.0837	.73491	18	.69088	.95562	1.0464	.72297
43	.67837	.92331	1.0831	.73472	17	.69109	.95618	1.0458	.72277
44	.67859	.92385	1.0824	.73452	16	.69130	.95673	1.0452	.72257
45	.67880	.92439	1.0818	.73432	15	.69151	.95729	1.0446	.72236
46	.67901	.92493	1.0812	.73413	14	.69172	.95785	1.0440	.72216
47	.67923	.92547	1.0805	.73393	13	.69193	.95841	1.0434	.72196
48	.67944	.92601	1.0799	.73373	12	.69214	.95897	1.0428	.72176
49	.67965	.92655	1.0793	.73353	11	.69235	.95952	1.0422	.72156
50	.67987	.92709	1.0786	.73333	10	.69256	.96008	1.0416	.72136
51	.68008	.92763	1.0780	.73314	9	.69277	.96064	1.0410	.72116
52	.68029	.92817	1.0774	.73294	8	.69298	.96120	1.0404	.72095
53	.68051	.92872	1.0768	.73274	7	.69319	.96176	1.0398	.72075
54	.68072	.92926	1.0761	.73254	6	.59340	.96232	1.0392	.72055
55	.68093	.92980	1.0755	.73234	5	.69361	.96288	1.0385	.72035
56	.68115	.93034	1.0749	.73215	4	.69382	.96344	1.0379	.72015
57	.68136	.93088	1.0742	.73195	3	.69403	.96400	1.0373	.71995
58	.68157	.93143	1.0736	.73175	2	.69424	.96457	1.0367	.71974
59	.68179	.93197	1.0730	.73155	1	.69445	.96513	1.0361	.71954
60	.68200	.93252	1.0724	.73135	0	.69466	.96569	1.0355	.71934
′	Cos	Ctn	Tan	Sin	′	Cos	Ctn	Tan	Sin
132° (312°)				(227°) 47°	133° (313°)				(226°) 46°

Table B-2 (*cont.*)

′	Sin	Tan	Ctn	Cos	′
0	.69466	.96569	1.0355	.71934	60
1	.69487	.96625	1.0349	.71914	59
2	.69508	.96681	1.0343	.71894	58
3	.69529	.96738	1.0337	.71873	57
4	.69549	.96794	1.0331	.71853	56
5	.69570	.96850	1.0325	.71833	55
6	.69591	.96907	1.0319	.71813	54
7	.69612	.96963	1.0313	.71792	53
8	.69633	.97020	1.0307	.71772	52
9	.69654	.97076	1.0301	.71752	51
10	.69675	.97133	1.0295	.71732	50
11	.69696	.97189	1.0289	.71711	49
12	.69717	.97246	1.0283	.71691	48
13	.69737	.97302	1.0277	.71671	47
14	.69758	.97359	1.0271	.71650	46
15	.69779	.97416	1.0265	.71630	45
16	.69800	.97472	1.0259	.71610	44
17	.69821	.97529	1.0253	.71590	43
18	.69842	.97586	1.0247	.71569	42
19	.69862	.97643	1.0241	.71549	41
20	.69883	.97700	1.0235	.71529	40
21	.69904	.97756	1.0230	.71508	39
22	.69925	.97813	1.0224	.71488	38
23	.69946	.97870	1.0218	.71468	37
24	.69966	.97927	1.0212	.71447	36
25	.69987	.97984	1.0206	.71427	35
26	.70008	.98041	1.0200	.71407	34
27	.70029	.98098	1.0194	.71386	33
28	.70049	.98155	1.0188	.71366	32
29	.70070	.98213	1.0182	.71345	31
30	.70091	.98270	1.0176	.71325	30
31	.70112	.98327	1.0170	.71305	29
32	.70132	.98384	1.0164	.71284	28
33	.70153	.98441	1.0158	.71264	27
34	.70174	.98499	1.0152	.71243	26
35	.70195	.98556	1.0147	.71223	25
36	.70215	.98613	1.0141	.71203	24
37	.70236	.98671	1.0135	.71182	23
38	.70257	.98728	1.0129	.71162	22
39	.70277	.98786	1.0123	.71141	21
40	.70298	.98843	1.0117	.71121	20
41	.70319	.98901	1.0111	.71100	19
42	.70339	.98958	1.0105	.71080	18
43	.70360	.99016	1.0099	.71059	17
44	.70381	.99073	1.0094	.71039	16
45	.70401	.99131	1.0088	.71019	15
46	.70422	.99189	1.0082	.70998	14
47	.70443	.99247	1.0076	.70978	13
48	.70463	.99304	1.0070	.70957	12
49	.70484	.99362	1.0064	.70937	11
50	.70505	.99420	1.0058	.70916	10
51	.70525	.99478	1.0052	.70896	9
52	.70546	.99536	1.0047	.70875	8
53	.70567	.99594	1.0041	.70855	7
54	.70587	.99652	1.0035	.70834	6
55	.70608	.99710	1.0029	.70813	5
56	.70628	.99768	1.0023	.70793	4
57	.70649	.99826	1.0017	.70772	3
58	.70670	.99884	1.0012	.70752	2
59	.70690	.99942	1.0006	.70731	1
60	.70711	1.0000	1.0000	.70711	0

′	Cos	Ctn	Tan	Sin	′

Table B-3 Natural Trigonometric Functions

Radians	Degrees	Sin	Tan	Cot	Cos		
.0000	0° 00′	.0000	.0000	∞	1.0000	90° 00′	1.5708
029	10	029	029	343.8	000	89° 50′	679
058	20	058	058	171.9	000	40	650
.0087	30	.0087	.0087	114.6	1.0000	30	1.5621
116	40	116	116	85.94	.9999	20	592
145	50	145	145	68.75	999	10	563
.0175	1° 00′	.0175	.0175	57.29	.9998	89° 00′	1.5533
204	10	204	204	49.10	998	88° 50′	504
233	20	233	233	42.96	997	40	475
.0262	30	.0262	.0262	38.19	.9997	30	1.5446
291	40	291	291	34.37	996	20	417
320	50	320	320	31.24	995	10	388
.0349	2° 00′	.0349	.0349	28.64	.9994	88° 00′	1.5359
378	10	378	378	26.43	993	87° 50′	330
407	20	407	407	24.54	992	40	301
.0436	30	.0436	.0436	22.90	.9990	30	1.5272
465	40	465	466	21.47	989	20	243
495	50	494	495	20.21	988	10	213
.0524	3° 00′	.0523	.0524	19.08	.9986	87° 00′	1.5184
553	10	552	553	18.07	985	86° 50′	155
582	20	581	582	17.17	983	40	126
.0611	30	.0610	.0612	16.35	.9981	30	1.5097
640	40	640	641	15.60	980	20	068
669	50	669	670	14.92	978	10	039
.0698	4° 00′	.0698	.0699	14.30	.9976	86° 00′	1.5010
727	10	727	729	13.73	974	85° 50′	981
756	20	756	758	13.20	971	40	952
.0785	30	.0785	.0787	12.71	.9969	30	1.4923
814	40	814	816	12.25	967	20	893
844	50	843	846	11.83	964	10	864
.0873	5° 00′	.0872	.0875	11.43	.9962	85° 00′	1.4835
902	10	901	904	11.06	959	84° 50′	806
931	20	929	934	10.71	957	40	777
.0960	30	.0958	.0963	10.39	.9954	30	1.4748
989	40	987	992	10.08	951	20	719
.1018	50	.1016	.1022	9.788	948	10	690
.1047	6° 00′	.1045	.1051	9.514	.9945	84° 00′	1.4661
076	10	074	080	9.255	942	83° 50′	632
105	20	103	110	9.010	939	40	603
.1134	30	.1132	.1139	8.777	.9936	30	1.4573
164	40	161	169	8.556	932	20	544
193	50	190	198	8.345	929	10	515
.1222	7° 00′	.1219	.1228	8.144	.9925	83° 00′	1.4486
251	10	248	257	7.953	922	82° 50′	457
280	20	276	287	7.770	918	40	428
.1309	30	.1305	.1317	7.596	.9914	30	1.4399
338	40	334	346	7.429	911	20	370
367	50	363	376	7.269	907	10	341
.1396	8° 00′	.1392	.1405	7.115	.9903	82° 00′	1.4312
425	10	421	435	6.968	899	81° 50′	283
454	20	449	465	6.827	894	40	254
.1484	30	.1478	.1495	6.691	.9890	30	1.4224
513	40	507	524	6.561	886	20	195
542	50	536	554	6.435	881	10	166
.1571	9° 00′	.1564	.1584	6.314	.9877	81° 00′	1.4137
600	10	593	614	197	872	80° 50′	108
629	20	622	644	084	868	40	079
.1658	30	.1650	.1673	5.976	.9863	30	1.4050
687	40	679	703	871	858	20	1.4021
716	50	708	733	769	853	10	992
		Cos	Cot	Tan	Sin	Degrees	Radians

Table B-3 (*cont.*)

Radians	Degrees	Sin	Tan	Cot	Cos		
.1745	10° 00′	.1736	.1763	5.671	.9848	80° 00′	1.3963
774	10	765	793	576	843	79° 50′	934
804	20	794	823	485	838	40	904
.1833	30	.1822	.1853	5.396	.9833	30	1.3875
862	40	851	883	309	827	20	846
891	50	880	914	226	822	10	817
.1920	11° 00′	.1908	.1944	5.145	.9816	79° 00′	1.3788
949	10	937	974	066	811	78° 50′	759
978	20	965	.2004	4.989	805	40	730
.2007	30	.1994	.2035	4.915	.9799	30	1.3701
036	40	.2022	065	843	793	20	672
065	50	051	095	773	787	10	643
.2094	12° 00′	.2079	.2126	4.705	.9781	78° 00′	1.3614
123	10	108	156	638	775	77° 50′	584
153	20	136	186	574	769	40	555
.2182	30	.2164	.2217	4.511	.9763	30	1.3526
211	40	193	247	449	757	20	497
240	50	221	278	390	750	10	468
.2269	13° 00′	.2250	.2309	4.331	.9744	77° 00′	1.3439
298	10	278	339	275	737	76° 50′	410
327	20	306	370	219	730	40	381
.2356	30	.2334	.2401	4.165	.9724	30	1.3352
385	40	363	432	113	717	20	323
414	50	391	462	061	710	10	294
.2443	14° 00′	.2419	.2493	4.011	.9703	76° 00′	1.3265
473	10	447	524	3.962	696	75° 50′	235
502	20	476	555	914	689	40	206
.2531	30	.2504	.2586	3.867	.9681	30	1.3177
560	40	532	617	821	674	20	148
589	50	560	648	776	667	10	119
.2618	15° 00′	.2588	.2679	3.732	.9659	75° 00′	1.3090
647	10	616	711	689	652	74° 50′	061
676	20	644	742	647	644	40	032
.2705	30	.2672	.2773	3.606	.9636	30	1.3003
734	40	700	805	566	628	20	974
763	50	728	836	526	621	10	945
.2793	16° 00′	.2756	.2867	3.487	.9613	74° 00′	1.2915
822	10	784	899	450	605	73° 50′	886
851	20	812	931	412	596	40	857
.2880	30	.2840	.2962	3.376	.9588	30	1.2828
909	40	868	994	340	580	20	799
938	50	896	.3026	305	572	10	770
.2967	17° 00′	.2924	.3057	3.271	.9563	73° 00′	1.2741
996	10	952	089	237	555	72° 50′	712
.3025	20	979	121	204	546	40	683
.3054	30	.3007	.3153	3.172	.9537	30	1.2654
083	40	035	185	140	528	20	625
113	50	062	217	108	520	10	595
.3142	18° 00′	.3090	.3249	3.078	.9511	72° 00′	1.2566
171	10	118	281	047	502	71° 50′	537
200	20	145	314	018	492	40	508
.3229	30	.3173	.3346	2.989	.9483	30	1.2479
258	40	201	378	960	474	20	450
287	50	228	411	932	465	10	421
.3316	19° 00′	.3256	.3443	2.904	.9455	71° 00′	1.2392
345	10	283	476	877	446	70° 50′	363
374	20	311	508	850	436	40	334
.3403	30	.3338	.3541	2.824	.9426	30	1.2305
432	40	365	574	798	417	20	275
462	50	393	607	773	407	10	246
		Cos	Cot	Tan	Sin	Degrees	Radians

Radians	Degrees	Sin	Tan	Cot	Cos		
.3491	20° 00′	.3420	.3640	2.747	.9397	70° 00′	1.2217
520	10	448	673	723	387	69° 50′	188
549	20	475	706	699	377	40	159
.3578	30	.3502	.3739	2.675	.9367	30	1.2130
607	40	529	772	651	356	20	101
636	50	557	805	628	346	10	072
.3665	21° 00′	.3584	.3839	2.605	.9336	69° 00′	1.2043
694	10	611	872	583	325	68° 50′	1.2014
723	20	638	906	560	315	40	985
.3752	30	.3665	.3939	2.539	.9304	30	1.1956
782	40	692	973	517	293	20	926
811	50	719	.4006	496	283	10	897
.3840	22° 00′	.3746	.4040	2.475	9.272	68° 00′	1.1868
869	10	773	074	455	261	67° 50′	839
898	20	800	108	434	250	40	810
.3927	30	.3827	.4142	2.414	.9239	30	1.1781
956	40	854	176	394	228	20	752
985	50	881	210	375	216	10	723
.4014	23° 00′	.3907	.4245	2.356	.9205	67° 00′	1.1694
043	10	934	279	337	194	66° 50′	665
072	20	961	314	318	182	40	636
.4102	30	.3987	.4348	2.300	9.171	30	1.1606
131	40	.4014	383	282	159	20	577
160	50	041	417	264	147	10	548
.4189	24° 00′	.4067	.4452	2.246	.9135	66° 00′	1.1519
218	10	094	487	229	124	65° 50′	490
247	20	120	522	211	112	40	461
.4276	30	.4147	.4557	2.194	.9100	30	1.1432
305	40	173	592	177	088	20	403
334	50	200	628	161	075	10	374
.4363	25° 00′	.4226	.4663	2.145	.9063	65° 00′	1.1345
392	10	253	699	128	051	64° 50′	316
422	20	279	734	112	038	40	286
.4451	30	.4305	.4770	2.097	.9026	30	1.1257
480	40	331	806	081	013	20	228
509	50	358	841	066	001	10	199
.4538	26° 00′	.4384	.4877	2.050	.8988	64° 00′	1.1170
567	10	410	913	035	975	63° 50′	141
596	20	436	950	020	962	40	112
.4625	30	.4462	.4986	2.006	.8949	30	1.1083
654	40	488	.5022	1.991	936	20	054
683	50	514	059	977	923	10	1.1025
.4712	27° 00′	.4540	.5095	1.963	.8910	63° 00′	1.0996
741	10	566	132	949	897	62° 50′	966
771	20	592	169	935	884	40	937
.4800	30	.4617	.5206	1.921	.8870	30	1.0908
829	40	643	243	907	857	20	879
858	50	669	280	894	843	10	850
.4887	28° 00′	.4695	.5317	1.881	.8829	62° 00′	1.0821
916	10	720	354	868	816	61° 50′	792
945	20	746	392	855	802	40	763
.4974	30	.4772	.5430	1.842	.8788	30	1.0734
.5003	40	797	467	829	774	20	705
032	50	823	505	816	760	10	676
.5061	29° 00′	.4848	.5543	1.804	.8746	61° 00′	1.0647
091	10	874	581	792	732	60° 50′	617
120	20	899	619	780	718	40	588
.5149	30	.4924	.5658	1.767	.8704	30	1.0559
178	40	950	696	756	689	20	530
207	50	975	735	744	675	10	501
		Cos	Cot	Tan	Sin	Degrees	Radians

Table B-3 (*cont.*)

Radians	Degrees	Sin	Tan	Cot	Cos		
.5236	30° 00′	.5000	.5774	1.732	.8660	60° 00′	1.0472
265	10	025	812	720	646	59° 50′	443
294	20	050	851	709	631	40	414
.5323	30	.5075	.5890	1.698	.8616	30	1.0385
352	40	100	930	686	601	20	356
381	50	125	969	675	587	10	327
.5411	31° 00′	.5150	.6009	1.664	.8572	59° 00′	1.0297
440	10	175	048	653	557	58° 50′	268
469	20	200	088	643	542	40	239
.5498	30	.5225	.6128	1.632	.8526	30	1.0210
527	40	250	168	621	511	20	181
556	50	275	208	611	496	10	152
.5585	32° 00′	.5299	.6249	1.600	.8480	58° 00′	1.0123
614	10	324	289	590	465	57° 50′	094
643	20	348	330	580	450	40	065
.5672	30	.5373	.6371	1.570	.8434	30	1.0036
701	40	398	412	560	418	20	1.0007
730	50	422	453	550	403	10	977
.5760	33° 00′	.5446	.6494	1.540	.8387	57° 00′	.9948
789	10	471	536	530	371	56° 50′	919
818	20	495	577	520	355	40	890
.5847	30	.5519	.6619	1.511	.8339	30	.9861
876	40	544	661	501	323	20	832
905	50	568	703	1.492	307	10	803
.5934	34° 00′	.5592	.6745	1.483	.8290	56° 00′	.9774
963	10	616	787	473	274	55° 50′	745
992	20	640	830	464	258	40	716
.6021	30	.5664	.6873	1.455	.8241	30	.9687
050	40	688	916	446	225	20	657
080	50	712	959	437	208	10	628
.6109	35° 00′	.5736	.7002	1.428	.8192	55° 00′	.9599
138	10	760	046	419	175	54° 50′	570
167	20	783	089	411	158	40	541
.6196	30	.5807	.7133	1.402	.8141	30	.9512
225	40	831	177	393	124	20	483
254	50	854	221	385	107	10	454
.6283	36° 00′	.5878	.7265	1.376	.8090	54° 00′	.9425
312	10	901	310	368	073	53° 50′	396
341	20	925	355	360	056	40	367
.6370	30	.5948	.7400	1.351	.8039	30	.9338
400	40	972	445	343	021	20	308
429	50	995	490	335	004	10	279
.6458	37° 00′	.6018	.7536	1.327	.7986	53° 00′	.9250
487	10	041	581	319	969	52° 50′	221
516	20	065	627	311	951	40	192
.6545	30	.6088	.7673	1.303	.7934	30	.9163
574	40	111	720	295	916	20	134
603	50	134	766	288	898	10	105
.6632	38° 00′	.6157	.7813	1.280	.7880	52° 00′	.9076
661	10	180	860	272	862	51° 50′	047
690	20	202	907	265	844	40	.9018
.6720	30	.6225	.7954	1.257	.7826	30	.8988
749	40	248	.8002	250	808	20	959
778	50	271	050	242	790	10	930
.6807	39° 00′	.6293	.8098	1.235	.7771	51° 00′	.8901
836	10	316	146	228	753	50° 50′	872
865	20	338	195	220	735	40	843
.6894	30	.6361	.8243	1.213	.7716	30	.8814
923	40	383	292	206	698	20	785
952	50	406	342	199	679	10	756
		Cos	Cot	Tan	Sin	Degrees	Radians

Table B-3 (*cont.*)

Radians	Degrees	Sin	Tan	Cot	Cos		
.6981	40° 00′	.6428	.8391	1.192	.7660	50° 00′	.8727
.7010	10	450	441	185	642	49° 50′	698
039	20	472	491	178	623	40	668
.7069	30	.6494	.8541	1.171	.7604	30	.8639
098	40	517	591	164	585	20	610
127	50	539	642	157	566	10	581
.7156	41° 00′	.6561	.8693	1.150	.7547	49° 00′	.8552
185	10	583	744	144	528	48° 50′	523
214	20	604	796	137	509	40	494
.7243	30	.6626	.8847	1.130	.7490	30	.8465
272	40	648	899	124	470	20	436
301	50	670	952	117	451	10	407
.7330	42° 00′	6691	.9004	1.111	.7431	48° 00′	.8378
359	10	713	057	104	412	47° 50′	348
389	20	734	110	098	392	40	319
.7418	30	.6756	.9163	1.091	.7373	30	.8290
447	40	777	217	085	353	20	261
476	50	799	271	079	333	10	232
.7505	43° 00′	.6820	.9325	1.072	.7314	47° 00′	.8203
534	10	841	380	066	294	46° 50′	174
563	20	862	435	060	274	40	145
.7592	30	.6884	.9490	1.054	.7254	30	.8116
621	40	905	545	048	234	20	087
650	50	926	601	042	214	10	058
.7679	44° 00′	.6947	.9657	1.036	.7193	46° 00′	.8029
709	10	967	713	030	173	45° 50′	999
738	20	988	770	024	153	40	970
.7767	30	.7009	.9827	1.018	.7133	30	.7941
796	40	030	884	012	112	20	912
825	50	050	942	006	092	10	883
.7854	45° 00′	.7071	1.000	1.000	.7071	45° 00′	.7854
		Cos	Cot	Tan	Sin	Degrees	Radians

N.		0	1	2	3	4	5	6	7	8	9
100	00	000	043	087	130	173	217	260	303	346	389
101		432	475	518	561	604	647	689	732	775	817
102		860	903	945	988	*030	*072	*115	*157	*199	*242
103	01	284	326	368	410	452	494	536	578	620	662
104		703	745	787	828	870	912	953	995	*036	*078
105	02	119	160	202	243	284	325	366	407	449	490
106		531	572	612	653	694	735	776	816	857	898
107		938	979	*019	*060	*100	*141	*181	*222	*262	*302
108	03	342	383	423	463	503	543	583	623	663	703
109		743	782	822	862	902	941	981	*021	*060	*100
110	04	139	179	218	258	297	336	376	415	454	493
111		532	571	610	650	689	727	766	805	844	883
112		922	961	999	*038	*077	*115	*154	*192	*231	*269
113	05	308	346	385	423	461	500	538	576	614	652
114		690	729	767	805	843	881	918	956	994	*032
115	06	070	108	145	183	221	258	296	333	371	408
116		446	483	521	558	595	633	670	707	744	781
117		819	856	893	930	967	*004	*041	*078	*115	*151
118	07	188	225	262	298	335	372	408	445	482	518
119		555	591	628	664	700	737	773	809	846	882
120		918	954	990	*027	*063	*099	*135	*171	*207	*243
121	08	279	314	350	386	422	458	493	529	565	600
122		636	672	707	743	778	814	849	884	920	955
123		991	*026	*061	*096	*132	*167	*202	*237	*272	*307
124	09	342	377	412	447	482	517	552	587	621	656
125		691	726	760	795	830	864	899	934	968	*003
126	10	037	072	106	140	175	209	243	278	312	346
127		380	415	449	483	517	551	585	619	653	687
128		721	755	789	823	857	890	924	958	992	*025
129	11	059	093	126	160	193	227	261	294	327	361
130		394	428	461	494	528	561	594	628	661	694
131		727	760	793	826	860	893	926	959	992	*024
132	12	057	090	123	156	189	222	254	287	320	352
133		385	418	450	483	516	548	581	613	646	678
134		710	743	775	808	840	872	905	937	969	*001
135	13	033	066	098	130	162	194	226	258	290	322
136		354	386	418	450	481	513	545	577	609	640
137		672	704	735	767	799	830	862	893	925	956
138		988	*019	*051	*082	*114	*145	*176	*208	*239	*270
139	14	301	333	364	395	426	457	489	520	551	582
140		613	644	675	706	737	768	799	829	860	891
141		922	953	983	*014	*045	*076	*106	*137	*168	*198
142	15	229	259	290	320	351	381	412	442	473	503
143		534	564	594	625	655	685	715	746	776	806
144		836	866	897	927	957	987	*017	*047	*077	*107
145	16	137	167	197	227	256	286	316	346	376	406
146		435	465	495	524	554	584	613	643	673	702
147		732	761	791	820	850	879	909	938	967	997
148	17	026	056	085	114	143	173	202	231	260	289
149		319	348	377	406	435	464	493	522	551	580
150		609	638	667	696	725	754	782	811	840	869

N.	0	1	2	3	4	5	6	7	8	9

Proportional parts

	44	43	42
1	4.4	4.3	4.2
2	8.8	8.6	8.4
3	13.2	12.9	12.6
4	17.6	17.2	16.8
5	22.0	21.5	21.0
6	26.4	25.8	25.2
7	30.8	30.1	29.4
8	35.2	34.4	33.6
9	39.6	38.7	37.8

	41	40	39
1	4.1	4.0	3.9
2	8.2	8.0	7.8
3	12.3	12.0	11.7
4	16.4	16.0	15.6
5	20.5	20.0	19.5
6	24.6	24.0	23.4
7	28.7	28.0	27.3
8	32.8	32.0	31.2
9	36.9	36.0	35.1

	38	37	36
1	3.8	3.7	3.6
2	7.6	7.4	7.2
3	11.4	11.1	10.8
4	15.2	14.8	14.4
5	19.0	18.5	18.0
6	22.8	22.2	21.6
7	26.6	25.9	25.2
8	30.4	29.6	28.8
9	34.2	33.3	32.4

	35	34	33
1	3.5	3.4	3.3
2	7.0	6.8	6.6
3	10.5	10.2	9.9
4	14.0	13.6	13.2
5	17.5	17.0	16.5
6	21.0	20.4	19.8
7	24.5	23.8	23.1
8	28.0	27.2	26.4
9	31.5	30.6	29.7

	32	31	30
1	3.2	3.1	3.0
2	6.4	6.2	6.0
3	9.6	9.3	9.0
4	12.8	12.4	12.0
5	16.0	15.5	15.0
6	19.2	18.6	18.0
7	22.4	21.7	21.0
8	25.6	24.8	24.0
9	28.8	27.9	27.0

.00 000–.17 869

Table B-4 (*cont.*)

150–200

N.		0	1	2	3	4	5	6	7	8	9
150	17	609	638	667	696	725	754	782	811	840	869
151		898	926	955	984	*013	*041	*070	*099	*127	*156
152	18	184	213	241	270	298	327	355	384	412	441
153		469	498	526	554	583	611	639	667	696	724
154		752	780	808	837	865	893	921	949	977	*005
155	19	033	061	089	117	145	173	201	229	257	285
156		312	340	368	396	424	451	479	507	535	562
157		590	618	645	673	700	728	756	783	811	838
158		866	893	921	948	976	*003	*030	*058	*085	*112
159	20	140	167	194	222	249	276	303	330	358	385
160		412	439	466	493	520	548	575	602	629	656
161		683	710	737	763	790	817	844	871	898	925
162		952	978	*005	*032	*059	*085	*112	*139	*165	*192
163	21	219	245	272	299	325	352	378	405	431	458
164		484	511	537	564	590	617	643	669	696	722
165		748	775	801	827	854	880	906	932	958	958
166	22	011	037	063	089	115	141	167	194	220	246
167		272	298	324	350	376	401	427	453	479	505
168		531	557	583	608	634	660	686	712	737	763
169		789	814	840	866	891	917	943	968	994	*019
170	23	045	070	096	121	147	172	198	223	249	274
171		300	325	350	376	401	426	452	477	502	528
172		553	578	603	629	654	679	704	729	754	779
173		805	830	855	880	905	930	955	980	*005	*030
174	24	055	080	105	130	155	180	204	229	254	279
175		304	329	353	378	403	428	452	477	502	527
176		551	576	601	625	650	674	699	724	748	773
177		797	822	846	871	895	920	944	969	993	*018
178	25	042	066	091	115	139	164	188	212	237	261
179		285	310	334	358	382	406	431	455	479	503
180		527	551	575	600	624	648	672	696	720	744
181		768	792	816	840	864	888	912	935	959	983
182	26	007	031	055	079	102	126	150	174	198	221
183		245	269	293	316	340	364	387	411	435	458
184		482	505	529	553	576	600	623	647	670	694
185		717	741	764	788	811	834	858	881	905	928
186		951	975	998	*021	*045	*068	*091	*114	*138	*161
187	27	184	207	231	254	277	300	323	346	370	393
188		416	439	462	485	508	531	554	577	600	623
189		646	669	692	715	738	761	784	807	830	852
190		875	898	921	944	967	989	*012	*035	*058	*081
191	28	103	126	149	171	194	217	240	262	285	307
192		330	353	375	398	421	443	466	488	511	533
193		556	578	601	623	646	668	691	713	735	758
194		780	803	825	847	870	892	914	937	959	981
195	29	003	026	048	070	092	115	137	159	181	203
196		226	248	270	292	314	336	358	380	403	425
197		447	469	491	513	535	557	579	601	623	645
198		667	688	710	732	754	776	798	820	842	863
199		885	907	929	951	973	994	*016	*038	*060	*081
200	30	103	125	146	168	190	211	233	255	276	298
N.		0	1	2	3	4	5	6	7	8	9

Proportional parts

	29	28
1	2.9	2.8
2	5.8	5.6
3	8.7	8.4
4	11.6	11.2
5	14.5	14.0
6	17.4	16.8
7	20.3	19.6
8	23.2	22.4
9	26.1	25.2

	27	26
1	2.7	2.6
2	5.4	5.2
3	8.1	7.8
4	10.8	10.4
5	13.5	13.0
6	16.2	15.6
7	18.9	18.2
8	21.6	20.8
9	24.3	23.4

	25
1	2.5
2	5.0
3	7.5
4	10.0
5	12.5
6	15.0
7	17.5
8	20.0
9	22.5

	24	23
1	2.4	2.3
2	4.8	4.6
3	7.2	6.9
4	9.6	9.2
5	12.0	11.5
6	14.4	13.8
7	16.8	16.1
8	19.2	18.4
9	21.6	20.7

	22	21
1	2.2	2.1
2	4.4	4.2
3	6.6	6.3
4	8.8	8.4
5	11.0	10.5
6	13.2	12.6
7	15.4	14.7
8	17.6	16.8
9	19.8	18.9

.17 609–.30 298

Table B-4 (*cont.*)

200–250

N.		0	1	2	3	4	5	6	7	8	9
200	30	103	125	146	168	190	211	233	255	276	298
201		320	341	363	384	406	428	449	471	492	514
202		535	557	578	600	621	643	664	685	707	728
203		750	771	792	814	835	856	878	899	920	942
204		963	984	*006	*027	*048	*069	*091	*112	*133	*154
205	31	175	197	218	239	260	281	302	323	345	366
206		387	408	429	450	471	492	513	534	555	576
207		597	618	639	660	681	702	723	744	765	785
208		806	827	848	869	890	911	931	952	973	994
209	32	015	035	056	077	098	118	139	160	181	201
210		222	243	263	284	305	325	346	366	387	408
211		428	449	469	490	510	531	552	572	593	613
212		634	654	675	695	715	736	756	777	797	818
213		838	858	879	899	919	940	960	980	*001	*021
214	33	041	062	082	102	122	143	163	183	203	224
215		244	264	284	304	325	345	365	385	405	425
216		445	465	486	506	526	546	566	586	606	626
217		646	666	686	706	726	746	766	786	806	826
218		846	866	885	905	925	945	965	985	*005	*025
219	34	044	064	084	104	124	143	163	183	203	223
220		242	262	282	301	321	341	361	380	400	420
221		439	459	479	498	518	537	557	577	596	616
222		635	655	674	694	713	733	753	772	792	811
223		830	850	869	889	908	928	947	967	986	*005
224	35	025	044	064	083	102	122	141	160	180	199
225		218	238	257	276	295	315	334	353	372	392
226		411	430	449	468	488	507	526	545	564	583
277		603	622	641	660	679	698	717	736	755	774
228		793	813	832	851	870	889	908	927	946	965
229		984	*003	*021	*040	*059	*078	*097	*116	*135	*154
230	36	173	192	211	229	248	267	286	305	324	342
231		361	380	399	418	436	455	474	493	511	530
232		549	568	586	605	624	642	661	680	698	717
233		736	754	773	791	810	829	847	866	884	903
234		922	940	959	977	996	*014	*033	*051	*070	*088
235	37	107	125	144	162	181	199	218	236	254	273
236		291	310	328	346	365	383	401	420	438	457
237		475	493	511	530	548	566	585	603	621	639
238		658	676	694	712	731	749	767	785	803	822
239		840	858	876	894	912	931	949	967	985	*003
240	38	021	039	057	075	093	112	130	148	166	184
241		202	220	238	256	274	292	310	328	346	364
242		382	399	417	435	453	471	489	507	525	543
243		561	578	596	614	632	650	668	686	703	721
244		739	757	775	792	810	828	846	863	881	899
245		917	934	952	970	987	*005	*023	*041	*058	*076
246	39	094	111	129	146	164	182	199	217	235	252
247		270	287	305	322	340	358	375	393	410	428
248		445	463	480	498	515	533	550	568	585	602
249		620	637	655	672	690	707	724	742	759	777
250		794	811	829	846	863	881	898	915	933	950

Proportional parts

	22	21
1	2.2	2.1
2	4.4	4.2
3	6.6	6.3
4	8.8	8.4
5	11.0	10.5
6	13.2	12.6
7	15.4	14.7
8	17.6	16.8
9	19.8	18.9

	20
1	2.0
2	4.0
3	6.0
4	8.0
5	10.0
6	12.0
7	14.0
8	16.0
9	18.0

	19
1	1.9
2	3.8
3	5.7
4	7.6
5	9.5
6	11.4
7	13.3
8	15.2
9	17.1

	18
1	1.8
2	3.6
3	5.4
4	7.2
5	9.0
6	10.8
7	12.6
8	14.4
9	16.2

	17
1	1.7
2	3.4
3	5.1
4	6.8
5	8.5
6	10.2
7	11.9
8	13.6
9	15.3

N.	0	1	2	3	4	5	6	7	8	9	Proportional parts

.30 103–.39 950

250–300

N.		0	1	2	3	4	5	6	7	8	9	Proportional parts
250	39	794	811	829	846	863	881	898	915	933	950	**18**
251		967	985	*002	*019	*037	*054	*071	*088	*106	*123	1 1.8
252	40	140	157	175	192	209	226	243	261	278	295	2 3.6
253		312	329	346	364	381	398	415	432	449	466	3 5.4
254		483	500	518	535	552	569	586	603	620	637	4 7.2
255		654	671	688	705	722	739	756	773	790	807	5 9.0
256		824	841	858	875	892	909	926	943	960	976	6 10.8
257		993	*010	*027	*044	*061	*078	*095	*111	*128	*145	7 12.6
258	41	162	179	196	212	229	246	263	280	296	313	8 14.4
259		330	347	363	380	397	414	430	447	464	481	9 16.2
260		497	514	531	547	564	581	597	614	631	647	**17**
261		664	681	697	714	731	747	764	780	797	814	1 1.7
262		830	847	863	880	896	913	929	946	963	979	2 3.4
263		996	*012	*029	*045	*062	*078	*095	*111	*127	*144	3 5.1
264	42	160	177	193	210	226	243	259	275	292	308	4 6.8
265		325	341	357	374	390	406	423	439	455	472	5 8.5
266		488	504	521	537	553	570	586	602	619	635	6 10.2
267		651	667	684	700	716	732	749	765	781	797	7 11.9
268		813	830	846	862	878	894	911	927	943	959	8 13.6
269		975	991	*008	*024	*040	*056	*072	*088	*104	*120	9 15.3
270	43	136	152	169	185	201	217	233	249	265	281	**16**
271		297	313	329	345	361	377	393	409	425	441	1 1.6
272		457	473	489	505	521	537	553	569	584	600	2 3.2
273		616	632	648	664	680	696	712	727	743	759	3 4.8
274		775	791	807	823	838	854	870	886	902	917	4 6.4
275		933	949	965	981	996	*012	*028	*044	*059	*075	5 8.0
276	44	091	107	122	138	154	170	185	201	217	232	6 9.6
277		248	264	279	295	311	326	342	358	373	389	7 11.2
278		404	420	436	451	467	483	498	514	529	545	8 12.8
279		560	576	592	607	623	638	654	669	685	700	9 14.4
280		716	731	747	762	778	793	809	824	840	855	**15**
281		871	886	902	917	932	948	963	979	994	*010	1 1.5
282	45	025	040	056	071	086	102	117	133	148	163	2 3.0
283		179	194	209	225	240	255	271	286	301	317	3 4.5
284		332	347	362	378	393	408	423	439	454	469	4 6.0
285		484	500	515	530	545	561	576	591	606	621	5 7.5
286		637	652	667	682	697	712	728	743	758	773	6 9.0
287		788	803	818	834	849	864	879	894	909	924	7 10.5
288		939	954	969	984	*000	*015	*030	*045	*060	*075	8 12.0
289	46	090	105	120	135	150	165	180	195	210	225	9 13.5
290		240	255	270	285	300	315	330	345	359	374	**14**
291		389	404	419	434	449	464	479	494	509	523	1 1.4
292		538	553	568	583	598	613	627	642	657	672	2 2.8
293		687	702	716	731	746	761	776	790	805	820	3 4.2
294		835	850	864	879	894	909	923	938	953	967	4 5.6
295		982	997	*012	*026	*041	*056	*070	*085	*100	*114	5 7.0
296	47	129	144	159	173	188	202	217	232	246	261	6 8.4
297		276	290	305	319	334	349	363	378	392	407	7 9.8
298		422	436	451	465	480	494	509	524	538	553	8 11.2
299		567	582	596	611	625	640	654	669	683	698	9 12.6
300		712	727	741	756	770	784	799	813	828	842	log e = .43429
N.		0	1	2	3	4	5	6	7	8	9	Proportional parts

.39 794–.47 842

Table B-4 (cont.)

300–350

N.		0	1	2	3	4	5	6	7	8	9
300	47	712	727	741	756	770	784	799	813	828	842
301		857	871	885	900	914	929	943	958	972	986
302	48	001	015	029	044	058	073	087	101	116	130
303		144	159	173	187	202	216	230	244	259	273
304		287	302	316	330	344	359	373	387	401	416
305		430	444	458	473	487	501	515	530	544	558
306		572	586	601	615	629	643	657	671	686	700
307		714	728	742	756	770	785	799	813	827	841
308		855	869	883	897	911	926	940	954	968	982
309		996	*010	*024	*038	*052	*066	*080	*094	*108	*122
310	49	136	150	164	178	192	206	220	234	248	262
311		276	290	304	318	332	346	360	374	388	402
312		415	429	443	457	471	485	499	513	527	541
313		554	568	582	596	610	624	638	651	665	679
314		693	707	721	734	748	762	776	790	803	817
315		831	845	859	872	886	900	914	927	941	955
316		969	982	996	*010	*024	*037	*051	*065	*079	*092
317	50	106	120	133	147	161	174	188	202	215	229
318		243	256	270	284	297	311	325	338	352	365
319		379	393	406	420	433	447	461	474	488	501
320		515	529	542	556	569	583	596	610	623	637
321		651	664	678	691	705	718	732	745	759	772
322		786	799	813	826	840	853	866	880	893	907
323		920	934	947	961	974	987	*001	*014	*028	*041
324	51	055	068	081	095	108	121	135	148	162	175
325		188	202	215	228	242	255	268	282	295	308
326		322	335	348	362	375	388	402	415	428	441
327		455	468	481	495	508	521	534	548	561	574
328		587	601	614	627	640	654	667	680	693	706
329		720	733	746	759	772	786	799	812	825	838
330		851	865	878	891	904	917	930	943	957	970
331		983	996	*009	*022	*035	*048	*061	*075	*088	*101
332	52	114	127	140	153	166	179	192	205	218	231
333		244	257	270	284	297	310	323	336	349	362
334		375	388	401	414	427	440	453	466	479	492
335		504	517	530	543	556	569	582	595	608	621
336		634	647	660	673	686	699	711	724	737	750
337		763	776	789	802	815	827	840	853	866	879
338		892	905	917	930	943	956	969	982	994	*007
339	53	020	033	046	058	071	084	097	110	123	135
340		148	161	173	186	199	212	224	237	250	263
341		275	288	301	314	326	339	352	364	377	390
342		403	415	428	441	453	466	479	491	504	517
343		529	542	555	567	580	593	605	618	631	643
344		656	668	681	694	706	719	732	744	757	769
345		782	794	807	820	832	845	857	870	882	895
346		908	920	933	945	958	970	983	995	*008	*020
347	54	033	045	058	070	083	095	108	120	133	145
348		158	170	183	195	208	220	233	245	258	270
349		283	295	307	320	332	345	357	370	382	394
350		407	419	432	444	456	469	481	494	506	518

$\log \pi = .49715$

N.	0	1	2	3	4	5	6	7	8	9	Proportional parts

.47 712–.54 518

Proportional parts

15		14		13		12	
1	1.5	1	1.4	1	1.3	1	1.2
2	3.0	2	2.8	2	2.6	2	2.4
3	4.5	3	4.2	3	3.9	3	3.6
4	6.0	4	5.6	4	5.2	4	4.8
5	7.5	5	7.0	5	6.5	5	6.0
6	9.0	6	8.4	6	7.8	6	7.2
7	10.5	7	9.8	7	9.1	7	8.4
8	12.0	8	11.2	8	10.4	8	9.6
9	13.5	9	12.6	9	11.7	9	10.8

350–400

N.		0	1	2	3	4	5	6	7	8	9	Proportional parts
350	54	407	419	432	444	456	469	481	494	506	518	
351		531	543	555	568	580	593	605	617	630	642	
352		654	667	679	691	704	716	728	741	753	765	
353		777	790	802	814	827	839	851	864	876	888	
354		900	913	925	937	949	962	974	986	998	*011	
355	55	023	035	047	060	072	084	096	108	121	133	
356		145	157	169	182	194	206	218	230	242	255	
357		267	279	291	303	315	328	340	352	364	376	
358		388	400	413	425	437	449	461	473	485	497	
359		509	522	534	546	558	570	582	594	606	618	
360		630	642	654	666	678	691	703	715	727	739	
361		751	763	775	787	799	811	823	835	847	859	
362		871	883	895	907	919	931	943	955	967	979	
363		991	*003	*015	*027	*038	*050	*062	*074	*086	*098	
364	56	110	122	134	146	158	170	182	194	205	217	
365		229	241	253	265	277	289	301	312	324	336	
366		348	360	372	384	396	407	419	431	443	455	
367		467	478	490	502	514	526	538	549	561	573	
368		585	597	608	620	632	644	656	667	679	691	
369		703	714	726	738	750	761	773	785	797	808	
370		820	832	844	855	867	879	891	902	914	926	
371		937	949	961	972	984	996	*008	*019	*031	*043	
372	57	054	066	078	089	101	113	124	136	148	159	
373		171	183	194	206	217	229	241	252	264	276	
374		287	299	310	322	334	345	357	368	380	392	
375		403	415	426	438	449	461	473	484	496	507	
376		519	530	542	553	565	576	588	600	611	623	
377		634	646	657	669	680	692	703	715	726	738	
378		749	761	772	784	795	807	818	830	841	852	
379		864	875	887	898	910	921	933	944	955	967	
380		978	990	*001	*013	*024	*035	*047	*058	*070	*081	
381	58	092	104	115	127	138	149	161	172	184	195	
382		206	218	229	240	252	263	274	286	297	309	
383		320	331	343	354	365	377	388	399	410	422	
384		433	444	456	467	478	490	501	512	524	535	
385		546	557	569	580	591	602	614	625	636	647	
386		659	670	681	692	704	715	726	737	749	760	
387		771	782	794	805	816	827	838	850	861	872	
388		883	894	906	917	928	939	950	961	973	984	
389		995	*006	*017	*028	*040	*051	*062	*073	*084	*095	
390	59	106	118	129	140	151	162	173	184	195	207	
391		218	229	240	251	262	273	284	295	306	318	
392		329	340	351	362	373	384	395	406	417	428	
393		439	450	461	472	483	494	506	517	528	539	
394		550	561	572	583	594	605	616	627	638	649	
395		660	671	682	693	704	715	726	737	748	759	
396		770	780	791	802	813	824	835	846	857	868	
397		879	890	901	912	923	934	945	956	966	977	
398		988	999	*010	*021	*032	*043	*054	*065	*076	*086	
399	60	097	108	119	130	141	152	163	173	184	195	
400		206	217	228	239	249	260	271	282	293	304	
N.		0	1	2	3	4	5	6	7	8	9	Proportional parts

Proportional parts:

	13
1	1.3
2	2.6
3	3.9
4	5.2
5	6.5
6	7.8
7	9.1
8	10.4
9	11.7

	12
1	1.2
2	2.4
3	3.6
4	4.8
5	6.0
6	7.2
7	8.4
8	9.6
9	10.8

	11
1	1.1
2	2.2
3	3.3
4	4.4
5	5.5
6	6.6
7	7.7
8	8.8
9	9.9

	10
1	1.0
2	2.0
3	3.0
4	4.0
5	5.0
6	6.0
7	7.0
8	8.0
9	9.0

.54 407–.60 304

Table B-4 (*cont.*)

400–450

N.		0	1	2	3	4	5	6	7	8	9
400	60	206	217	228	239	249	260	271	282	293	304
401		314	325	336	347	358	369	379	390	401	412
402		423	433	444	455	466	477	487	498	509	520
403		531	541	552	563	574	584	595	606	617	627
404		638	649	660	670	681	692	703	713	724	735
405		746	756	767	778	788	799	810	821	831	842
406		853	863	874	885	895	906	917	927	938	949
407		959	970	981	991	*002	*013	*023	*034	*045	*055
408	61	066	077	087	098	109	119	130	140	151	162
409		172	183	194	204	215	225	236	247	257	268
410		278	289	300	310	321	331	342	352	363	374
411		384	395	405	416	426	437	448	458	469	479
412		490	500	511	521	532	542	553	563	574	584
413		595	606	616	627	637	648	658	669	679	690
414		700	711	721	731	742	752	763	773	784	794
415		805	815	826	836	847	857	868	878	888	899
416		909	920	930	941	951	962	972	982	993	*003
417	62	014	024	034	045	055	066	076	086	097	107
418		118	128	138	149	159	17?	180	190	201	211
419		221	232	242	252	263	273	284	294	304	315
420		325	335	346	356	366	377	387	397	408	418
421		428	439	449	459	469	480	490	500	511	521
422		531	542	552	562	572	583	593	603	613	624
423		634	644	655	665	675	685	696	706	716	726
424		737	747	757	767	778	788	798	808	818	829
425		839	849	859	870	880	890	900	910	921	931
426		941	951	961	972	982	992	*002	*012	*022	*033
427	63	043	053	063	073	083	094	104	114	124	134
428		144	155	165	175	185	195	205	215	225	236
429		246	256	266	276	286	296	306	317	327	337
430		347	357	367	377	387	397	407	417	428	438
431		448	458	468	478	488	498	508	518	528	538
432		548	558	568	579	589	599	609	619	629	639
433		649	659	669	679	689	699	709	719	729	739
434		749	759	769	779	789	799	809	819	829	839
435		849	859	869	879	889	899	909	919	929	939
436		949	959	969	979	988	998	*008	*018	*028	*038
437	64	048	058	068	078	088	098	108	118	128	137
438		147	157	167	177	187	197	207	217	227	237
439		246	256	266	276	286	296	306	316	326	335
440		345	355	365	375	385	395	404	414	424	434
441		444	454	464	473	483	493	503	513	523	532
442		542	552	562	572	582	591	601	611	621	631
443		640	650	660	670	680	689	699	709	719	729
444		738	748	758	768	777	787	797	807	816	826
445		836	846	856	865	875	885	895	904	914	924
446		933	943	953	963	972	982	992	*002	*011	*021
447	65	031	040	050	060	070	079	089	099	108	118
448		128	137	147	157	167	176	186	196	205	215
449		225	234	244	254	263	273	283	292	302	312
450		321	331	341	350	360	369	379	389	398	408

N.	0	1	2	3	4	5	6	7	8	9	Proportional parts

Proportional parts

	11		10		9
1	1.1	1	1.0	1	.9
2	2.2	2	2.0	2	1.8
3	3.3	3	3.0	3	2.7
4	4.4	4	4.0	4	3.6
5	5.5	5	5.0	5	4.5
6	6.6	6	6.0	6	5.4
7	7.7	7	7.0	7	6.3
8	8.8	8	8.0	8	7.2
9	9.9	9	9.0	9	8.1

.60 206–.65 408

450–500

N.		0	1	2	3	4	5	6	7	8	9	Proportional parts
450	65	321	331	341	350	360	369	379	389	398	408	
451		418	427	437	447	456	466	475	485	495	504	
452		514	523	533	543	552	562	571	581	591	600	
453		610	619	629	639	648	658	667	677	686	696	
454		706	715	725	734	744	753	763	772	782	792	
455		801	811	820	830	839	849	858	868	877	887	
456		896	906	916	925	935	944	954	963	973	982	
457		992	*001	*011	*020	*030	*039	*049	*058	*068	*077	
458	66	087	096	106	115	124	134	143	153	162	172	
459		181	191	200	210	219	229	238	247	257	266	
460		276	285	295	304	314	323	332	342	351	361	
461		370	380	389	398	408	417	427	436	445	455	
462		464	474	483	492	502	511	521	530	539	549	
463		558	567	577	586	596	605	614	624	633	642	
464		652	661	671	680	689	699	708	717	727	736	
465		745	755	764	773	783	792	801	811	820	829	
466		839	848	857	867	876	885	894	904	913	922	
467		932	941	950	960	969	978	987	997	*006	*015	
468	67	025	034	043	052	062	071	080	089	099	108	
469		117	127	136	145	154	164	173	182	191	201	
470		210	219	228	237	247	256	265	274	284	293	
471		302	311	321	330	339	348	357	367	376	385	
472		394	403	413	422	431	440	449	459	468	477	
473		486	495	504	514	523	532	541	550	560	569	
474		578	587	596	605	614	624	633	642	651	660	
475		669	679	688	697	706	715	724	733	742	752	
476		761	770	779	788	797	806	815	825	834	843	
477		852	861	870	879	888	897	906	916	925	934	
478		943	952	961	970	979	988	997	*006	*015	*024	
479	68	034	043	052	061	070	079	088	097	106	115	
480		124	133	142	151	160	169	178	187	196	205	
481		215	224	233	242	251	260	269	278	287	296	
482		305	314	323	332	341	350	359	368	377	386	
483		395	404	413	422	431	440	449	458	467	476	
484		485	494	502	511	520	529	538	547	556	565	
485		574	583	592	601	610	619	728	637	646	655	
486		664	673	681	690	699	708	717	726	735	744	
487		753	762	771	780	789	797	806	815	824	833	
488		842	851	860	869	878	886	895	904	913	922	
489		931	940	949	958	966	975	984	993	*002	*011	
490	69	020	028	037	046	055	064	073	082	090	099	
491		108	117	126	135	144	152	161	170	179	188	
492		197	205	214	223	232	241	249	258	267	276	
493		285	294	302	311	320	329	338	346	355	364	
494		373	381	390	399	408	417	425	434	443	452	
495		461	469	478	487	496	504	513	522	531	539	
496		548	557	566	574	583	592	601	609	618	627	
497		636	644	653	662	671	679	688	697	705	714	
498		723	732	740	749	758	767	775	784	793	801	
499		810	819	827	836	845	854	862	871	880	888	
500		897	906	914	923	932	940	949	958	966	975	
N.		0	1	2	3	4	5	6	7	8	9	Proportional parts

Proportional parts:

	10
1	1.0
2	2.0
3	3.0
4	4.0
5	5.0
6	6.0
7	7.0
8	8.0
9	9.0

	9
1	.9
2	1.8
3	2.7
4	3.6
5	4.5
6	5.4
7	6.3
8	7.2
9	8.1

	8
1	.8
2	1.6
3	2.4
4	3.2
5	4.0
6	4.8
7	5.6
8	6.4
9	7.2

.65 321–.69 975

Table B-4 (*cont.*)

500–550

N.		0	1	2	3	4	5	6	7	8	9
500	69	897	906	914	923	932	940	949	958	966	975
501		984	992	*001	*010	*018	*027	*036	*044	*053	*062
502	70	070	079	088	096	105	114	122	131	140	148
503		157	165	174	183	191	200	209	217	226	234
504		243	252	260	269	278	286	295	303	312	321
505		329	338	346	355	364	372	381	389	398	406
506		415	424	432	441	449	458	467	475	484	492
507		501	509	518	526	535	544	552	561	569	578
508		586	595	603	612	621	629	638	646	655	663
509		672	680	689	697	706	714	723	731	740	749
510		757	766	774	783	791	800	808	817	825	834
511		842	851	859	868	876	885	893	902	910	919
512		927	935	944	952	961	969	978	986	995	*003
513	71	012	020	029	037	046	054	063	071	079	088
514		096	105	113	122	130	139	147	155	164	172
515		181	189	198	206	214	223	231	240	248	257
516		265	273	282	290	299	307	315	324	332	341
517		349	357	366	374	383	391	399	408	416	425
518		433	441	450	458	466	475	483	492	500	508
519		517	525	533	542	550	559	567	575	584	592
520		600	609	617	625	634	642	650	659	667	675
521		684	692	700	709	717	725	734	742	750	759
522		767	775	784	792	800	809	817	825	834	842
523		850	858	867	875	883	892	900	908	917	925
524		933	941	950	958	966	975	983	991	999	*008
525	72	016	024	032	041	049	057	066	074	082	090
526		099	107	115	123	132	140	148	156	165	173
527		181	189	198	206	214	222	230	239	247	255
528		263	272	280	288	296	304	313	321	329	337
529		346	354	362	370	378	387	395	403	411	419
530		428	436	444	452	460	469	477	485	493	501
531		509	518	526	534	542	550	558	567	575	583
532		591	599	607	616	624	632	640	648	656	665
533		673	681	689	697	705	713	722	730	738	746
534		754	762	770	779	787	795	803	811	819	827
535		835	843	852	860	868	876	884	892	900	908
536		916	925	933	941	949	957	965	973	981	989
537		997	*006	*014	*022	*030	*038	*046	*054	*062	*070
538	73	078	086	094	102	111	119	127	135	143	151
539		159	167	175	183	191	199	207	215	223	231
540		239	247	255	263	272	280	288	296	304	312
541		320	328	336	344	352	360	368	376	384	392
542		400	408	416	424	432	440	448	456	464	472
543		480	488	496	504	512	520	528	536	544	552
544		560	568	576	584	592	600	608	616	624	632
545		640	648	656	664	672	679	687	695	703	711
546		719	727	735	743	751	759	767	775	783	791
547		799	807	815	823	830	838	846	854	862	870
548		878	886	894	902	910	918	926	933	941	949
549		957	965	973	981	989	997	*005	*013	*020	*028
550	74	036	044	052	060	068	076	084	092	099	107

Proportional parts

	9
1	.9
2	1.8
3	2.7
4	3.6
5	4.5
6	5.4
7	6.3
8	7.2
9	8.1

	8
1	.8
2	1.6
3	2.4
4	3.2
5	4.0
6	4.8
7	5.6
8	6.4
9	7.2

	7
1	.7
2	1.4
3	2.1
4	2.8
5	3.5
6	4.2
7	4.9
8	5.6
9	6.3

| N. | 0 | 1 | 2 | 3 | 4 | 5 | 6 | 7 | 8 | 9 | Proportional parts |

.69 897–.74 107

550–600

N.		0	1	2	3	4	5	6	7	8	9	Proportional parts
550	74	036	044	052	060	068	076	084	092	099	107	
551		115	123	131	139	147	155	162	170	178	186	
552		194	202	210	218	225	233	241	249	257	265	
553		273	280	288	296	304	312	320	327	335	343	
554		351	359	367	374	382	390	398	406	414	421	
555		429	437	445	453	461	468	476	484	492	500	
556		507	515	523	531	539	547	554	562	570	578	
557		586	593	601	609	617	624	632	640	648	656	
558		663	671	679	687	695	702	710	718	726	733	
559		741	749	757	764	772	780	788	796	803	811	
560		819	827	834	842	850	858	865	873	881	889	
561		896	904	912	920	927	935	943	950	958	966	
562		974	981	989	997	*005	*012	*020	*028	*035	*043	
563	75	051	059	066	074	082	089	097	105	113	120	
564		128	136	143	151	159	166	174	182	189	197	
565		205	213	220	228	236	243	251	259	266	274	
566		282	289	297	305	312	320	328	335	343	351	
567		358	366	374	381	389	397	404	412	420	427	
568		435	442	450	458	465	473	481	488	496	504	
569		511	519	526	534	542	549	557	565	572	580	
570		587	595	603	610	618	626	633	641	648	656	
571		664	671	679	686	694	702	709	717	724	732	
572		740	747	755	762	770	778	785	793	800	808	
573		815	823	831	838	846	853	861	868	876	884	
574		891	899	906	914	921	929	937	944	952	959	
575		967	974	982	989	997	*005	*012	*020	*027	*035	
576	76	042	050	057	065	072	080	087	095	103	110	
577		118	125	133	140	148	155	163	170	178	185	
578		193	200	208	215	223	230	238	245	253	260	
579		268	275	283	290	298	305	313	320	328	335	
580		343	350	358	365	373	380	388	395	403	410	
581		418	425	433	440	448	455	462	470	477	485	
582		492	500	507	515	522	530	537	545	552	559	
583		567	574	582	589	597	604	612	619	626	634	
584		641	649	656	664	671	678	686	693	701	708	
585		716	723	730	738	745	753	760	768	775	782	
586		790	797	805	812	819	827	834	842	849	856	
587		864	871	879	886	893	901	908	916	923	930	
588		938	945	953	960	967	975	982	989	997	*004	
589	77	012	019	026	034	041	048	056	063	070	078	
590		085	093	100	107	115	122	129	137	144	151	
591		159	166	173	181	188	195	203	210	217	225	
592		232	240	247	254	262	269	276	283	291	298	
593		305	313	320	327	335	342	349	357	364	371	
594		379	386	393	401	408	415	422	430	437	444	
595		452	459	466	474	481	488	495	503	510	517	
596		525	532	539	546	554	561	568	576	583	590	
597		597	605	612	619	627	634	641	648	656	663	
598		670	677	685	692	699	706	714	721	728	735	
599		743	750	757	764	772	779	786	793	801	808	
600		815	822	830	837	844	851	859	866	873	880	
N.		0	1	2	3	4	5	6	7	8	9	Proportional parts

Proportional parts (right column):

	8
1	.8
2	1.6
3	2.4
4	3.2
5	4.0
6	4.8
7	5.6
8	6.4
9	7.2

	7
1	.7
2	1.4
3	2.1
4	2.8
5	3.5
6	4.2
7	4.9
8	5.6
9	6.3

.74 036–.77 880

Table B-4 (*cont.*)

600–650

N.		0	1	2	3	4	5	6	7	8	9	Proportional parts
600	77	815	822	830	837	844	851	859	866	873	880	
601		887	895	902	909	916	924	931	938	945	952	
602		960	967	974	981	988	996	*003	*010	*017	*025	
603	78	032	039	046	053	061	068	075	082	089	097	
604		104	111	118	125	132	140	147	154	161	168	
605		176	183	190	197	204	211	219	226	233	240	
606		247	254	262	269	276	283	290	297	305	312	
607		319	326	333	340	347	355	362	369	376	383	
608		390	398	405	412	419	426	433	440	447	455	
609		462	469	476	483	490	497	504	512	519	526	
610		533	540	547	554	561	569	576	583	590	597	
611		604	611	618	625	633	640	647	654	661	668	
612		675	682	689	696	704	711	718	725	732	739	
613		746	753	760	767	774	781	789	796	803	810	
614		817	824	831	838	845	852	859	866	873	880	
615		888	895	902	909	916	923	930	937	944	951	
616		958	965	972	979	986	993	*000	*007	*014	*021	
617	79	029	036	043	050	057	064	071	078	085	092	
618		099	106	113	120	127	134	141	148	155	162	
619		169	176	183	190	197	204	211	218	225	232	
620		239	246	253	260	267	274	281	288	295	302	
621		309	316	323	330	337	344	351	358	365	372	
622		379	386	393	400	407	414	421	428	435	442	
623		449	456	463	470	477	484	491	498	505	511	
624		518	525	532	539	546	553	560	567	574	581	
625		588	595	602	609	616	623	630	637	644	650	
626		657	664	671	678	685	692	699	706	713	720	
627		727	734	741	748	754	761	768	775	782	789	
628		796	803	810	817	824	831	837	844	851	858	
629		865	872	879	886	893	900	906	913	920	927	
630		934	941	948	955	962	969	975	982	989	996	
631	80	003	010	017	024	030	037	044	051	058	065	
632		072	079	085	092	099	106	113	120	127	134	
633		140	147	154	161	168	175	182	188	195	202	
634		209	216	223	229	236	243	250	257	264	271	
635		277	284	291	298	305	312	318	325	332	339	
636		346	353	359	366	373	380	387	393	400	407	
637		414	421	428	434	441	448	455	462	468	475	
638		482	489	496	502	509	516	523	530	536	543	
639		550	557	564	570	577	584	591	598	604	611	
640		618	625	632	638	645	652	659	665	672	679	
641		686	693	699	706	713	720	726	733	740	747	
642		754	760	767	774	781	787	794	801	808	814	
643		821	828	835	841	848	855	862	868	875	882	
644		889	895	902	909	916	922	929	936	943	949	
645		956	963	969	976	983	990	996	*003	*010	*017	
646	81	023	030	037	043	050	057	064	070	077	084	
647		090	097	104	111	117	124	131	137	144	151	
648		158	164	171	178	184	191	198	204	211	218	
649		224	231	238	245	251	258	265	271	278	285	
650		291	298	305	311	318	325	331	338	345	351	
N.		0	1	2	3	4	5	6	7	8	9	Proportional parts

Proportional parts:

	8
1	.8
2	1.6
3	2.4
4	3.2
5	4.0
6	4.8
7	5.6
8	6.4
9	7.2

	7
1	.7
2	1.4
3	2.1
4	2.8
5	3.5
6	4.2
7	4.9
8	5.6
9	6.3

	6
1	.6
2	1.2
3	1.8
4	2.4
5	3.0
6	3.6
7	4.2
8	4.3
9	5.4

.77 815–.81 351

650–700

N.		0	1	2	3	4	5	6	7	8	9
650	81	291	298	305	311	318	325	331	338	345	351
651		358	365	371	378	385	391	398	405	411	418
652		425	431	438	445	451	458	465	471	478	485
653		491	498	505	511	518	525	531	538	544	551
654		558	564	571	578	584	591	598	604	611	617
655		624	631	637	644	651	657	664	671	677	684
656		690	697	704	710	717	723	730	737	743	750
657		757	763	770	776	783	790	796	803	809	816
658		823	829	836	842	849	856	862	869	875	882
659		889	895	902	908	915	921	928	935	941	948
660		954	961	968	974	981	987	994	*000	*007	*014
661	82	020	027	033	040	046	053	060	066	073	079
662		086	092	099	105	112	119	125	132	138	145
663		151	158	164	171	178	184	191	197	204	210
664		217	223	230	236	243	249	256	263	269	276
665		282	289	295	302	308	315	321	328	334	341
666		347	354	360	367	373	380	387	393	400	406
667		413	419	426	432	439	445	452	458	465	471
668		478	484	491	497	504	510	517	523	530	536
669		543	549	556	562	569	575	582	588	595	601
670		607	614	620	627	633	640	646	653	659	666
671		672	679	685	692	698	705	711	718	724	730
672		737	743	750	756	763	769	776	782	789	795
673		802	808	814	821	827	834	840	847	853	860
674		866	872	879	885	892	898	905	911	918	924
675		930	937	943	950	956	963	969	975	982	988
676		995	*001	*008	*014	*020	*027	*033	*040	*046	*052
677	83	059	065	072	078	085	091	097	104	110	117
678		123	129	136	142	149	155	161	168	174	181
679		187	193	200	206	213	219	225	232	238	245
680		251	257	264	270	276	283	289	296	302	308
681		315	321	327	334	340	347	353	359	366	372
682		378	385	391	398	404	410	417	423	429	436
683		442	448	455	461	467	474	480	487	493	499
684		506	512	518	525	531	537	544	550	556	563
685		569	575	582	588	594	601	607	613	620	626
686		632	639	645	651	658	664	670	677	683	689
687		696	702	708	715	721	727	734	740	746	753
688		759	765	771	778	784	790	797	803	809	816
689		822	828	835	841	847	853	860	866	872	879
690		885	891	897	904	910	916	923	929	935	942
691		948	954	960	967	973	979	985	992	998	*004
692	84	011	017	023	029	036	042	048	055	061	067
693		073	080	086	092	098	105	111	117	123	130
694		136	142	148	155	161	167	173	180	186	192
695		198	205	211	217	223	230	236	242	248	255
696		261	267	273	280	286	292	298	305	311	317
697		323	330	336	342	348	354	361	367	373	379
698		386	392	398	404	410	417	423	429	435	442
699		448	454	460	466	473	479	485	491	497	504
700		510	516	522	528	535	541	547	553	559	566

N.	0	1	2	3	4	5	6	7	8	9	Proportional parts

Proportional parts

	7
1	.7
2	1.4
3	2.1
4	2.8
5	3.5
6	4.2
7	4.9
8	5.6
9	6.3

	6
1	.6
2	1.2
3	1.8
4	2.4
5	3.0
6	3.6
7	4.2
8	4.8
9	5.4

.81 291–.84 566

Table B-4 (*cont.*)

700–750

N.		0	1	2	3	4	5	6	7	8	9	Proportional parts
700	84	510	516	522	528	535	541	547	553	559	566	
701		572	578	584	590	597	603	609	615	621	628	
702		634	640	646	652	658	665	671	677	683	689	
703		696	702	708	714	720	726	733	739	745	751	
704		757	763	770	776	782	788	794	800	807	813	
705		819	825	831	837	844	850	856	862	868	874	
706		880	887	893	899	905	911	917	924	930	936	
707		942	948	954	960	967	973	979	985	991	997	
708	85	003	009	016	022	028	034	040	046	052	058	
709		065	071	077	083	089	095	101	107	114	120	
710		126	132	138	144	150	156	163	169	175	181	
711		187	193	199	205	211	217	224	230	236	242	
712		248	254	260	266	272	278	285	291	297	303	
713		309	315	321	327	333	339	345	352	358	364	
714		370	376	382	388	394	400	406	412	418	425	
715		431	437	443	449	455	461	467	473	479	485	
716		491	497	503	509	516	522	528	534	540	546	
717		552	558	564	570	576	582	588	594	600	606	
718		612	618	625	631	637	643	649	655	661	667	
719		673	679	685	691	697	703	709	715	721	727	
720		733	739	745	751	757	763	769	775	781	788	
721		794	800	806	812	818	824	830	836	842	848	
722		854	860	866	872	878	884	890	896	902	908	
723		914	920	926	932	938	944	950	956	962	968	
724		974	980	986	992	998	*004	*010	*016	*022	*028	
725	86	034	040	046	052	058	064	070	076	082	088	
726		094	100	106	112	118	124	130	136	141	147	
727		153	159	165	171	177	183	189	195	201	207	
728		213	219	225	231	237	243	249	255	261	267	
729		273	279	285	291	297	303	308	314	320	326	
730		332	338	344	350	356	362	368	374	380	386	
731		392	398	404	410	415	421	427	433	439	445	
732		451	457	463	469	475	481	487	493	499	504	
733		510	516	522	528	534	540	546	552	558	564	
734		570	576	581	587	593	599	605	611	617	623	
735		629	635	641	646	652	658	664	670	676	682	
736		688	694	700	705	711	717	723	729	735	741	
737		747	753	759	764	770	776	782	788	794	800	
738		806	812	817	823	829	835	841	847	853	859	
739		864	870	876	882	888	894	900	906	911	917	
740		923	929	935	941	947	953	958	964	970	976	
741		982	988	994	999	*005	*011	*017	*023	*029	*035	
742	87	040	046	052	058	064	070	075	081	087	093	
743		099	105	111	116	122	128	134	140	146	151	
744		157	163	169	175	181	186	192	198	204	210	
745		216	221	227	233	239	245	251	256	262	268	
746		274	280	286	291	297	303	309	315	320	326	
747		332	338	344	349	355	361	367	373	379	384	
748		390	396	402	408	413	419	425	431	437	442	
749		448	454	460	466	471	477	483	489	495	500	
750		506	512	518	523	529	535	541	547	552	558	

N.	0	1	2	3	4	5	6	7	8	9	Proportional parts

Proportional parts:

	7		6		5
1	.7	1	.6	1	.5
2	1.4	2	1.2	2	1.0
3	2.1	3	1.8	3	1.5
4	2.8	4	2.4	4	2.0
5	3.5	5	3.0	5	2.5
6	4.2	6	3.6	6	3.0
7	4.9	7	4.2	7	3.5
8	5.6	8	4.8	8	4.0
9	6.3	9	5.4	9	4.5

.84 510–.87 558

Table B-4 (*cont.*)

750–800

N.		0	1	2	3	4	5	6	7	8	9	Proportional parts
750	87	506	512	518	523	529	535	541	547	552	558	
751		564	570	576	581	587	593	599	604	610	616	
752		622	628	633	639	645	651	656	662	668	674	
753		679	685	691	697	703	708	714	720	726	731	
754		737	743	749	754	760	766	772	777	783	789	
755		795	800	806	812	818	823	829	835	841	846	
756		852	858	864	869	875	881	887	892	898	904	
757		910	915	921	927	933	938	944	950	955	961	
758		967	973	978	984	990	996	*001	*007	*013	*018	
759	88	024	030	036	041	047	053	058	064	070	076	
760		081	087	093	098	104	110	116	121	127	133	
761		138	144	150	156	161	167	173	178	184	190	
762		195	201	207	213	218	224	230	235	241	247	
763		252	258	264	270	275	281	287	292	298	304	
764		309	315	321	326	332	338	343	349	355	360	
765		366	372	377	383	389	395	400	406	412	417	
766		423	429	434	440	446	451	457	463	468	474	
767		480	485	491	497	502	508	513	519	525	530	
768		536	542	547	553	559	564	570	576	581	587	
769		593	598	604	610	615	621	627	632	638	643	
770		649	655	660	666	672	677	683	689	694	700	
771		705	711	717	722	728	734	739	745	750	756	
772		762	767	773	779	784	790	795	801	807	812	
773		818	824	829	835	840	846	852	857	863	868	
774		874	880	885	891	897	902	908	913	919	925	
775		930	936	941	947	953	958	964	969	975	981	
776		986	992	997	*003	*009	*014	*020	*025	*031	*037	
777	89	042	048	053	059	064	070	076	081	087	092	
778		098	104	109	115	120	126	131	137	143	148	
779		154	159	165	170	176	182	187	193	198	204	
780		209	215	221	226	232	237	243	248	254	260	
781		265	271	276	282	287	293	298	304	310	315	
782		321	326	332	337	343	348	354	360	365	371	
783		376	382	387	393	398	404	409	415	421	426	
784		432	437	443	448	454	459	465	470	476	481	
785		487	492	498	504	509	515	520	526	531	537	
786		542	548	553	559	564	570	575	581	586	592	
787		597	603	609	614	620	625	631	636	642	647	
788		653	658	664	669	675	680	686	691	697	702	
789		708	713	719	724	730	735	741	746	752	757	
790		763	768	774	779	785	790	796	801	807	812	
791		818	823	829	834	840	845	851	856	862	867	
792		873	878	883	889	894	900	905	911	916	922	
793		927	933	938	944	949	955	960	966	971	977	
794		982	988	993	998	*004	*009	*015	*020	*026	*031	
795	90	037	042	048	053	059	064	069	075	080	086	
796		091	097	102	108	113	119	124	129	135	140	
797		146	151	157	162	168	173	179	184	189	195	
798		200	206	211	217	222	227	233	238	244	249	
799		255	260	266	271	276	282	287	293	298	304	
800		309	314	320	325	331	336	342	347	352	358	
N.		0	1	2	3	4	5	6	7	8	9	Proportional parts

Proportional parts

	6
1	.6
2	1.2
3	1.8
4	2.4
5	3.0
6	3.6
7	4.2
8	4.8
9	5.4

	5
1	.5
2	1.0
3	1.5
4	2.0
5	2.5
6	3.0
7	3.5
8	4.0
9	4.5

.87 506–.90 358

Table B-4 (*cont.*)

N.		0	1	2	3	4	5	6	7	8	9	Proportional parts		
800	90	309	314	320	325	331	336	342	347	352	358			
801		363	369	374	380	385	390	396	401	407	412			
802		417	423	428	434	439	445	450	455	461	466			
803		472	477	482	488	493	499	504	509	515	520			
804		526	531	536	542	547	553	558	563	569	574			
805		580	585	590	596	601	607	612	617	623	628			
806		634	639	644	650	655	660	666	671	677	682			
807		687	693	698	703	709	714	720	725	730	736			
808		741	747	752	757	763	768	773	779	784	789			
809		795	800	806	811	816	822	827	832	838	843			
810		849	854	859	865	870	875	881	886	891	897			6
811		902	907	913	918	924	929	934	940	945	950			
812		956	961	966	972	977	982	988	993	998	*004	1	.6	
813	91	009	014	020	025	030	036	041	046	052	057	2	1.2	
814		062	068	073	078	084	089	094	100	105	110	3	1.8	
815		116	121	126	132	137	142	148	153	158	164	4	2.4	
816		169	174	180	185	190	196	201	206	212	217	5	3.0	
817		222	228	233	238	243	249	254	259	265	270	6	3.6	
818		275	281	286	291	297	302	307	312	318	323	7	4.2	
819		328	334	339	344	350	355	360	365	371	376	8	4.8	
												9	5.4	
820		381	387	392	397	403	408	413	418	424	429			
821		434	440	445	450	455	461	466	471	477	482			
822		487	492	498	503	508	514	519	524	529	535			
823		540	545	551	556	561	566	572	577	582	587			
824		593	598	603	609	614	619	624	630	635	640			
825		645	651	656	661	666	672	677	682	687	693			
826		698	703	709	714	719	724	730	735	740	745			
827		751	756	761	766	772	777	782	787	793	798			
828		803	808	814	819	824	829	834	840	845	850			
829		855	861	866	871	876	882	887	892	897	903			
830		908	913	918	924	929	934	939	944	950	955			5
831		960	965	971	976	981	986	991	997	*002	*007			
832	92	012	018	023	028	033	038	044	049	054	059	1	.5	
833		065	070	075	080	085	091	096	101	106	111	2	1.0	
834		117	122	127	132	137	143	148	153	158	163	3	1.5	
835		169	174	179	184	189	195	200	205	210	215	4	2.0	
836		221	226	231	236	241	247	252	257	262	267	5	2.5	
837		273	278	283	288	293	298	304	309	314	319	6	3.0	
838		324	330	335	340	345	350	355	361	366	371	7	3.5	
839		376	381	387	392	397	402	407	412	418	423	8	4.0	
												9	4.5	
840		428	433	438	443	449	454	459	464	469	474			
841		480	485	490	495	500	505	511	516	521	526			
842		531	536	542	547	552	557	562	567	572	578			
843		583	588	593	598	603	609	614	619	624	629			
844		634	639	645	650	655	660	665	670	675	681			
845		686	691	696	701	706	711	716	722	727	732			
846		737	742	747	752	758	763	768	773	778	783			
847		788	793	799	804	809	814	819	824	829	834			
848		840	845	850	855	860	865	870	875	881	886			
849		891	896	901	906	911	916	921	927	932	937			
850		942	947	952	957	962	967	973	978	983	988			

N.	0	1	2	3	4	5	6	7	8	9	Proportional parts

.90 309–.92 988

Table B-4 (*cont.*)

850–900

N.		0	1	2	3	4	5	6	7	8	9
850	92	942	947	952	957	962	967	973	978	983	988
851		993	998	*003	*008	*013	*018	*024	*029	*034	*039
852	93	044	049	054	059	064	069	075	080	085	090
853		095	100	105	110	115	120	125	131	136	141
854		146	151	156	161	166	171	176	181	186	192
855		197	202	207	212	217	222	227	232	237	242
856		247	252	258	263	268	273	278	283	288	293
857		298	303	308	313	318	323	328	334	339	344
858		349	354	359	364	369	374	379	384	389	394
859		399	404	409	414	420	425	430	435	440	445
860		450	455	460	465	470	475	480	485	490	495
861		500	505	510	515	520	526	531	536	541	546
862		551	556	561	566	571	576	581	586	591	596
863		601	606	611	616	621	626	631	636	641	646
864		651	656	661	666	671	676	682	687	692	697
865		702	707	712	717	722	727	732	737	742	747
866		752	757	762	767	772	777	782	787	792	797
867		802	807	812	817	822	827	832	837	842	847
868		852	857	862	867	872	877	882	887	892	897
869		902	907	912	917	922	927	932	937	942	947
870		952	957	962	967	972	977	982	987	992	997
871	94	002	007	012	017	022	027	032	037	042	047
872		052	057	062	067	072	077	082	086	091	096
873		101	106	111	116	121	126	131	136	141	146
874		151	156	161	166	171	176	181	186	191	196
875		201	206	211	216	221	226	231	236	240	245
876		250	255	260	265	270	275	280	285	290	295
877		300	305	310	315	320	325	330	335	340	345
878		349	354	359	364	369	374	379	384	389	394
879		399	404	409	414	419	424	429	433	438	443
880		448	453	458	463	468	473	478	483	488	493
881		498	503	507	512	517	522	527	532	537	542
882		547	552	557	562	567	571	576	581	586	591
883		596	601	606	611	616	621	626	630	635	640
884		645	650	655	660	665	670	675	680	685	689
885		694	699	704	709	714	719	724	729	734	738
886		743	748	753	758	763	768	773	778	783	787
887		792	797	802	807	812	817	822	827	832	836
888		841	846	851	856	861	866	871	876	880	885
889		890	895	900	905	910	915	919	924	929	934
890		939	944	949	954	959	963	968	973	978	983
891		988	993	998	*002	*007	*012	*017	*022	*027	*032
892	95	036	041	046	051	056	061	066	071	075	080
893		085	090	095	100	105	109	114	119	124	129
894		134	139	143	148	153	158	163	168	173	177
895		182	187	192	197	202	207	211	216	221	226
896		231	236	240	245	250	255	260	265	270	274
897		279	284	289	294	299	303	308	313	318	323
898		328	332	337	342	347	352	357	361	366	371
899		376	381	386	390	395	400	405	410	415	419
900		424	429	434	439	444	448	453	458	463	468

Proportional parts

	6
1	.6
2	1.2
3	1.8
4	2.4
5	3.0
6	3.6
7	4.2
8	4.8
9	5.4

	5
1	.5
2	1.0
3	1.5
4	2.0
5	2.5
6	3.0
7	3.5
8	4.0
9	4.5

	4
1	.4
2	.8
3	1.2
4	1.6
5	2.0
6	2.4
7	2.8
8	3.2
9	3.6

.92 942–.95 468

Table B-4 (*cont.*)

N.		0	1	2	3	4	5	6	7	8	9	Proportional parts
900	95	424	429	434	439	444	448	453	458	463	468	
901		472	477	482	487	492	497	501	506	511	516	
902		521	525	530	535	540	545	550	554	559	564	
903		569	574	578	583	588	593	598	602	607	612	
904		617	622	626	631	636	641	646	650	655	660	
905		665	670	674	679	684	689	694	698	703	708	
906		713	718	722	727	732	737	742	746	751	756	
907		761	766	770	775	780	785	789	794	799	804	
908		809	813	818	823	828	832	837	842	847	852	
909		856	861	866	871	875	880	885	890	895	899	
910		904	909	914	918	923	928	933	938	942	947	
911		952	957	961	966	971	976	980	985	990	995	**5**
912		999	*004	*009	*014	*019	*023	*028	*033	*038	*042	1 .5
913	96	047	052	057	061	066	071	076	080	085	090	2 1.0
914		095	099	104	109	114	118	123	128	133	137	3 1.5
915		142	147	152	156	161	166	171	175	180	185	4 2.0
916		190	194	199	204	209	213	218	223	227	232	5 2.5
917		237	242	246	251	256	261	265	270	275	280	6 3.0
918		284	289	294	298	303	308	313	317	322	327	7 3.5
919		332	336	341	346	350	355	360	365	369	374	8 4.0
920		379	384	388	393	398	402	407	412	417	421	9 4.5
921		426	431	435	440	445	450	454	459	464	468	
922		473	478	483	487	492	497	501	506	511	515	
923		520	525	530	534	539	544	548	553	558	562	
924		567	572	577	581	586	591	595	600	605	609	
925		614	619	624	628	633	638	642	647	652	656	
926		661	666	670	675	680	685	689	694	699	703	
927		708	713	717	722	727	731	736	741	745	750	
928		755	759	764	769	774	778	783	788	792	797	
929		802	806	811	816	820	825	830	834	839	844	
930		848	853	858	862	867	872	876	881	886	890	
931		895	900	904	909	914	918	923	928	932	937	**4**
932		942	946	951	956	960	965	970	974	979	984	1 .4
933		988	993	997	*002	*007	*011	*016	*021	*025	*030	2 .8
934	97	035	039	044	049	053	058	063	067	072	077	3 1.2
935		081	086	090	095	100	104	109	114	118	123	4 1.6
936		128	132	137	142	146	151	155	160	165	169	5 2.0
937		174	179	183	188	192	197	202	206	211	216	6 2.4
938		220	225	230	234	239	243	248	253	257	262	7 2.8
939		267	271	276	280	285	290	294	299	304	308	8 3.2
940		313	317	322	327	331	336	340	345	350	354	9 3.6
941		359	364	368	373	377	382	387	391	396	400	
942		405	410	414	419	424	428	433	437	442	447	
943		451	456	460	465	470	474	479	483	488	493	
944		497	502	506	511	516	520	525	529	534	539	
945		543	548	552	557	562	566	571	575	580	585	
946		589	594	598	603	607	612	617	621	626	630	
947		635	640	644	649	653	658	663	667	672	676	
948		681	685	690	695	699	704	708	713	717	722	
949		727	731	736	740	745	749	754	759	763	768	
950		772	777	782	786	791	795	800	804	809	813	
N.		0	1	2	3	4	5	6	7	8	9	Proportional parts

.95 424–.97 813

Table B-4 (cont.)

N.		0	1	2	3	4	5	6	7	8	9
950	97	772	777	782	786	791	795	800	804	809	813
951		818	823	827	832	836	841	845	850	855	859
952		864	868	873	877	882	886	891	896	900	905
953		909	914	918	923	928	932	937	941	946	950
954		955	959	964	968	973	978	982	987	991	996
955	98	000	005	009	014	019	023	028	032	037	041
956		046	050	055	059	064	068	073	078	082	087
957		091	096	100	105	109	114	118	123	127	132
958		137	141	146	150	155	159	164	168	173	177
959		182	186	191	195	200	204	209	214	218	223
960		227	232	236	241	245	250	254	259	263	268
961		272	277	281	286	290	295	299	304	308	313
962		318	322	327	331	336	340	345	349	354	358
963		363	367	372	376	381	385	390	394	399	403
964		408	412	417	421	426	430	435	439	444	448
965		453	457	462	466	471	475	480	484	489	493
966		498	502	507	511	516	520	525	529	534	538
967		543	547	552	556	561	565	570	574	579	583
968		588	592	597	601	605	610	614	619	623	628
969		632	637	641	646	650	655	659	664	668	673
970		677	682	686	691	695	700	704	709	713	717
971		722	726	731	735	740	744	749	753	758	762
972		767	771	776	780	784	789	793	798	802	807
973		811	816	820	825	829	834	838	843	847	851
974		856	860	865	869	874	878	883	887	892	896
975		900	905	909	914	918	923	927	932	936	941
976		945	949	954	958	963	967	972	976	981	985
977		989	994	998	*003	*007	*012	*016	*021	*025	*029
978	99	034	038	043	047	052	056	061	065	069	074
979		078	083	087	092	096	100	105	109	114	118
980		123	127	131	136	140	145	149	154	158	162
981		167	171	176	180	185	189	193	198	202	207
982		211	216	220	224	229	233	238	242	247	251
983		255	260	264	269	273	277	282	286	291	295
984		300	304	308	313	317	322	326	330	335	339
985		344	348	352	357	361	366	370	374	379	383
986		388	392	396	401	405	410	414	419	423	427
987		432	436	441	445	449	454	458	463	467	471
988		476	480	484	489	493	498	502	506	511	515
989		520	524	528	533	537	542	546	550	555	559
990		564	568	572	577	581	585	590	594	599	603
991		607	612	616	621	625	629	634	638	642	647
992		651	656	660	664	669	673	677	682	686	691
993		695	699	704	708	712	717	721	726	730	734
994		739	743	747	752	756	760	765	769	774	778
995		782	787	791	795	800	804	808	813	817	822
996		826	830	835	839	843	848	852	856	861	865
997		870	874	878	883	887	891	896	900	904	909
998		913	917	922	926	930	935	939	944	948	952
999		957	961	965	970	974	978	983	987	991	996
1000	00	000	004	009	013	017	022	026	030	035	039

N.	0	1	2	3	4	5	6	7	8	9	Proportional parts

Proportional parts

	5
1	.5
2	1.0
3	1.5
4	2.0
5	2.5
6	3.0
7	3.5
8	4.0
9	4.5

	4
1	.4
2	.8
3	1.2
4	1.6
5	2.0
6	2.4
7	2.8
8	3.2
9	3.6

.97 772–.99 996

Table B-4 (*cont.*)

N.		0	1	2	3	4	5	6	7	8	9	d.
1000	000	0000	0434	0869	1303	1737	2171	2605	3039	3473	3907	434
1001		4341	4775	5208	5642	6076	6510	6943	7377	7810	8244	434
1002		8677	9111	9544	9977	*0411	*0844	*1277	*1710	*2143	*2576	433
1003	001	3009	3442	3875	4308	4741	5174	5607	6039	6472	6905	433
1004		7337	7770	8202	8635	9067	9499	9932	*0364	*0796	*1228	432
1005	002	1661	2093	2525	2957	3389	3821	4253	4685	5116	5548	432
1006		5980	6411	6843	7275	7706	8138	8569	9001	9432	9863	431
1007	003	0295	0726	1157	1588	2019	2451	2882	3313	3744	4174	431
1008		4605	5036	5467	5898	6328	6759	7190	7620	8051	8481	431
1009		8912	9342	9772	*0203	*0633	*1063	*1493	*1924	*2354	*2784	430
1010	004	3214	3644	4074	4504	4933	5363	5793	6223	6652	7082	430
1011		7512	7941	8371	8800	9229	9659	*0088	*0517	*0947	*1376	429
1012	005	1805	2234	2663	3092	3521	3950	4379	4808	5237	5666	429
1013		6094	6523	6952	7380	7809	8238	8666	9094	9523	9951	429
1014	006	0380	0808	1236	1664	2092	2521	2949	3377	3805	4233	428
1015		4660	5088	5516	5944	6372	6799	7227	7655	8082	8510	428
1016		8937	9365	9792	*0219	*0647	*1074	*1501	*1928	*2355	*2782	427
1017	007	3210	3637	4064	4490	4917	5344	5771	6198	6624	7051	427
1018		7478	7904	8331	8757	9184	9610	*0037	*0463	*0889	*1316	426
1019	008	1742	2168	2594	3020	3446	3872	4298	4724	5150	5576	426
1020		6002	6427	6853	7279	7704	8130	8556	8981	9407	9832	426
1021	009	0257	0683	1108	1533	1959	2384	2809	3234	3659	4084	425
1022		4509	4934	5359	5784	6208	6633	7058	7483	7907	8332	425
1023		8756	9181	9605	*0030	*0454	*0878	*1303	*1727	*2151	*2575	424
1024	010	3000	3424	3848	4272	4696	5120	5544	5967	6391	6815	424
1025		7239	7662	8086	8510	8933	9357	9780	*0204	*0627	*1050	424
1026	011	1474	1897	2320	2743	3166	3590	4013	4436	4859	5282	423
1027		5704	6127	6550	6973	7396	7818	8241	8664	9086	9509	423
1028		9931	*0354	*0776	*1198	*1621	*2043	*2465	*2887	*3310	*3732	422
1029	012	4154	4576	4998	5420	5842	6264	6685	7107	7529	7951	422
1030		8372	8794	9215	9637	*0059	*0480	*0901	*1323	*1744	*2165	422
1031	013	2587	3008	3429	3850	4271	4692	5113	5534	5955	6376	421
1032		6797	7218	7639	8059	8480	8901	9321	9742	*0162	*0583	421
1033	014	1003	1424	1844	2264	2685	3105	3525	3945	4365	4785	420
1034		5205	5625	6045	6465	6885	7305	7725	8144	8564	8984	420
1035		9403	9823	*0243	*0662	*1082	*1501	*1920	*2340	*2759	*3178	420
1036	015	3598	4017	4436	4855	5274	5693	6112	6531	6950	7369	419
1037		7788	8206	8625	9044	9462	9881	*0300	*0718	*1137	*1555	419
1038	016	1974	2392	2810	3229	3647	4065	4483	4901	5319	5737	418
1039		6155	6573	6991	7409	7827	8245	8663	9080	9498	9916	418
1040	017	0333	0751	1168	1586	2003	2421	2838	3256	3673	4090	417
1041		4507	4924	5342	5759	6176	6593	7010	7427	7844	8260	417
1042		8677	9094	9511	9927	*0344	*0761	*1177	*1594	*2010	*2427	417
1043	018	2843	3259	3676	4092	4508	4925	5341	5757	6173	6589	416
1044		7005	7421	7837	8253	8669	9084	9500	9916	*0332	*0747	416
1045	019	1163	1578	1994	2410	2835	3240	3656	4071	4486	4902	415
1046		5317	5732	6147	6562	6977	7392	7807	8222	8637	9052	415
1047		9467	9882	*0296	*0711	*1126	*1540	*1955	*2369	*2784	*3198	415
1048	020	3613	4027	4442	4856	5270	5684	6099	6513	6927	7341	414
1049		7755	8169	8583	8997	9411	9824	*0238	*0652	*1066	*1479	414
1050	021	1893	2307	2720	3134	3547	3961	4374	4787	5201	5614	413
N.		0	1	2	3	4	5	6	7	8	9	d.

.000 0000–.021 5614

1050–1100

N.		0	1	2	3	4	5	6	7	8	9	d.
1050	021	1893	2307	2720	3134	3547	3961	4374	4787	5201	5614	413
1051		6027	6440	6854	7267	7680	8093	8506	8919	9332	9745	413
1052	022	0157	0570	0983	1396	1808	2221	2634	3046	3459	3871	413
1053		4284	4696	5109	5521	5933	6345	6758	7170	7582	7994	412
1054		8406	8818	9230	9642	*0054	*0466	*0878	*1289	*1701	*2113	412
1055	023	2525	2936	3348	3759	4171	4582	4994	5405	5817	6228	411
1056		6639	7050	7462	7873	8284	8695	9106	9517	9928	*0339	411
1057	024	0750	1161	1572	1982	2393	2804	3214	3625	4036	4446	411
1058		4857	5267	5678	6088	6498	6909	7319	7729	8139	8549	410
1059		8960	9370	9780	*0190	*0600	*1010	*1419	*1829	*2239	*2649	410
1060	025	3059	3468	3878	4288	4697	5107	5516	5926	6335	6744	410
1061		7154	7563	7972	8382	8791	9200	9609	*0018	*0427	*0836	409
1062	026	1245	1654	2063	2472	2881	3289	3698	4107	4515	4924	409
1063		5333	5741	6150	6558	6967	7375	7783	8192	8600	9008	408
1064		9416	9824	*0233	*0641	*1049	*1457	*1865	*2273	*2680	*3088	408
1065	027	3496	3904	4312	4719	5127	5535	5942	6350	6757	7165	408
1066		7572	7979	8387	8794	9201	9609	*0016	*0423	*0830	*1237	407
1067	028	1644	2051	2458	2865	3272	3679	4086	4492	4899	5306	407
1068		5713	6119	6526	6932	7339	7745	8152	8558	8964	9371	406
1069		9777	*0183	*0590	*0996	*1402	*1808	*2214	*2620	*3026	*3432	406
1070	029	3838	4244	4649	5055	5461	5867	6272	6678	7084	7489	406
1071		7895	8300	8706	9111	9516	9922	*0327	*0732	*1138	*1543	405
1072	030	1948	2353	2758	3163	3568	3973	4378	4783	5188	5592	405
1073		5997	6402	6807	7211	7616	8020	8425	8830	9234	9638	405
1074	031	0043	0447	0851	1256	1660	2064	2468	2872	3277	3681	404
1075		4085	4489	4893	5296	5700	6104	6508	6912	7315	7719	404
1076		8123	8526	8930	9333	9737	*0140	*0544	*0947	*1350	*1754	403
1077	032	2157	2560	2963	3367	3770	4173	4576	4979	5382	5785	403
1078		6188	6590	6993	7396	7799	8201	8604	9007	9409	9812	403
1079	033	0214	0617	1019	1422	1824	2226	2629	3031	3433	3835	402
1080		4238	4640	5042	5444	5846	6248	6650	7052	7453	7855	402
1081		8257	8659	9060	9462	9864	*0265	*0667	*1068	*1470	*1871	402
1082	034	2273	2674	3075	3477	3878	4279	4680	5081	5482	5884	401
1083		6285	6686	7087	7487	7888	8289	8690	9091	9491	9892	401
1084	035	0293	0693	1094	1495	1895	2296	2696	3096	3497	3897	400
1085		4297	4698	5098	5498	5898	6298	6698	7098	7498	7898	400
1086		8298	8698	9098	9498	9898	*0297	*0697	*1097	*1496	*1896	400
1087	036	2295	2695	3094	3494	3893	4293	4692	5091	5491	5890	399
1088		6289	6688	7087	7486	7885	8284	8683	9082	9481	9880	399
1089	037	0279	0678	1076	1475	1874	2272	2671	3070	3468	3867	399
1090		4265	4663	5062	5460	5858	6257	6655	7053	7451	7849	398
1091		8248	8646	9044	9442	9839	*0237	*0635	*1033	*1431	*1829	398
1092	038	2226	2624	3022	3419	3817	4214	4612	5009	5407	5804	398
1093		6202	6599	6996	7393	7791	8188	8585	8982	9379	9776	397
1094	039	0173	0570	0967	1364	1761	2158	2554	2951	3348	3745	397
1095		4141	4538	4934	5331	5727	6124	6520	6917	7313	7709	397
1096		8106	8502	8898	9294	9690	*0086	*0482	*0878	*1274	*1670	396
1097	040	2066	2462	2858	3254	3650	4045	4441	4837	5232	5628	396
1098		6023	6419	6814	7210	7605	8001	8396	8791	9187	9582	395
1099		9977	*0372	*0767	*1162	*1557	*1952	*2347	*2742	*3137	*3532	395
1100	041	3927	4322	4716	5111	5506	5900	6295	6690	7084	7479	395
N.		0	1	2	3	4	5	6	7	8	9	d.

.021 1893–.041 7479

Table B-4 (*cont.*)

1100–1150

N.		0	1	2	3	4	5	6	7	8	9	d.
1100	041	3927	4322	4716	5111	5506	5900	6295	6690	7084	7479	395
1101		7873	8268	8662	9056	9451	9845	*0239	*0633	*1028	*1422	394
1102	042	1816	2210	2604	2998	3392	3786	4180	4574	4968	5361	394
1103		5755	6149	6543	6936	7330	7723	8117	8510	8904	9297	394
1104		9691	*0084	*0477	*0871	*1264	*1657	*2050	*2444	*2837	*3230	393
1105	043	3623	4016	4409	4802	5195	5587	5980	6373	6766	7159	393
1106		7551	7944	8337	8729	9122	9514	9907	*0299	*0692	*1084	393
1107	044	1476	1869	2261	2653	3045	3437	3829	4222	4614	5006	392
1108		5398	5790	6181	6573	6965	7357	7749	8140	8532	8924	392
1109		9315	9707	*0099	*0490	*0882	*1273	*1664	*2056	*2447	*2839	392
1110	045	3230	3621	4012	4403	4795	5186	5577	5968	6359	6750	391
1111		7141	7531	7922	8313	8704	9095	9485	9876	*0267	*0657	391
1112	046	1048	1438	1829	2219	2610	3000	3391	3781	4171	4561	390
1113		4952	5342	5732	6122	6512	6902	7292	7682	8072	8462	390
1114		8852	9242	9632	*0021	*0411	*0801	*1190	*1580	*1970	*2359	390
1115	047	2749	3138	3528	3917	4306	4696	5085	5474	5864	6253	389
1116		6642	7031	7420	7809	8198	8587	8976	9365	9754	*0143	389
1117	048	0532	0921	1309	1698	2087	2475	2864	3253	3641	4030	389
1118		4418	4806	5195	5583	5972	6360	6748	7136	7525	7913	388
1119		8301	8689	9077	9465	9853	*0241	*0629	*1017	*1405	*1792	388
1120	049	2180	2568	2956	3343	3731	4119	4506	4894	5281	5669	388
1121		6056	6444	6831	7218	7606	7993	8380	8767	9154	9541	387
1122		9929	*0316	*0703	*1090	*1477	*1863	*2250	*2637	*3024	*3411	387
1123	050	3798	4184	4571	4958	5344	5731	6117	6504	6890	7277	387
1124		7663	8049	8436	8822	9208	9595	9981	*0367	*0753	*1139	386
1125	051	1525	1911	2297	2683	3069	3455	3841	4227	4612	4998	386
1126		5384	5770	6155	6541	6926	7312	7697	8083	8468	8854	386
1127		9239	9624	*0010	*0395	*0780	*1166	*1551	*1936	*2321	*2706	385
1128	052	3091	3476	3861	4246	4631	5016	5400	5785	6170	6555	385
1129		6939	7324	7709	8093	8478	8862	9247	9631	*0016	*0400	385
1130	053	0784	1169	1553	1937	2321	2706	3090	3474	3858	4242	384
1131		4626	5010	5394	5778	6162	6546	6929	7313	7697	8081	384
1132		8464	8848	9232	9615	9999	*0382	*0766	*1149	*1532	*1916	384
1133	054	2299	2682	3066	3449	3832	4215	4598	4981	5365	5748	383
1134		6131	6514	6896	7279	7662	8045	8428	8811	9193	9576	383
1135		9959	*0341	*0724	*1106	*1489	*1871	*2254	*2636	*3019	*3401	382
1136	055	3783	4166	4548	4930	5312	5694	6077	6459	6841	7223	382
1137		7605	7987	8369	8750	9132	9514	9896	*0278	*0659	*1041	382
1138	056	1423	1804	2186	2567	2949	3330	3712	4093	4475	4856	381
1139		5237	5619	6000	6381	6762	7143	7524	7905	8287	8668	381
1140		9049	9429	9810	*0191	*0572	*0953	*1334	*1714	*2095	*2476	381
1141	057	2856	3237	3618	3998	4379	4759	5140	5520	5900	6181	381
1142		6661	7041	7422	7802	8182	8562	8942	9322	9702	*0082	380
1143	058	0462	0842	1222	1602	1982	2362	2741	3121	3501	3881	380
1144		4260	4640	5019	5399	5778	6158	6537	6917	7296	7676	380
1145		8055	8434	8813	9193	9572	9951	*0330	*0709	*1088	*1467	379
1146	059	1846	2225	2604	2983	3362	3741	4119	4498	4877	5256	379
1147		5634	6013	6391	6770	7148	7527	7905	8284	8662	9041	379
1148		9419	9797	*0175	*0554	*0932	*1310	*1688	*2066	*2444	*2822	378
1149	060	3200	3578	3956	4334	4712	5090	5468	5845	6223	6601	378
1150		6978	7356	7734	8111	8489	8866	9244	9621	9999	*0376	378
N.		0	1	2	3	4	5	6	7	8	9	d.

.041 3927–.061 0376

Table B-4 (*cont.*)

N.		0	1	2	3	4	5	6	7	8	9	d.
1150	060	6978	7356	7734	8111	8489	8866	9244	9621	9999	*0376	378
1151	061	0753	1131	1508	1885	2262	2639	3017	3394	3771	4148	377
1152		4525	4902	5279	5656	6032	6409	6786	7163	7540	7916	377
1153		8293	8670	9046	9423	9799	*0176	*0552	*0929	*1305	*1682	377
1154	062	2058	2434	2811	3187	3563	3939	4316	4692	5068	5444	376
1155		5820	6196	6572	6948	7324	7699	8075	8451	8827	9203	376
1156		9578	9954	*0330	*0705	*1081	*1456	*1832	*2207	*2583	*2958	376
1157	063	3334	3709	4084	4460	4835	5210	5585	5960	6335	6711	375
1158		7086	7461	7836	8211	8585	8960	9335	9710	*0085	*0460	375
1159	064	0834	1209	1584	1958	2333	2708	3082	3457	3831	4205	375
1160		4580	4954	5329	5703	6077	6451	6826	7200	7574	7948	374
1161		8322	8696	9070	9444	9818	*0192	*0566	*0940	*1314	*1688	374
1162	065	2061	2435	2809	3182	3556	3930	4303	4677	5050	5424	374
1163		5797	6171	6544	6917	7291	7664	8037	8410	8784	9157	373
1164		9530	9903	*0276	*0649	*1022	*1395	*1768	*2141	*2514	*2886	373
1165	066	3259	3632	4005	4377	4750	5123	5495	5868	6241	6613	373
1166		6986	7358	7730	8103	8475	8847	9220	9592	9964	*0336	372
1167	067	0709	1081	1453	1825	2197	2569	2941	3313	3685	4057	372
1168		4428	4800	5172	5544	5915	6287	6659	7030	7402	7774	372
1169		8145	8517	8888	9259	9631	*0002	*0374	*0745	*1116	*1487	371
1170	068	1859	2230	2601	2972	3343	3714	4085	4456	4827	5198	371
1171		5569	5940	6311	6681	7052	7423	7794	8164	8535	8906	371
1172		9276	9647	*0017	*0388	*0758	*1129	*1499	*1869	*2240	*2610	370
1173	069	2980	3350	3721	4091	4461	4831	5201	5571	5941	6311	370
1174		6681	7051	7421	7791	8160	8530	8900	9270	9639	*0009	370
1175	070	0379	0748	1118	1487	1857	2226	2596	2965	3335	3704	369
1176		4073	4442	4812	5181	5550	5919	6288	6658	7027	7396	369
1177		7765	8134	8503	8871	9240	9609	9978	*0347	*0715	*1084	369
1178	071	1453	1822	2190	2559	2927	3296	3664	4033	4401	4770	369
1179		5138	5506	5875	6243	6611	6979	7348	7716	8084	8452	368
1180		8820	9188	9556	9924	*0292	*0660	*1028	*1396	*1763	*2131	368
1181	072	2499	2867	3234	3602	3970	4337	4705	5072	5440	5807	368
1182		6175	6542	6910	7277	7644	8011	8379	8746	9113	9480	367
1183		9847	*0215	*0582	*0949	*1316	*1683	*2050	*2416	*2783	*3150	367
1184	073	3517	3884	4251	4617	4984	5351	5717	6084	6450	6817	367
1185		7184	7550	7916	8283	8649	9016	9382	9748	*0114	*0481	366
1186	074	0847	1213	1579	1945	2311	2677	3043	3409	3775	4141	366
1187		4507	4873	5239	5605	5970	6336	6702	7068	7433	7799	366
1188		8164	8530	8895	9261	9626	9992	*0357	*0723	*1088	*1453	365
1189	075	1819	2184	2549	2914	3279	3644	4010	4375	4740	5105	365
1190		5470	5835	6199	6564	6929	7294	7659	8024	8388	8753	365
1191		9118	9482	9847	*0211	*0576	*0940	*1305	*1669	*2034	*2398	364
1192	076	2763	3127	3491	3855	4220	4584	4948	5312	5676	6040	364
1193		6404	6768	7132	7496	7860	8224	8588	8952	9316	9680	364
1194	077	0043	0407	0771	1134	1498	1862	2225	2589	2952	3316	364
1195		3679	4042	4406	4769	5133	5496	5859	6222	6585	6949	363
1196		7312	7675	8038	8401	8764	9127	9490	9853	*0216	*0579	363
1197	078	0942	1304	1667	2030	2393	2755	3118	3480	3843	4206	363
1198		4568	4931	5293	5656	6018	6380	6743	7105	7467	7830	362
1199		8192	8554	8916	9278	9640	*0003	*0365	*0727	*1089	*1451	362
1200	079	1812	2174	2536	2898	3260	3622	3983	4345	4707	5068	362
N.		0	1	2	3	4	5	6	7	8	9	d.

.060 6978–.079 5068

Table B-5　Natural Logarithms of Numbers

n	$\log_e n$	n	$\log_e n$	n	$\log_e n$
	*	4.5	1.5041	9.0	2.1972
.1	7.6974	4.6	1.5261	9.1	2.2083
.2	8.3906	4.7	1.5476	9.2	2.2192
.3	8.7960	4.8	1.5686	9.3	2.2300
.4	9.0837	4.9	1.5892	9.4	2.2407
.5	9.3069	5.0	1.6094	9.5	2.2513
.6	9.4892	5.1	1.6292	9.6	2.2618
.7	9.6433	5.2	1.6487	9.7	2.2721
.8	9.7769	5.3	1.6677	9.8	2.2824
.9	9.8946	5.4	1.6864	9.9	2.2925
1.0	.0000	5.5	1.7047	10	2.3026
1.1	.0953	5.6	1.7228	11	2.3979
1.2	.1823	5.7	1.7405	12	2.4849
1.3	.2624	5.8	1.7579	13	2.5649
1.4	.3365	5.9	1.7750	14	2.6391
1.5	.4055	6.0	1.7918	15	2.7081
1.6	.4700	6.1	1.8083	16	2.7726
1.7	.5306	6.2	1.8245	17	2.8332
1.8	.5878	6.3	1.8405	18	2.8904
1.9	.6419	6.4	1.8563	19	2.9444
2.0	.6931	6.5	1.8718	20	2.9957
2.1	.7419	6.6	1.8871	25	3.2189
2.2	.7885	6.7	1.9021	30	3.4012
2.3	.8329	6.8	1.9169	35	3.5553
2.4	.8755	6.9	1.9315	40	3.6889
2.5	.9163	7.0	1.9459	45	3.8067
2.6	.9555	7.1	1.9601	50	3.9120
2.7	.9933	7.2	1.9741	55	4.0073
2.8	1.0296	7.3	1.9879	60	4.0943
2.9	1.0647	7.4	2.0015	65	4.1744
3.0	1.0986	7.5	2.0149	70	4.2485
3.1	1.1314	7.6	2.0281	75	4.3175
3.2	1.1632	7.7	2.0412	80	4.3820
3.3	1.1939	7.8	2.0541	85	4.4427
3.4	1.2238	7.9	2.0669	90	4.4998
3.5	1.2528	8.0	2.0794	100	4.6052
3.6	1.2809	8.1	2.0919	110	4.7005
3.7	1.3083	8.2	2.1041	120	4.7875
3.8	1.3350	8.3	2.1163	130	4.8676
3.9	1.3610	8.4	2.1282	140	4.9416
4.0	1.3863	8.5	2.1401	150	5.0106
4.1	1.4110	8.6	2.1518	160	5.0752
4.2	1.4351	8.7	2.1633	170	5.1358
4.3	1.4586	8.8	2.1748	180	5.1930
4.4	1.4816	8.9	2.1861	190	5.2470

* Subtract 10 for $n < 1$. Thus $\log_e .1 = 7.6974 - 10 = -2.3026$.

Table B-6   Exponential Functions

x	e^x	e^{-x}	x	e^x	e^{-x}
.00	1.0000	1.0000	1.5	4.4817	.2231
.01	1.0101	.9901	1.6	4.9530	.2019
.02	1.0202	.9802	1.7	5.4739	.1827
.03	1.0305	.9705	1.8	6.0496	.1653
.04	1.0408	.9608	1.9	6.6859	.1496
.05	1.0513	.9512	2.0	7.3891	.1353
.06	1.0618	.9418	2.1	8.1662	.1225
.07	1.0725	.9324	2.2	9.0250	.1108
.08	1.0833	.9331	2.3	9.9742	.1003
.09	1.0942	.9139	2.4	11.023	.0907
.10	1.1052	.9048	2.5	12.182	.0821
.11	1.1163	.8958	2.6	13.464	.0743
.12	1.1275	.8869	2.7	14.880	.0672
.13	1.1388	.8781	2.8	16.445	.0608
.14	1.1503	.8694	2.9	18.174	.0550
.15	1.1618	.8607	3.0	20.086	.0498
.16	1.1735	.8521	3.1	22.198	.0450
.17	1.1853	.8437	3.2	24.533	.0408
.18	1.1972	.8353	3.3	27.113	.0369
.19	1.2092	.8270	3.4	29.964	.0334
.20	1.2214	.8187	3.5	33.115	.0302
.21	1.2337	.8106	3.6	36.598	.0273
.22	1.2461	.8025	3.7	40.447	.0247
.23	1.2586	.7945	3.8	44.701	.0224
.24	1.2712	.7866	3.9	49.402	.0202
.25	1.2840	.7788	4.0	54.598	.0183
.30	1.3499	.7408	4.1	60.340	.0166
.35	1.4191	.7047	4.2	66.686	.0150
.40	1.4918	.6703	4.3	73.700	.0136
.45	1.5683	.6376	4.4	81.451	.0123
.50	1.6487	.6065	4.5	90.017	.0111
.55	1.7333	.5769	4.6	99.484	.0101
.60	1.8221	.5488	4.7	109.95	.0091
.65	1.9155	.5220	4.8	121.51	.0082
.70	2.0138	.4966	4.9	134.29	.0074
.75	2.1170	.4724	5.0	148.41	.0067
.80	2.2255	.4493	5.5	244.69	.0041
.85	2.3396	.4274	6.0	403.43	.0025
.90	2.4596	.4066	6.5	665.14	.0015
.95	2.5857	.3867	7.0	1096.6	.0009
1.0	2.7183	.3679	7.5	1808.0	.0006
1.1	3.0042	.3329	8.0	2981.0	.0003
1.2	3.3201	.3012	8.5	4914.8	.0002
1.3	3.6693	.2725	9.0	8103.1	.0001
1.4	4.0552	.2466	10.0	22026	.00005

Appendix C

PROOF OF THE LAW OF COSINES

The following equations are called the *law of cosines*.

$$a^2 = b^2 + c^2 - 2bc \cos A \qquad (1)$$

$$b^2 = a^2 + c^2 - 2ac \cos B \qquad (2)$$

$$c^2 = a^2 + b^2 - 2ab \cos C \qquad (3)$$

To prove equation (1) consider the drawing at the left.

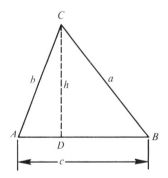

In $\triangle CBD$, $a^2 = h^2 + (BD)^2$. Now $(BD) = [c - (AD)]$, hence

$$a^2 = h^2 + [c - (AD)]^2$$
$$= h^2 + c^2 - 2c(AD) + (AD)^2$$
$$= h^2 + (AD)^2 + c^2 - 2c(AD)$$

but $h^2 + (AD)^2 = b^2$ and $(AD) = b \cos A$, hence

$$a^2 = b^2 + c^2 - 2c(b \cos A) \quad \text{or} \quad a^2 = b^2 + c^2 - 2bc \cos A \qquad (1)$$

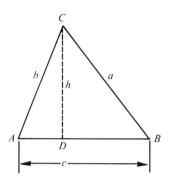

In $\triangle ACD$, $b^2 = h^2 + (AD)^2$. Now $(AD) = [c - (BD)]$, hence

$$b^2 = h^2 + [c - (BD)]^2$$
$$= h^2 + c^2 - 2c(BD) + (BD)^2$$
$$= h^2 + (BD)^2 + c^2 - 2c(BD)$$

but $h^2 + (BD)^2 = a^2$ and $(BD) = a \cos B$, hence

$$b^2 = a^2 + c^2 - 2c(a \cos B) \quad \text{or} \quad b^2 = a^2 + c^2 - 2ac \cos B \qquad (2)$$

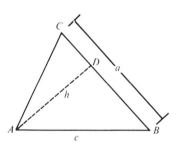

In $\triangle ABD$, $c^2 = h^2 + (BD)^2$. Now $(BD) = [a - (CD)]$, hence

$$c^2 = h^2 + [a - (CD)]^2$$
$$= h^2 + a^2 - 2a(CD) + (CD)^2$$
$$= h^2 + (CD)^2 + a^2 - 2a(CD)$$

but, $h^2 + (CD)^2 = b^2$ and $(CD) = b \cos C$, hence

$$c^2 = b^2 + a^2 - 2a(b \cos C) \quad \text{or} \quad c^2 = a^2 + b^2 - 2ab \cos C \qquad (3)$$

Answers to Selected Problems

Chapter 1

Set 1 1. 5×10^2 3. 9.524×10^4 5. 5.726371×10^6 7. 8.5×10^{-2} 9. 7.5×10^{-6} 11. 8.941×10^5 13. 5.21724398×10^8 15. 2.9873×10^{-3}

Set 2 1. 2.9979×10^8 3. 3.16×10^7 5. 1.6×10^{-19} 7. 1.49×10^{-10} 9. 5.4×10^5 11. 6.371×10^6 13. 5.486×10^{-4} 15. 6.328×10^{-7}

Set 3 1. 1×10^{-1} 3. 6.75×10^2 5. 1.238896×10^2 7. 1.0×10^{11} 9. 1.836×10^8

Set 4 1. 1.0×10^5 3. 1.424×10^5 5. 2.25×10 7. 5.6×10^{18} 9. 6.29×10^{-3}

Chapter 2

Set 1 1. comm. for mul. 3. comm. for add. 5. Distributive 7. comm. for mul. 9. assoc. for mul.

Set 2 1. Integer, rational, real 3. Integer, rational, real 5. rational, real 7. irrational, real 9. irrational, real

Set 3 1. 13 3. -2 5. 44 7. -7 9. -4 11. -4 13. 0 15. 0 17. 9

Set 4 1. 10 3. 20 5. 8 7. 33 9. 112

Set 5 1. a^7 3. x^4 5. z^{14} 7. 1,000,000 9. 7 11. $\dfrac{5}{8}$ 13. $3\,a^3b^2c^4$ 15. 1 17. $\dfrac{724}{11}$ 19. a^4b^5

Set 6 1. $8a$ 3. $5x^2$ 5. $-\theta$ 7. $7\,ax^2$ 9. $-5xyz^2$ 11. $3abc$ 13. $13p$ 15. $7+4y$ 17. $7\,cd$ 19. $4x-y+1$

Set 7. 1. 3 3. 4 5. 12 7. 675 9. 63; 1. $\dfrac{\sqrt{5}}{5}$ 3. $\dfrac{\sqrt{10}}{4}$ 5. $\dfrac{\sqrt{2}}{4}$ 7. $\dfrac{\sqrt{10}}{5}$ 9. $\dfrac{\sqrt{15xy}}{5y}$

Set 8. 1. $\dfrac{1}{2}$, (1) numerator, (2) denominator 3. $\dfrac{3}{5}$, (3) numerator, (5) denominator 5. $\dfrac{13}{8}$, (13) numerator, (8) denominator

Set 9 1. proper 3. decimal 5. mixed 7. proper 9. decimal 11. decimal 13. improper 15. mixed

439

Set 10   1. $\frac{1}{2}$ 3. $\frac{1}{5}$ 5. $\frac{2}{5}$ 7. $\frac{1}{5}$ 9. $\frac{2a}{3}$ 11. $\frac{9}{10w}$

Set 11   1. yes 3. no 5. yes 7. no 9. yes 11. no

Set 12   1. 4 3. 16 5. 3 7. 64 9. 2 11. 1 13. 3 15. 3 17. 56 19. 6

Set 13   1. $\frac{1}{2}$ 3. $\frac{5}{6}$ 5. $\frac{7}{6x}$ 7. $\frac{37}{32a}$ 9. $4\frac{2}{3}$ 11. $\frac{3}{10}$ 13. $\frac{51}{40e}$ 15. $13\frac{11}{15}$

Set 14   1. $\frac{6^2}{4}$ 3. $\frac{3ei}{2}$ 5. $\frac{4}{15y^4}$ 7. $\frac{4L^2}{5k}$

Set 15   1. $\frac{2bx}{9}$ 3. $\frac{4}{6e^4k^4}$ 5. $\frac{2b^4}{c^2}$ 7. $\frac{1}{2}$

Set 16   1. 14 3. 7 5. 7 7. 3 9. 4 11. 8 13. 2 15. -3

Set 17   1. 4 3. 4000 5. 2; 1. 4 3. 18 5. 300; 1. 360 3. 250 5. 125

Set 18   1. 12, 48 3. 44, 54 5. 21, 42, 84 7. 150, 175, 450 9. 4, 11

11. 20, 16, 8 13. 550 15. 1000, 800, 1200 17. 18, 72 19. $1\frac{1}{5}$

Set 19   1. $\frac{V-Vo}{t}$ 3. $P_1 - QR$ 5. $\frac{1}{2\pi F^x c}$ 7. $\frac{2(hf - e\,Ew)}{v^2}$ 9. $\frac{A + dh}{P_2 - p_1 - dh}$

11. $\frac{2\pi d^3 DS}{16\,D^2 + 8Dd}$ 13. $\frac{Rr^2}{r_2 - R}$ 15. $\frac{5}{9}(F - 32)$ 17. $\frac{eo - Aes}{ABeo}$ 19. $\frac{fl\,(V - V_a) - fs}{fs}$

21. $\frac{e}{\sin wt}$ 23. $\frac{F}{QUB}$ 25. $\frac{x(m + a) - a\sin\theta\,(x - 2a - m)}{a\sin\theta}$

Set 20   1. I 3. II 5. I 7. II 9. I

Set 22   1. 5,0 3. 0,6 5. 2,0 7. 0,4 9. 3,0

Set 23   1. 1 3. $-\frac{3}{2}$ 5. $\frac{11}{3}$ 7. 8 9. $-\frac{2}{5}$

Set 24   1. $m=2$ $b=0$ 3. $m=2$ $b=-4$ 5. $m=2$ $b=1$ 7. $m=-2$ $b=1$

9. $m=-\frac{1}{18}$ $b=-\frac{2}{3}$ 11. $m=2$ $b=6$ 13. $m=\frac{6}{5}$ 15. $m=1$ $b=-1$

Set 25   1. $y=x+2$ 3. $y=4x-1$ 5. $3y=2x-6$ 7. $y=3x-6$ 9. $8x+2y=7$

Set 26   1. $x-y=0$ 3. $y=x-2$ 5. $y=3x+1$ 7. $6x+y+2=0$ 9. $x+2y=-5$

Set 27   1. perpendicular 3. perpendicular 5. perpendicular 7. perpendicular
9. parallel

Set 29   1. (6,2) 3. $(1,-2)$ 5. (1,1) 7. $(0,-2)$ 9. $\left(\frac{34}{11}, -\frac{5}{11}\right)$

Set 30   1. (1,4) 3. $(-3,-3)$ 5. (4,3) 7. $\left(\frac{1}{2}, -3\right)$ 9. $\left(\frac{40}{7}, \frac{37}{7}\right)$

Set 31   1. 24, 72 3. 14, 55 5. $I_1 = \frac{2}{55}$, $I_2 = \frac{7}{55}$ 7. 50, 25 9. 20,20

11. $32\frac{1}{2}, 57\frac{1}{2}$ 13. 2, 14   15. $1\frac{1}{2}, 3\frac{1}{2}$

Set 32   1. -1 3. 17 5. 10 7. -29

Set 33   1. $(-1,3)$ 3. $\left(-1\frac{1}{2}, 3\right)$ 5. $(3,-1)$ 7. $\left(\frac{69}{11}, \frac{57}{11}\right)$

Set 34   1. 13 3. 1 5. -2

Set 35   1. $x=1$ 3. $r=1$ 5. $x=-\dfrac{2}{17}$ 7. $x=-3$ 9. $x=3$

$\qquad\qquad y=2$ $s=\frac{1}{2}$ $y=10\dfrac{7}{17}$ $y=2$ $y=-2$

$\qquad\qquad z=3$ $t=3/2$ $z=-\dfrac{1}{17}$ $z=4$ $z=-4$

Chapter 3

Set 1   1. $\angle ABC$, $\angle CBA$, $\angle B$, $\angle b$

Set 2   1. segment 3. ray 5. ray 7. ray 9. complementary angles 11. right angle 13. right angle 15. straight angle

Set 3   3. (a) alternate interior (b) adjacent (c) vertical (d) corresponding (e) vertical (f) alternate interior (g) corresponding (h) alternate exterior

Set 4   (a) acute (b) obtuse (c) right (d) right (e) obtuse (f) acute

Set 5   1. $\triangle$RST; $\triangle$STR; $\triangle$TRS; $\triangle$TSR

Set 6   1. 33 in. 3. 7 in. 5. $11\dfrac{1}{2}$ in.

Set 7   1. 60 3. 14 in. 5. 71.5 sq ft. 7. 68.4 ft.

Set 8   1. 5 3. 14.4 5. 12.04 7. 12 9. 18.02

Set 9   1. yes 3. yes 5. yes 7. no 9. no

Set 10 $a=6.33, b=2.67$ 3. $s=1\dfrac{1}{2}, r=18$

Set 11   1. 40 ft. 3. 108 ft. 5. 3333.33 ft.

Set 12   1. 167° 3. rectangle, square, rhombus

Set 13   1. 94.08 sq in. 3. $A=772.84$ sq in., $P=111.2$ in. 5. 722.4 sq in. 7. 11.07 in. 9. 127.5 ft. 11. $975

Set 15   1. 1962.5 sq in. 3. 71.7 sq ft. 5. No, only one-half as much 7. 11.07 in. 9. 908.57 ft. 11. 800

Set 16   1. 1563 sq in. 3. 2 gal.

Set 17   1. $h=10.48, V=100.78, s=110$ sq. in., $T=148.47$ sq in. 3. 17.28 5. 9.42 in.

Set 18   1. 90.5 3. 44.36 5. 12 qts.

Set 19   1. 443.34, 597.2, 1025.73 3. 14.99 5. 1 : 8

Set 20   1. 314, 523.33 3. 9 : 121, 27 : 1331 5. 4090

Set 21   1. 1809.12, 994.2 3. 1192, 739.6 5. 474.39, 151.07 7. 920; 665.7

Chapter 4

Set 1   1. $1:3$ 3. $\dfrac{16}{1}$ 5. $1:24$ 7. $16:1$ 9. $\dfrac{500}{1}$

Set 2   1. 2 3. $\dfrac{49}{50}$ 5. 1 7. 35 9. 400,000

Set 3   1. $d=kv$ 3. $a=\dfrac{k}{t^2}$ 5. $w=k\left(\dfrac{w-}{LA}\right)$

Set 4   1. 75 3. 480 5. 16

Set 5   1. 560 3. 525 5. $73\dfrac{1}{3}$ ft

Set 6   1. 1.4 in. 3. 2 sec. 5. 50°

Set 7   1. $3\dfrac{1}{3}$ 3. 220 gm

Set 8   1. 350 rpm 3. 870 rpm, 1740 rpm, 3480 rpm 5. $11,433\dfrac{1}{3}$ rpm

Set 9   1. 4.39 lbs 3. 360 lbs

Chapter 5

Set 1   1. B, b, ABC 3. P, Y, QPO 5. A, a, BAC 7. Y, ZUX, XUZ
9. ϕ, ZOX, XOZ

Set 2   1. 2π 3. $\dfrac{\pi}{180}$ 5. $\dfrac{\pi}{2}$ 7. $\dfrac{\pi}{6}$ 9. 3π 11. $\dfrac{11\pi}{90}$ 13. $\dfrac{31\pi}{18}$ 15. $\dfrac{\pi}{36}$

Set 3   1. 57.3 3. 18.2 5. 22.9 7. 4.6° 9. 258

Set 4   1. 45 3. 5 5. 220 7. 180 9. 342

Set 5   1. 5 3. 6.08 5. 23.2 7. 84 9. 8.062

Set 6

	1.	3.	5.	7.	9.
$\sin\phi=$	.60	.707	.98	.394	.316
$\cos\phi=$	.80	.707	.196	.921	.949
$\tan\phi=$	.75	1.0	5.0	.428	.333
$\cot\phi=$	1.33	1.0	.20	2.33	3.0
$\sec\phi=$	1.25	1.41	5.1	1.08	1.05
$\csc\phi=$	1.66	1.41	1.02	2.53	3.16

Set 7

	1.	3.	5.	7.	9.	11.
$\sin\phi=$	.707	.984	.390	.999	.275	.224
$\cos\phi=$	.707	.173	.920	.034	.961	.974
$\tan\phi=$	1.00	5.67	.424	28.6	.286	.230
$\cot\phi=$	1.00	.176	2.35	.034	3.48	4.33

Set 8   1. 30° 3. 45° 5. 10° 7. 90° 9. 3° 11. 30° 13. 45° 15. 22°

Set 9   1. .1808 3. .5604 5. .7038 7. 1.929 9. .9728

Set 10   1. 14°31′ 3. 40°10′ 5. 26°21′ 7. 21°57′ 9. 37°52′ 11. 23°12′
13. 58°50′ 15. 68°15′

Set 11 1. 5 3. 35 5. 213.6 7. 0.472 9. 151.328 11. 813.9 13. 72.8
15. $16.1 \times 9.27 \times 7.49$

Set 12 1. 4.77 3. 5 5. 10.61 7. 8.12

Set 13 1. $B=27.3°, C=90°, b=4.31, c=9.40$ 3. $A=69.7, C=90°, a=42.49$
5. $A=36°52', B=53°8', C=90°, b=80$ 7. $A=30°, B=60°, C=90°, a=29.93$
9. $A=42.1°, C=90°, a=4789.08, c=7142.86$

Set 14 1. 111.80 3. 15 5. 11.62 7. 488.78 9. 1020 11. 80 13. 4.02
15. 33.04

Set 15 1. 3.999, 3.007 3. 64.26, 415.044 5. .106, .020 7. 86.6, 50
9. 3917.334, 3917.334

Set 19 1. 2.09 3. 6°52' 5. 915.83 7. 0.0000026

Set 20 1. $\sin \theta = 12/13, \cos \theta = 5/13, \tan \theta = 12/5, \cot \theta = 5/12$
3. $\sin \theta = 2/2, \cos = 2/2, \tan \theta = -1, \cot \theta = -1$
5. $\sin \theta = 1/2, \cos \theta = 3/2, \tan \theta = 3/3, \cot \theta = 3$
7. $\sin \theta = -1/2, \cos \theta = 3/2, \tan \theta = 3/3, \cot \theta = -3$

Set 22 1. $b=69.5, c=76.6, C=87°$ 3. $a=70.5, B=49°, C=59.5°$
5. $a=288, B=49°, C=59.5°$ 7. $b=35.7, A=64°, C=76°$
9. $A=56.4°, B=33.6°, C=90°$

Chapter 6

Set 1 1. $4j$ 3. $2\sqrt{2}j$ 5. $3\sqrt{5j}/5$ 7. $E\sqrt{Rj}/R$ 9. $-\sqrt{2Bj}$

Set 2 1. $5+3j$ 3. $-1+2\sqrt{2}j$ 5. $-2-j$ 7. $5+3\sqrt{3}j$ 9. $-4-2\sqrt{2}j$
      $5-3j$ $-1-2\sqrt{2}j$ $-2+j$ $5-3\sqrt{3}j$ $-4+2\sqrt{2}j$

1. $\dfrac{12-24j}{5}$ 3. $\dfrac{3-9j}{10}$ 5. j 7. $2-8j$ 9. 0

Set 5 5.84∠31° 3. 9.48∠58° 5. 29.1∠116° 7. 510∠117° 9. 1265∠235°

Set 6 1. 51∠26° 3. 920∠325° 5. 1180∠183° 7. .327∠44°

Set 7 1. $6(\cos 75° + j\sin 75°)$ 3. $7(\cos 140° + j\sin 140°)$
5. $12(\cos 7° + j\sin 7°)$ 7. $5(\cos 62° + j\sin 62°)$
9. $1/7(-\cos 37° - j\sin 37°)$

Set 8 1. $8∠2(\cos 315° + J\sin 315°)$ or $8-j8$
3. $64(\cos 180° + J\sin 180°)$ or -64
5. $3125(\cos 1530° + J\sin 1530°$ or $3125j$
7. $r_1=1.74(\cos 40° + J\sin 40°)$ or $1.33+1.20j$
   $r_2=1.74(\cos 220° + J\sin 220°)$ or $-1.33-1.20j$
9. $2\sqrt{2}(\cos 225° + j\sin 225°)$ or $-1.99-1.99j$

Chapter 7

Set 1 1. $10^3=1000$ 3. $4^3=64$ 5. $25^{0.5}=5$ 7. $8^3=512$ 9. $e^x=y$

Set 2 1. $\log_{10} 100 = 2$ 3. $\log_2 32 = 5$ 5. $\log_4 1 = 0$ 7. $\log_{16} 4 = 0.5$ 9. $\log_2 256 = 8$

Set 3 1. 3 3. 4 5. 4 7. 64 9. 10

Set 4 1. 0.30103 3. 2.43136 5. 2.72916 7. 3.77641 9. 4.39314
11. $2.60206 - 10$ 13. 6.88326 15. $6.44091 - 10$ 17. $6.43712 - 10$
19. $7.83442 - 10$

Set 5 1. 1.9999 3. 700 5. 150,000 7. 52,000 9. 340,000,000

Set 6 1. 1.140 3. 19.719 5. 0.000015913 7. 856,500 9. 342.33

Chapter 8

Set 1 1. $15a^2 + 20a$ 3. $14rs + 21rt + 35r$ 5. $24m^2n + 32m^3 + 72m^3n$
7. $mpv^2 + 3mqv^2 + 4mv^2$ 9. $14x^3y^3 + 7x^2y^3 + 7x^2y^2$
11. $144e^3i^2 + 18e^2i^3$ 13. $45x^3y^2z + 20xy^3z$

Set 2 1. $7xy(2x + 3y + 5xy)$ 3. $3a^2bc(14 - 18bc - 21c)$
5. $mnp(39n^2 + 52mp - 61m^2n^2p^2)$ 7. $11x^3y^2(3x^2 + 11y^2 - 11xy^4 + 11x^4y)$
9. $8x^2(9x^3y^4z^7 + 5w^3 - 7xy^2 + 12x^2y^3z^5)$

Set 3 1. $9x^2 - y^2$ 3. $81e^2 - 25i^2$ 5. $225m^2 - 121n^2$ 7. $9x^2 - 25y^2$ 9. $5x^4y^2 - 16z^2$

Set 4 1. $(a - b)(a + b)$ 3. $(4u - 5v)(4u + 5v)$ 5. $(9x^2y + 7z)(9x^2y + 7z)$
7. $(8r^2s^3 - 4t)(8r^2s^3 + 4t)$ 9. $(15n^5 + 20p^4)(15n^5 - 20p^4)$

Set 5 1. $4x^2 + 20x + 25$ 3. $5c^2d^2 + 70cd + 49$ 5. $81u^2 - 54uv + 9v^2$
7. $36z^2 - 60zw + 25w^2$ 9. $16x^4 - 72x^2y^2 + 81y^4$

Set 6 1. $(3x + 1)^2$ 3. $(2e - 3)^2$ 5. $(5t - s)^2$ 7. $(9b + c^2)^2$ 9. $(x^2 + 4y)^2$

Set 7 1. $x^2 + 8x + 15$ 3. $^2e + 5e + 4$ 5. $x^2 + 12x + 27$ 7. $7n^2 + 16n - 15$
9. $105s^2 - 429s + 276$

Set 8 $(x + 4)(x + 2)$ 3. $(e - 5)(e - 1)$ 5. $(r + 3)(r + 7)$ 7. $(u + 15)(u - 4)$
9. $(2y - 3)(y + 2)$ 11. $(4u + 1)(3u + 2)$ 13. $(5a - 16b)(2a + b)$
15. $3D(D - s)(D + 2)$ 17. $(V - 18)(V + 3)$ 19. $(P + 12)(P - 8)$
21 $(2D - 1)(D + 8)$ 23. $2(2n^2 - 6n - 4)$ 25. $(3w - i)(4w + i)$
27. $(r - 7)(r - 5)$ 29. $(8h + 9)(6h - 5)$ 31. $(2r + 3)(r - 9)$

Set 9 1. $(x + 4)(x^2 - 4x + 16)$ 3. $(v - 5)(v^2 + 5v + 25)$ 5. $(a - 6)(a^2 + 6a + 36)$
7. $(s - 9)(s^2 + 9s + 81)$ 9. $(4r - 8)(16r^2 + 32r + 64)$

Set 10 1. $(x - 4)(x + 6y)$ 3. $(x + 2)(x^2 - 5)$ 5. $(h^2 + 6)(h + 3)$
7. $(5i^3 - 4)(e^2 + 6i^2)$

Set 11 1. $5y + 3$ 3. $2x^3 + x^2 - 3x + 4$ 7. $25z^2 + 10zw + 4w^2$ 9. $3V - 2$
11. $2h^2 - 2h + 5$

Set 12 1. $\dfrac{7E}{3R}$ 3. $\dfrac{120\lambda^5\beta^5}{\phi^4}$ 5. $\dfrac{5n^4m^8}{7n^2m^3 + 9}$ 7. $\dfrac{6A - 5B}{A}$ 9. $\dfrac{K + H}{-2K - H}$

Set 13 1. $\dfrac{9a}{4(a - b)}$ 3. $\dfrac{x^2(x^2 + 1)}{2(x^2 - 1)}$ 5. $\dfrac{-(a + 1)}{2}$ 7. $\dfrac{-4l^2(K - 2)}{K(5 + K)}$

Set 14 1. $\dfrac{2B}{9A}$ 3. $\dfrac{4}{6K^4}$ 5. $\dfrac{x}{x - 1}$ 7. $-\dfrac{4x + 5}{5x + 6}$

Set 15   1. $\dfrac{2M}{3}$ 3. 0 5. $\dfrac{-79e+198}{8e-20}$ 7. $\dfrac{-2x^2+3x}{32-2x^2}$ 9. $\dfrac{V^2-V+3}{(V-2)(V+3)(V-1)}$

Set 16   1. $\dfrac{3^x}{2y^2}$ 3. $\dfrac{8\phi^2y}{3}$ 5. $\dfrac{\lambda(\lambda-1)}{\lambda+1}$ 7. $\dfrac{B+5}{B+4}$

Set 17   1. $\dfrac{1}{5}$ 3. 2 5. -39 7. $\dfrac{7}{11}$

Set 18   1. $3, -3$ 3. $2\sqrt{2}, -2\sqrt{2}$ 5. $\dfrac{5}{2}, -\dfrac{5}{2}$ 7. $5, -5$ 9. $\dfrac{2}{5}, -\dfrac{2}{5}$

Set 19   1. $\dfrac{3}{5}, -1$ 3. $\dfrac{4}{3}, -5$ 5. $6, 3$ 7. $\dfrac{8}{3}$

1. $12, -2$ 3. $3+\sqrt{2}, 3-\sqrt{2}$ 5. $-\dfrac{3}{7}, 2$ 7. $-5+\sqrt{6}, -5-\sqrt{6}$

Set 20   1. $\dfrac{3}{2}, 1$ 3. $\dfrac{3}{5}, 2$ 5. $-\dfrac{2}{3}, \dfrac{1}{4}$ 7. $\dfrac{3+\sqrt{47i}}{14}, \dfrac{3-\sqrt{47i}}{14}$

9. $-\dfrac{1}{2}+\dfrac{\sqrt{14}}{8}, -\dfrac{1}{2}-\dfrac{\sqrt{14}}{8}$

Set 22   1. $x^3+3x^2y+3xy^2+y^3$ 3. $c^5+5c^4d+10c^3d^2+10c^2d^3+5cd^4+d^5$
5. $a^5-10a^4+40a^3-80a^2+80a-32$
7. $x^7+7x^6y+21x^5y^2+35x^4y^3+35x^3y^4+21x^2y^5+7xy^6+y^7$
9. $x^7+7x^6y+21x^5y^2+35x^4y^3+35x^3y^4+21x^2y^5+7xy^6+y^7$

Chapter 9

Set 1   1. 6 3. 12 5. -3 7. 0 9. $\dfrac{7}{2}$ 11. 2

Set 2   1. 3 3. $3-2x$ 5. $-\dfrac{2}{x^2}$ 7. $\dfrac{-2}{x-4x+4}$ 9. $3x^2$ 11. 158 mph

Set 3   1. 0 3. 2 5. $8x^3$ 7. $8x-4$ 9. $5(3x^2-2x+1)^4(6x-2)$ 11. $-2x^{-3}$

13. $\dfrac{1^{-\frac{2}{3}}}{3x}$ 15. $2x-\dfrac{1^{-\frac{3}{2}}}{2x}$ 17. 201 units; 49 units/sec

Set 4   1. (a) 56 ft/sec, -8 ft/sec (b) 225 ft (c) 7.5 sec
3. (a) 100 ft/sec (b) -4 ft/sec^2 (c) 1,250 ft
5. $80\sqrt{6}$ ft/sec. 7. (a) $16\sqrt{3}$ ft/sec 9. 45,792 Joules/sec
             (b) $\sqrt{3}$ sec

Set 5   1. 20 3. 160 5. $-\dfrac{1}{50\sqrt{2}\,i}$ 7. $\dfrac{-5\sqrt{2}\sqrt[3]{3}}{72}$ 9. $2x+2$ 11. $2x$ 13. $2x+1$

15. 32 in^2/sec 17. $\dfrac{5\sqrt{3}}{2}$ in^2/sec

Set 6   1. 7.07 mph 3. 1,256 in^3/min. 5. 9.42 ft^3/min 7.$\dfrac{5}{72\pi}$ in./min

9. $\dfrac{-2Q_1Q_2}{3\pi et^4}$

Set 7   1. $2\cos(2x)$ 3. $\sec^2 x$ 5. $-2\sin x \cos x$ 7. $6x\tan^2 x^2 \sec^2 x^2$

9. $6x\cos(3x^2)\sec^2(3x^2)-6x\sin(3x^2)\tan(3x^2)$ 11. $\dfrac{\sin x}{\cos^2 x}$

13. $\dfrac{-3\sec^2(3x)}{\tan^2(3x)}$ 15. $\dfrac{1}{5}\cos\left(\dfrac{1}{5x}\right)$ 17. $4\sec^2(4x)$

19. $6 \tan (3x) \sec^2 (3x) + \sin x$ 21. $\dfrac{2}{25}$ rad/sec

23. $43,200 \sin (60t) \cos (60t)$

Set 8   1. $1/x$ 3. $2 \sec \theta \csc \theta$ 5. $2 + \log_e x^2$ 7. $\dfrac{1}{x \log x}$ 9. $2e^{2x}$ 11. $\dfrac{e^x}{x} + e^x \log_e x$

13. $\dfrac{2e^{-x}}{x} - e^{-x} \log_e x^2$ 15. $0.69315(2x-3)\ 2^{(x^2-3x)}$ 17. $-\dfrac{1}{5}e^{-4}$ 19. A/P

Set 9   1. $y^1 = 3x^3 + 9x^2 - 2x + 1$, $y^{11} = 12x^2 + 18x - 2$, $y^{111} = 18x + 18$

3. $y^1 = 3/x$, $y^{11} = -3/x^2$, $y^{111} = 6/x^3$

5. $y^1 = 6x (x^2+3)^2$, $y^{11} = 24x^2 (x^2+3) + 6(x^2+3)^2$, $y^{111} = 120x^3 + 216x$

7. $y^1 = -\dfrac{\cos x}{\sin^2 x}$, $y^{11} = \dfrac{\sin x + 2 \cos^2 x}{\sin^3 x}$, $y^{111} = \dfrac{-5 \sin^2 \cos x - 6 \cos^3 x}{\sin^4 x}$

9. $y^1 = -x^{-2} - 2x^{-3}$, $y^{11} = 2x^{-3} + 6x^{-4}$ $y^{111} = -6x^{-4} - 24x^{-5}$

Set 10   1. 12, 12   3. $6 \times 3 \times 8$ 5. $10 \times 10 \times 5$ 9. $1\dfrac{1}{5}$ hrs.

Chapter 10

Set 1   1. $x^3 + c$ 3. $x^3/3 - 3x^2/2 + x + c$ 5. $1/3 \sin 3x + c$ 7. $\dfrac{1}{2} \log_e (x^2 - 1) + c$

9. $\frac{1}{4} (x^2 - 1)^2 + c$

Set 2   1. $x^3/3 + c$ 3. $e^x + c$ 5. $x^4/4 + x^3/3 + c$ 7. $-\cos x + c$

9. $x^3/3 + x^6/6 + \sin x + c$ 11. $s = \dfrac{-32.2\ t^2}{2} + c_1 t + c_2$, $v = -32.2t + c_1$

Set 3   1. $\dfrac{(x^2+1)^4}{8} + c$ 3. $-1/3(x^3+2)^{-1} + c$ 5. $-1/2 \cos (x^2) + c$

7. $1/3 \dfrac{a^{x^3}}{\log a} + c$ 9. $1/3 \sin^3 x + c$ 11. $\dfrac{(x^3 + x^2 + 2)^4}{4} + c$

Set 4   1. 60   3. 36.75   5. 328/3   7. 0   9. 2

Set 5   1. $V = -32\ t + 80$, $x = -16\ t^2 + 80t + 50$, 150 ft.

3. $V = 5 - 5\varepsilon^{\frac{-2t}{m}}$, $s = 5t + 10 \left(\dfrac{m}{4}\right) \left(\dfrac{-2t}{\varepsilon^m}\right) - \dfrac{10m}{4}$ 5. 2250

Set 6   1. 216 in·lb   3. 18750 ft·lbs   5. $\dfrac{416\pi}{3}$ 7. $-2/L$

Chapter 11

Set 1   1. $y = .143\ x - .039$ 3. $I = .256e + .023$ 5. $V = .36t + 100$, $t = -277.7$

Set 2   1. $y = 3.375\ x^2 - 4.125x + 5.218$ 3. $y = -0.000032x^2 + 0.999x - 8.80$

Set 3   1. $y = 3.23x^{-1.25}$ 3. $y = 686x^{-1.976}$ 5. $p = 412.5\ v^{-.815}$

1. $y = 0.517(10)^{0.215x}$ or $y = 0.517(e)^{0.495x}$

3. $y = 0.303(10)^{0.058x}$ or $y = 0.303(e)^{0.135x}$

5. $y = 9.888(10)^{-0.296x}$ or $y = 9.888(e)^{0.681x}$ or 0.000000013

Index